Lecture Notes in Mechanical Engineering

Lecture Notes in Mechanical Engineering (LNME) publishes the latest developments in Mechanical Engineering–quickly, informally and with high quality. Original research reported in proceedings and post-proceedings represents the core of LNME. Volumes published in LNME embrace all aspects, subfields and new challenges of mechanical engineering. Topics in the series include:

- Engineering Design
- Machinery and Machine Elements
- Mechanical Structures and Stress Analysis
- Automotive Engineering
- Engine Technology
- Aerospace Technology and Astronautics
- Nanotechnology and Microengineering
- Control, Robotics, Mechatronics
- MEMS
- Theoretical and Applied Mechanics
- Dynamical Systems, Control
- Fluid Mechanics
- Engineering Thermodynamics, Heat and Mass Transfer
- Manufacturing
- Precision Engineering, Instrumentation, Measurement
- Materials Engineering
- Tribology and Surface Technology

To submit a proposal or request further information, please contact the Springer Editor of your location:

China: Dr. Mengchu Huang at mengchu.huang@springer.com
India: Priya Vyas at priya.vyas@springer.com
Rest of Asia, Australia, New Zealand: Swati Meherishi at swati.meherishi@springer.com
All other countries: Dr. Leontina Di Cecco at Leontina.dicecco@springer.com

To submit a proposal for a monograph, please check our Springer Tracts in Mechanical Engineering at http://www.springer.com/series/11693 or contact Leontina.dicecco@springer.com

Indexed by SCOPUS. The books of the series are submitted for indexing to Web of Science.

More information about this series at http://www.springer.com/series/11236

Dagmar Klichová · Libor Sitek ·
Sergej Hloch · Joško Valentinčič
Editors

Advances in Water Jetting

Selected Papers from the International Conference on Water Jet 2019 - Research, Development, Applications, November 20–22, 2019, Čeladná, Czech Republic

Springer

Editors
Dagmar Klichová
Institute of Geonics of the Czech Academy of Sciences
Ostrava, Czech Republic

Libor Sitek
Institute of Geonics of the Czech Academy of Sciences
Ostrava, Czech Republic

Sergej Hloch
Faculty of Manufacturing Technologies TUKE with a seat in Prešov
Prešov, Slovakia

Joško Valentinčič
Faculty of Mechanical Engineering
University of Ljubljana
Ljubljana, Slovenia

ISSN 2195-4356 ISSN 2195-4364 (electronic)
Lecture Notes in Mechanical Engineering
ISBN 978-3-030-53490-5 ISBN 978-3-030-53491-2 (eBook)
https://doi.org/10.1007/978-3-030-53491-2

This Springer imprint is published by the registered company Springer Nature Switzerland AG
The registered company address is: Gewerbestrasse 11, 6330 Cham, Switzerland

Preface

Due to its diversity, the technology of high-speed water jets, to which this volume of the Lecture Notes in Mechanical Engineering (LNME) is dedicated, has long ceased to be a marginal and exceptional scientific and research area of study and has become an integral part of common industrial methods of machining and production. Thanks to its specific properties, it is spreading to industries where the water jet application would have been unthinkable a few years ago. In addition to the growing importance of this technology, the attention is paid to the research and development activities and obtained results directly related to the water jet technology.

The major changes that have recently taken place to an increasing extent, not only in industry but also in the economy, are closely linked to the massive digitalization and automation in production, whether we call this process the Fourth Industrial Revolution, Industry 4.0 or something else. These changes are, of course, reflected in manufacturing technologies, including the high-speed water jets. The demand for energy savings encourages professionals to create more efficient cutting, machining and cleaning machines and tools, thus improving the existing production processes. The continuing global trend towards complete automation is of fundamental importance for the further development of the water jet technology. Combining robotic and control systems with the so-called intelligent water jet nozzles is becoming a great challenge in the near future. In addition, the prospects of the use of water jets is very promising in other areas. The disposal of unexploded ammunition, micro-machining, car manufacturing, maintenance and disposal of underwater constructions, objects and ocean-going vessels, but also the joint replacement surgery, are only some of the sectors of possible water jet application. In synergy with modern and advanced decision-making and control methods, such as the fuzzy logic, machine learning or artificial intelligence, the water jet is becoming an effective, safe and reliable tool fulfilling, at the same time, strict environmental requirements. However, in order to meet all current and future challenges, big efforts will have to be made and scientific and engineering approaches will have to be effectively combined.

One of the platforms, effectively combing various approaches, is a series of conferences on the water jet technology organized every two years in attractive venues in the Czech Republic by the Department of Material Disintegration of the Institute of Geonics of the Czech Academy of Sciences in Ostrava. Over the years, initially the local meeting of experts from the Czech Republic and the surrounding areas has become a respected event of worldwide importance. Thanks to the generous financial support of both domestic and foreign sponsors; a rich accompanying programme and interesting industrial visits are the traditional parts of the conference. Members of the International Advisory Board of the conference regularly select the three best presentations which are subsequently evaluated financially.

This volume of Lecture Notes in Mechanical Engineering contains selected papers presented at the 6th international conference on water jetting **Water Jet 2019-Research, Development, Application** (WJ2019) held in Čeladná (Czechia) on 20–22 November 2019. A total of 39 scientific, technical or commercial contributions from 11 countries were presented. Based on the review procedure, 26 articles, the focus and quality of which meet the traditional high-level requirements of this publication, were accepted for publishing.

In conclusion, dear readers, allow me to thank all those who participated in the smooth running of the conference, namely the participants, speakers, members of the International Advisory Board and the organizational committee. In addition, I am very thankful to editors and reviewers for their hard work and, of course, the whole team at Springer Nature Switzerland AG, who has always been ready to help and give advice during the preparation of this book. It would not have been possible without their interest and dedication.

April 2020 Libor Sitek

Organization

International Advisory Board

Chair

Josef Foldyna	Institute of Geonics of the CAS, Czechia

Members

Lenka Bodnárová	Brno University of Technology, Czechia
Sergej Hloch	Faculty of Manufacturing Technologies TUKE with a seat in Prešov, Slovakia
Zdenko Krajný	Slovak University of Technology in Bratislava, Slovakia
Michael Jarchau	Hammelmann GmbH, Germany
Marco Linde	ANT Applied New Technologies AG, Germany
Guoyi Peng	Nihon University, College of Engineering, Japan
Franz Perndorfer	Perndorfer Maschinenbau KG, Austria
Frank Pude	INSPIRE AG, Switzerland
Franz Trieb	BFT GmbH, Austria
Joško Valentinčič	University of Ljubljana, Slovenia

Organizing Board

Chair

Libor Sitek	Institute of Geonics of the CAS, Czechia

Members

Eva Dudková	Institute of Geonics of the CAS, Czechia
Lucie Gurková	Institute of Geonics of the CAS, Czechia
Dagmar Klichová	Institute of Geonics of the CAS, Czechia

Daria Nováková	Institute of Geonics of the CAS, Czechia
Jiří Starý	Institute of Geonics of the CAS, Czechia

External Reviewers

Lenka Bodnárová	Brno University of Technology, Czechia
Augusto Bortolussi	Institute of Environmental Geology and Geoengineering of the Italian National Research Council, Italy
František Botko	Faculty of Manufacturing Technologies TUKE with a seat in Prešov, Slovakia
Axel Henning	OMAX Corporation, USA
Sergej Hloch	Faculty of Manufacturing Technologies TUKE with a seat in Prešov, Slovakia
Michael Jarchau	Hammelmann GmbH, Germany
Marko Jerman	University of Ljubljana, Slovenia
Krzysztof Kotwica	AGH University of Science and Technology, Poland
Zdenko Krajný	Slovak University of Technology in Bratislava, Slovakia
Andrej Lebar	University of Ljubljana, Slovenia
Dominika Lehocká	Faculty of Manufacturing Technologies TUKE with a seat in Prešov, Slovakia
Guoyi Peng	Nihon University, College of Engineering, Japan
Andrzej Perec	Jacob of Paradies University, Poland
Frank Pude	INSPIRE AG, Switzerland
Izidor Sabotin	University of Ljubljana, Slovenia
Franz Trieb	BFT GmbH, Austria
Joško Valentinčič	University of Ljubljana, Slovenia

Contents

Keynote Lectures

Effect of Particle Fragmentation on Cutting Performance in Abrasive Waterjets

Axel Henning(✉), Michael Lo, Ernst Schubert, and Peter Miles

OMAX Corporation, 21409 72nd Avenue S, Kent, WA 98032, USA
axel.henning@omax.com

Abstract. Abrasive waterjets have become a standard tool for machining a wide variety of materials. Today's applications of abrasive waterjet cutting can be found in many different industries that range from producing very small high precision parts to making rough separation cuts of thick steel plates. Advancements in understanding of the physics of the abrasive waterjet cutting process continues to further advance the state of the art in predictive modeling and motion control software of the abrasive waterjet cutting process. Considerable research has been devoted towards studying how the various abrasive waterjet process parameters affect one another and their influence on the final cutting results. While several approaches have evaluated the effect of hydraulic energy from the water jet itself, in this paper the focus is laid on the effect of the particles that are interacting with the material. In this paper the kinetic behavior and fragmentation of particles has been analyzed in a wide range of different conditions. New insights into fragmentation of the particles especially at higher pressures can explain the behavior of experimental cutting studies where higher pressures did not provide a significant improvement and in many cases a decrease of cutting performance. Studies like this will not only help us understand the very processes that are involved in the cutting operation but also help us to optimize the overall process to increase performance and reliability and also be competitive against other cutting methods while providing ample new applications in the future.

Keywords: Abrasive water jet · Particle erosion · Metal cutting · Fragmentation

1 Introduction

Abrasive waterjets have been well established in industrial application since its first introduction in the 1980s. Today, applications of abrasive waterjet cutting can be found in many different industries that range from producing very small high precision parts to making rough separation cuts of 150 mm thick steel plates, from separating tiny electronic components to medical surgery research. Advancements in understanding of the physics of the abrasive waterjet cutting process continues to further advance the state of the art in predictive modeling and motion control software of the abrasive waterjet cutting process [1, 2].

D. Klichová et al. (Eds.): Water Jet 2019, LNME, pp. 3–14, 2021.
https://doi.org/10.1007/978-3-030-53491-2_1

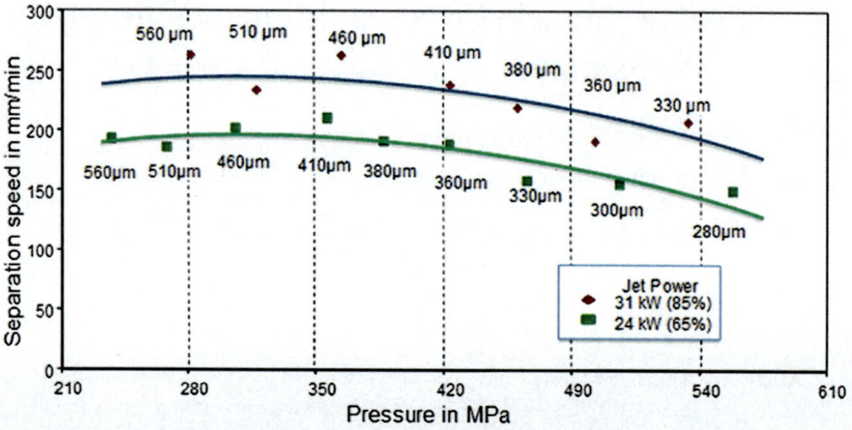

a) Constant abrasive load ratio (R=14%)

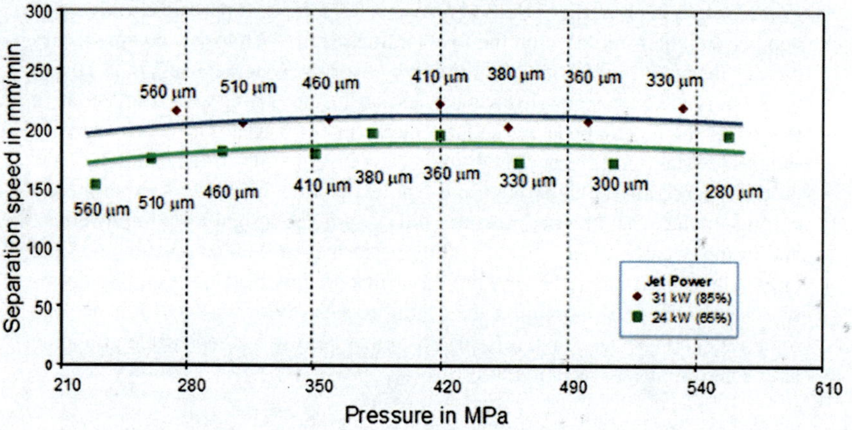

b) Constant abrasive feed rate (360g/min)

Fig. 1. Effect of pressure on separation speed for 25.4 mm mild steel (A36) with constant hydraulic power [6]. (Orifice diameter as indicated).

Currently, the most common parameter that is used to evaluate abrasive waterjet cutting performance is the operating pressure of the pump because it is the easiest parameter to adjust by simply varying either the pump speed or adjusting pressure regulators. This is a fallacy, as jet pressure is only a partial and indirect measure of the overall hydraulic power being delivered to the work piece for removing material. Hydraulic power incorporates the product of two variables, not just one: pressure and flow rate. For a constant input power rating, any increase in pressure requires a proportional decrease in flow rates. Higher pressures may be desirable because it drives the velocity of the abrasive particles higher which increases the kinetic energy of each particle. But the resultant decrease in flow rates, at a constant input power, decrease the ability to carry and accelerate more abrasive particles which decreases the abrasive kinetic power [3–5].

When evaluating the effect of operating pressure on cutting performance the experimental results show that, even when varying the pressure widely, its effect is negligible or even detrimental (see Fig. 1). While keeping the applied hydraulic power the same the cutting performance increased slightly when using the same amount of abrasive feed rate (Fig. 1b). It decreased slightly when using the same abrasive load ratio (Fig. 1a) [5, 6]. The effect of higher hydraulic power through better pump efficiency largely outshined the effect of pressure (65% - Intensifier design, 85% - Direct Drive design). In the end it is the abrasive particle that performs the erosion process by impacting on the workpiece. In this paper the focus is on how particles can be accelerated and what effect that has on the kinetic power they can obtain.

2 Fragmentation of Particles

The particle size has a large impact on speed and acceleration of the particles and on the available kinetic power to perform the cutting task. Therefore the question at hand was which distribution of particle sizes we should expect to impact on the material. The original abrasive material that is fed into the cutting head already consists of a wide distribution of particle sizes (see [7]) and previous evaluations [8, 9] have shown that fragmentation of particles occurs in the cutting head.

2.1 Experimental Setup and Procedure

This project goal was to characterize the amount of fragmentation of the abrasive particles that have passed through an abrasive water jet (AWJ) cutting head. Experimental data was collected to characterize levels of abrasive fragmentation based on perturbations of certain operating conditions. The data was characterized by altering one variable at a time while holding others constant in an effort to correlate direct cause and effect of each parameter.

The experimental setup consisted of a 1.5 m tall, 250 mm diameter PVC pipe standing on end. The pipe was fitted with a close-fitting cover and a removable plug at the bottom that had an integral drain. AWJ cutting heads were mounted to the top cover by means of an adaptor that ensured the head was aligned to fire axially into the tube. The tube itself was filled with water to decelerate and collect all the fired abrasives while preventing impact with any solid surface (Fig. 2).

Fig. 2. Abrasive fragmentation test apparatus.

For each test the nozzle was fired into the catcher at different pressures and abrasive feed rates without hitting a workpiece. After the test the abrasive material was collected by fully draining and flushing the catcher. The liquid and abrasives that were collected were passed through a filter in a dewatering process and then subsequently dried in an electric convection oven. Once dried, the particles were separated from the filter and classified by a vibratory sieve. The individual sieve pans were weighed, and the resulting weight and screen size data recorded. The filter was also weighed, and the additional weight of the trapped abrasive added to the fines. Particles smaller than the finest sieve, 63 μm, were collected and weighted together and considered 'fines' or 'pan'.

2.2 Experiments

In Fig. 3 the particle size distribution, by mass, of original garnet abrasive Barton Mines HPX80 is compared against the particle size distribution at the exit of the mixing tube. While the largest portion (35%) of the original abrasive was classified at 250 μm, the accelerated garnet showed a rather flat distribution with a maximum of 18% at 150 μm. Notably, approximately 18% by weight portion of fines were discovered. Due to the fines being smaller than 50 μm this 18% portion will most likely not result in a significant erosion on the workpiece.

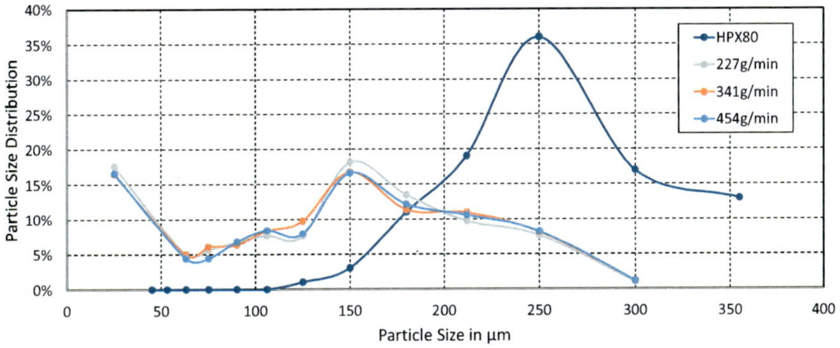

Fig. 3. Particle size distribution at different abrasive feed rates (pressure 345 MPa).

Also, it is very notable that changing the abrasive feed rate from 227 g/min to 454 g/min had no impact on the fragmentation of the particles. Changing the abrasive feed rate would increase the number of particles inside the cutting head at any given time and therefore the likelihood of particle interaction. Since this did not have a significant impact one can conclude that particle to particle interaction is not a significant factor in the fragmentation process.

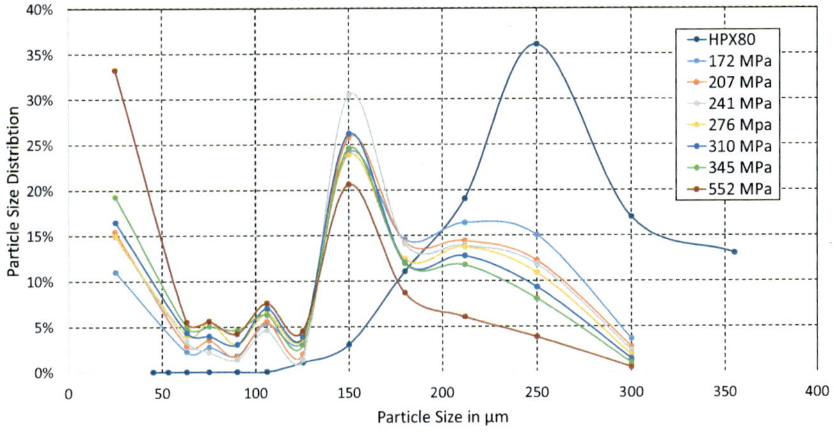

Fig. 4. Particle size distribution at different pressures with different hydraulic power (Orifice 250 μm mixing tube 760 μm, abrasive 340 g/min)

A second set of experiments was conducted where the size of the orifice and mixing tube were kept constant while changing the operating pressure from 186 MPa to 553 MPa. This results in a significant increase in hydraulic power with higher pressure. As can be seen in Fig. 4 the mass portion of fines significantly increased with higher pressures while the portion of large particles decreased with pressure.

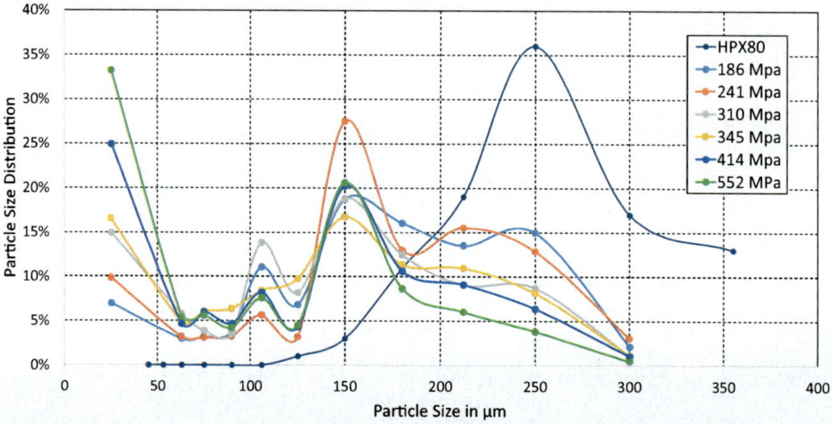

Fig. 5. Particle size distribution after firing at different pressures (see legend) with same hydraulic power.

In the third set of experiments the pressure was varied from 186 MPa to 553 MPa. But this time the orifice size was varied to keep the hydraulic power the same. Even though the hydraulic conditions are very different from the previous set, the result is fairly similar: The mass portion of fines increases with higher pressures and the mass portion of large particles decreases with higher pressure (Fig. 5).

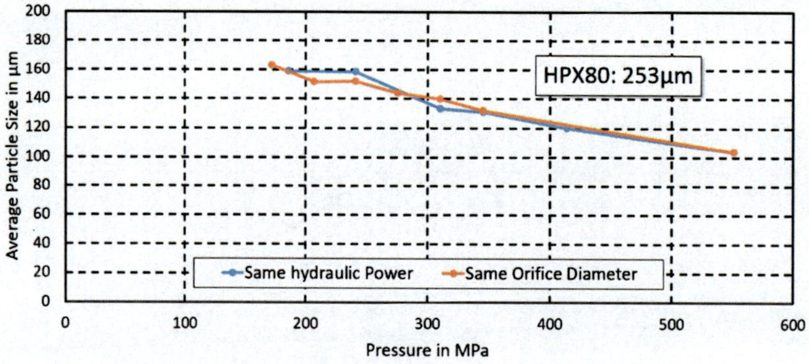

Fig. 6. Average particle size after firing at different pressures.

One way of comparing different fragmentation results is to calculate a weighted average particle size. The original garnet that was used in this test had an average particle size of 253 µm. In Fig. 6 the weighted average of the fragmented particle sizes is shown for both sets of experiments with very different conditions. One with constant hydraulic power with varying orifice size and one with constant orifice and therefore varying hydraulic power. It is very notable that both experiments resulted in a similar fragmentation behavior for a given pressure. It can therefore be noted that the average particle size and therefore the fragmentation is mainly affected by operating pressure.

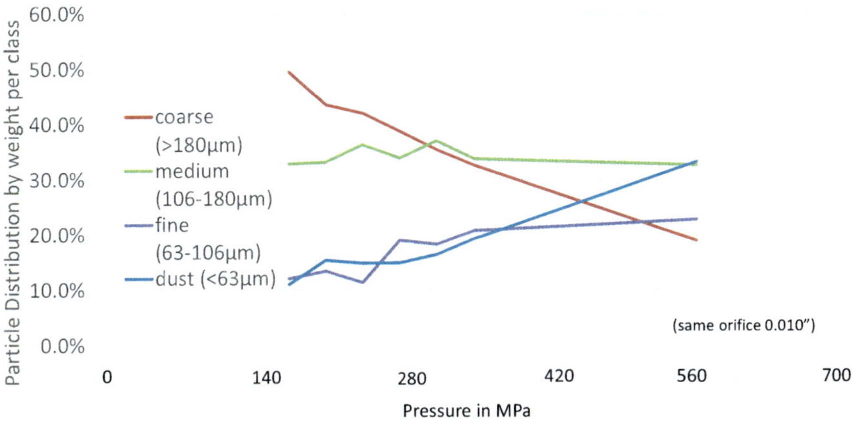

Fig. 7. Particle size distribution after firing at different pressures with same hydraulic power.

In Fig. 7 the particles are classified in broader classes: coarse (>180 µm), medium (106–180 µm), fine (63–106 µm), and dust (<63 µm). This representation clearly shows the effect of pressure on size distribution. The portion of coarse particles is significantly reduced with higher pressures from 50% down to 20% while the portion of dust increases significantly from 10% to over 30%.

3 Kinetic Power of Particles

3.1 Varying Hydraulic Power

While it is an important first step to analyze the speed and size of the particles that are hitting the workpiece and eventually perform the cutting operation, the true measure for cutting performance is the kinetic power of the particles. Each particle obtains a certain amount of kinetic energy.

$$E_{P,kin} = \frac{1}{2} m_a v_a^2 \tag{1}$$

The overall kinetic power, $P_{P,kin}$, of the cutting jet can be represented as a function of the particle's mass flow rate, $\dot{m}_a$, and its velocity

$$P_{P,kin} = \frac{1}{2} \dot{m}_a v_a^2 \tag{2}$$

In Fig. 8 the power of each particle size portion is displayed for different conditions. All tests were performed with the same orifice diameter (250 µm) and the same abrasive feed rate (340 g/min) while the pressure was varied from 186 MPa to 553 MPa. This means that the available hydraulic power changed from 3 kW to 19 kW. It is therefore no surprise that the test with the highest hydraulic power also showed the highest kinetic power. All different tests showed a peak kinetic power at the particle size cluster of 150 µm. It is very notable, though, that the portion of fines, that do not have a significant

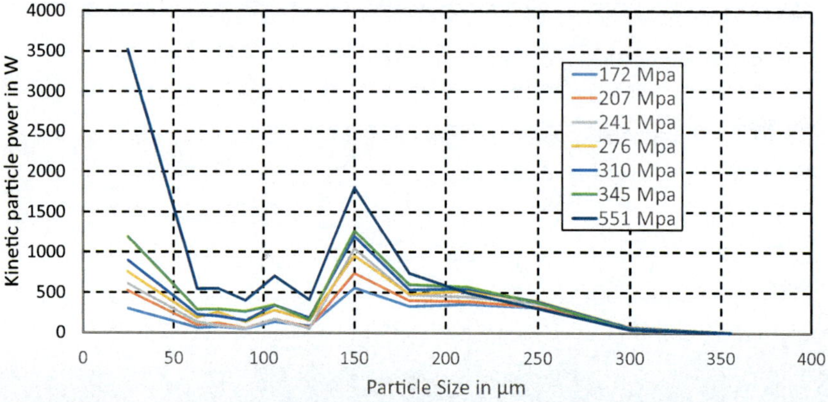

Fig. 8. Total kinetic power at different pressures (see legend) with the same orifice (250 µm) and the same abrasive feed rate (340 g/min).

impact on erosion increased with pressure and proved to be a major portion of the overall kinetic power especially at higher pressures.

By comparing the hydraulic power of the jet with the total kinetic power of effective particles it becomes clear that the kinetic efficiency (hydraulic power divided by the kinetic power) is linearly decreasing with higher pressures (Fig. 9). At 172 MPa the kinetic efficiency is about 60% whereas at 553 MPa the efficiency has dropped down to 30%. This will have a significant impact on cutting performance of the abrasive waterjet and might explain the behavior that was experimentally shown in Fig. 1 where higher pressures actually were seen to have a detrimental effect on cutting performance.

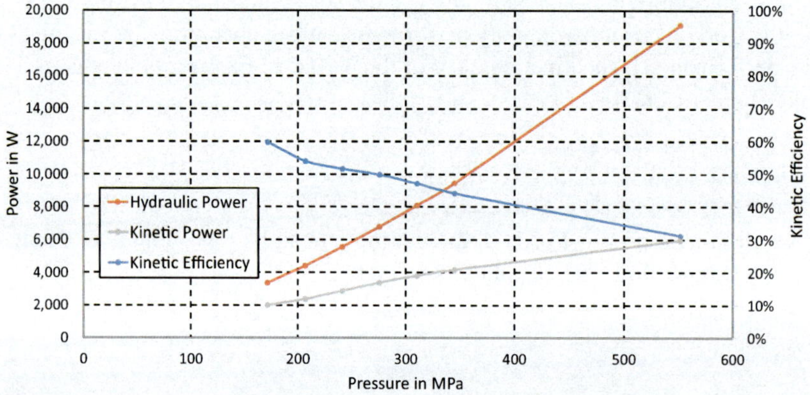

Fig. 9. Efficiency of converting hydraulic power into kinetic particle power (same orifice).

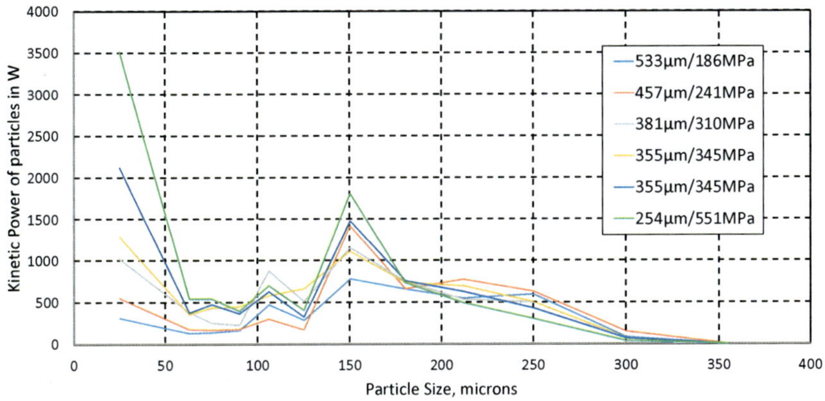

Fig. 10. Kinetic power of particles at different pressures, same hydraulic power (18 kW) and same abrasive feed rate (340 g/min) (orifice diameter and pressure as indicated in legend).

3.2 Constant Hydraulic Power

The next set of experiments were performed with the same hydraulic power while changing pressure. This was accomplished by using different orifice sizes to maintain a constant hydraulic power. In the first batch of tests the same abrasive feed rate (340 g/min) was used with all experiments. This means that as the hydraulic pressure increases the resulting water flowrate decreases, and the abrasive load ratio (abrasive mass flow rate divided by the waterjet mass flow rate) is increases. In Fig. 10 the kinetic power is displayed for each portion of fragmented particle sizes. As can be seen in Fig. 10 the distribution of kinetic power shows a general trend of increasing as the abrasive load ratio increases. But also here as in the previous test the amount of kinetic power within the fines is significantly increasing with higher pressures.

In the typical application of abrasive waterjet nozzle configuration's using a low water flow rate typically uses a lower abrasive feed rate. There are many practical reasons for this setting with the most important being that the capability of reliably delivering particles into the cutting head depends on the amount of air delivered through the venturi process for that orifice. In Fig. 11 the experiments were carried out with an abrasive load ratio of 11%. This means that the abrasive feed rate is larger at lower pressures due to the larger water flowrate at the same pressure. With those conditions it becomes clear that the experiments with the higher pressures show lower kinetic power of different portions of particle sizes. Also, as we have seen before, the amount of power that goes into fines, which are not effective in the cutting process, is the largest at higher pressures.

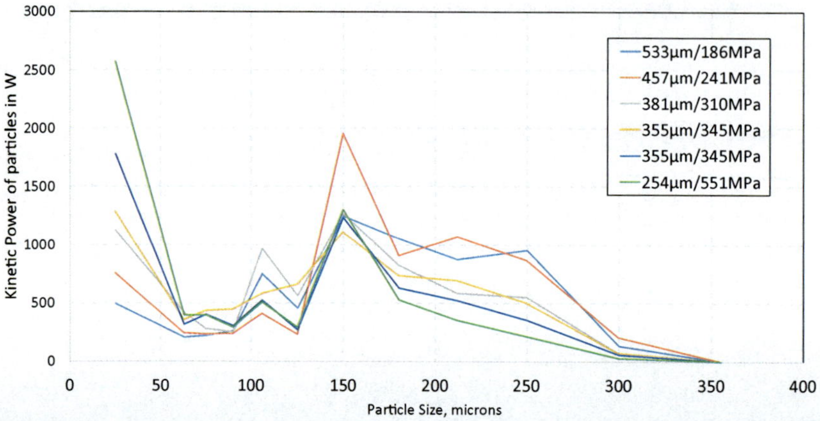

Fig. 11. Kinetic power of particles at different pressures, same hydraulic power (18 kW) and same abrasive load ratio (11%).

Figure 12 shows the total abrasive kinetic power for the tests that were carried out using the same hydraulic power, where in one case the abrasive flow rate was held constant (344 g/min), and the other case the abrasive loading was held constant (11%). Two distinct trends can be seen. First, the constant abrasive feed rate shows the total kinetic power increases at the lower pressures but then appears to find a maximum level after around 400 MPa where there is no significant change at higher pressures. Second, when the abrasive load ratio was kept constant the trend clearly shows a decrease in total abrasive kinetic power. The main reasons here are probably an insufficient supply of particles that could be accelerated and the large portion of fines that do not have a significant impact in the erosion process.

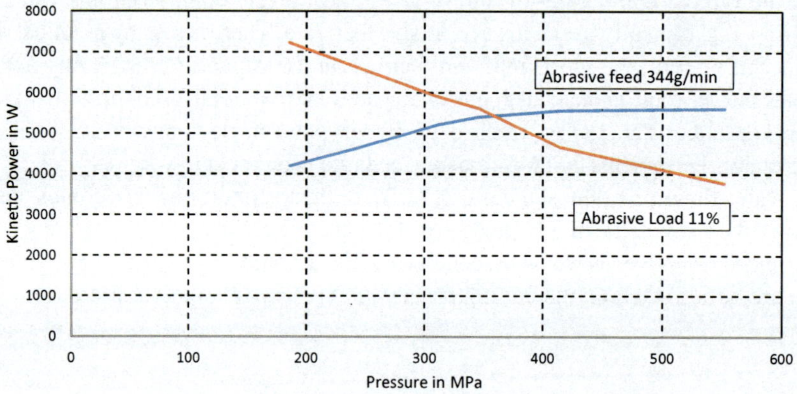

Fig. 12. Total kinetic power of particles over pressure at different abrasive feed rates.

4 Conclusions

Abrasive waterjet cutting has found large application in different industrial applications. As its usage widens the abrasive waterjet can be applied in areas that traditionally have been occupied by competing cutting methods. It is therefore of essential importance to understand its potential and opportunities for optimizing the cutting process. Over the years since its inception in the 1970s, innovations in pumping technology have not only increased reliability but also hydraulic power and pressure. In pure waterjets higher pressures have shown some benefits that users have not been able to match in cutting with abrasive waterjets. Experimental data actually showed minimal effect and even detrimental effects of higher pressures. By focusing on the actual particles, this paper tries to analyze the reason why nominally faster water jets do not translate in higher abrasive cutting performance.

In this paper we have been analyzing the behavior of abrasive particles during the acceleration process in the cutting head. For this we have looked at different conditions where either the hydraulic power of the process was kept constant or the orifice size was kept constant, which results in changing hydraulic power. Also, the effect of different abrasive feed rate conditions were taken into account.

In the first section the size distribution of particles exiting the mixing tube was analyzed. These are the particles that impact on the workpiece to perform the cutting operation. It was shown that pressure has the most significant effect on the particle distribution/fragmentation. The portion of larger particles decreases significantly with higher pressures and the amount of fines that have no significant impact in the cutting process increases substantially. Even though we were applying very different operating conditions, the average particle size showed the same behavior of linearly decreasing with increasing pressure. Massive fragmentation of particles into small ineffective fines would be a first indication why the abrasive waterjet is not as effective at higher pressures.

The second section then focused on the kinetic power of the abrasive water jet. Again it showed that the higher rate of fragmentation at higher pressures limited the capability of the abrasive jet to perform the cutting process. Only increasing the abrasive load at higher pressures helped to overcome the losses caused by massive fragmentation to maintain and eventually increase the total kinetic power. This is limited by the ability to feed the abrasives into the cutting head, though. With the same abrasive load, the kinetic power was significantly reduced at higher pressures.

In this paper the kinetic behavior and fragmentation of particles has been analyzed in a wide range of different conditions. With new insights into fragmentation of the particles especially at higher pressures we were able to explain the behavior of experimental cutting studies where higher pressures did not provide a significant improvement and in many cases a decrease of cutting performance. Studies like this will not only help us understand the very processes that are involved in the cutting operation but also help us to optimize the overall process to increase performance and reliability and also be competitive against other cutting methods while providing ample new applications in the future.

References

1. Henning, A.: Modellierung der Schnittgeometrie beim Schneiden mit dem Wasserabrasivstrahl, Dissertation, Universitaet Stuttgart. ISBN 978-3-939890-28-7, Jost Jetter Verlag, Heinsheim
2. Henning, A., Miles, P.: Efficiency drives cutting performance. In: Proceedings of the 22nd International Conference on Water Jetting, BHR Group, Seattle, USA, pp. 16–18, November 2016
3. Westkämper, E., Henning, A., Radons, G., Friedrich, R., Ditzinger, T.: Cutting edge quality through process modeling of the abrasive waterjet. In: Teti, R. (Hrsg.): CIRP; Proceedings of 2nd CIRP International, Seminar, 21–23 June 2000, pp. 179–188 (2000)
4. Henning, A., Liu, H.T., Olsen, C.: Economic and technical efficiency of high performance abrasive waterjet cutting. In: Proceedings of the 18th International Conference on Pressure Vessels, PVP2010, ASME, 18–22 July, Bellevue, Washington (2010)
5. Henning, A., Miles, P.: Effect of pressure on abrasive waterjet cutting performance. In: 22nd International Conference on Water Jetting, BHR Group, Haarlem, The Netherlands, 3–5 September 2014
6. Miles, P., Henning, A.: Abrasive kinetic cutting power: a function of power conversion. In: Proceedings of the 2015 WJTA-IMCA Conference and Expo, 2–4 November, New Orleans, Louisiana (2015)
7. https://www.barton.com/wp-content/uploads/2012/02/HPX_PSD_Table.pdf
8. Ohlsen, J.: Recycling von Feststoffen beim Wasserabrasivestrahlschneiden. Fortschr. Ber. VDI Reihe 15 Nr. 175 Düsseldorf: VDI Verlag (1997)
9. Zaremba, D., Heese, P., Bauer, M., Maier, H.J., Hassel, T.: Particle: disintegration in the abrasive water injection jet. In: Proceedings of the 2015 WJTA-IMCA Conference and Expo, 2–4 November, New Orleans, Louisiana (2015)

Measurement of the Effective Waterjet Diameter by Means of Force Signals

Mikhail Kliuev[1,2], Frank Pude[1(✉)], Josef Stirnimann[1], and Konrad Wegener[1,3]

[1] inspire AG, ETH Zurich, Technoparkstrasse 1, 8005 Zurich, Switzerland
pude@inspire.ethz.ch

[2] Moscow State Technological University STANKIN, Vadkovsky per. 1, Moscow 127994, Russia

[3] Institute of Machine Tools and Manufacturing, ETH Zurich, Leonhardstrasse 21, 8092 Zurich, Switzerland

Abstract. The application of both pure and abrasive waterjets is well established in manufacturing and related processes like surface preparation. Particle free surface processing in order to avoid any particle residuals on the machined workpiece surfaces was addressed as core requirement by quality engineers who had to fight against induced failure of functionalized surfaces after surface preparation by e.g. sand blasting.

Due to the need of multi-path processing for generating surfaces it is necessary to monitor the effective geometry of the impact zone of the waterjet and the related impact forces. By getting these data it becomes possible to calculate the necessary grade of overlapping of the neighbouring traces. This becomes more ambitious by permanently changing waterjet shapes caused by very little and not preventable changes of the micro structure inside the used nozzle. A nozzle change mostly induces a change of the jet shape and related geometry of the impact zone on the hit surface.

In this paper an apparatus and method for waterjet shape measurement are presented. By using this method it is possible to generate necessary information on nozzle wear related jet shape changes or new jet shapes after nozzle change which enables the operator to recalibrate the process parameters. The effect of machining velocity on waterjet characteristics is studied. Also the capability of sorting of not acceptable nozzles becomes possible and can avoid damage of surfaces to be processed.

Keywords: Production engineering · Waterjet · Jet geometry · Force signals

1 Introduction

The application of both pure and abrasive waterjets is well established in manufacturing and related processes like surface preparation. The pure waterjet type is mostly used for cutting of soft material like foamed plastic or paper webs. Quite recently in [1] particle free surface processing in order to avoid any residuals on the machined workpiece surface

D. Klichová et al. (Eds.): Water Jet 2019, LNME, pp. 15–27, 2021.
https://doi.org/10.1007/978-3-030-53491-2_2

was addressed as core requirement. This came up especially from quality engineers who had to fight against induced failure of coatings after surface preparation by e.g. sand blasting.

The potential of pure waterjets to generate different grades of roughness on the hit surfaces was shown by various research groups round the world. The paper [2] presented the utilisation of pure waterjets for roughening of Titanium surfaces.

Due to the need of multi-path processing for generating surfaces it is necessary to monitor the effective geometry of the impact zone of the waterjet and the related impact forces. These parameters depend on the distance of individually generated droplets to the jet axis. Only this way it is possible to get information about the penetration depth and also the path-width of the processed trace. By getting this data it becomes possible to calculate the necessary grade of overlapping of the neighbouring traces. This becomes more ambitious by permanently changing waterjet shapes caused by very little and not preventable changes of the micro structure of the effective tearing edge at the point of jet formation inside the used nozzle. A nozzle change automatically induces a change of the jet shape and related geometry of the impact zone on the hit surface.

It was shown by several research groups that the waterjet diameter can be measured by using different technological approaches. The application of an optical measurement system which uses images made by a photo camera was introduced by [8]. For the measurement of abrasive waterjet diameters it is necessary to understand the wear mechanism of the focusing tube [5, 6]. The measurement principles which were applied in this case using optical principles. The experimental set-up was equipped with a photo camera [4] or a laser transmitter/receiver system [5].

In this paper an apparatus and method for waterjet shape measurement is presented. By using this method it will be possible to generate necessary information on new jet shape after nozzle change which enables the operator to recalibrate the process parameters. Also the capability of sorting of not acceptable nozzles becomes possible and can avoid damage of surfaces.

2 Experimental Setup

The principle configuration of the measuring device is shown in Fig. 1. A steel plate is mounted on a type 9327C $F_x = 7.694$pC/N, $F_y = 7.672$pC/N, $F_z = 3.763$pC/N Kistler piezoelectric sensor. V_f is the feed rate of the jet. While the waterjet moves over the steel plate force signals are detected by the piezo electrical sensor. Figure 2 shows a typical shape of the kind of signal detected. X_E is the entrance distance; X_O is the jet output distance and X_{FC} the distance during which the whole jet cross-section of the jet is within the workpiece. S is the width of the plate plus twice the radius of the waterjet.

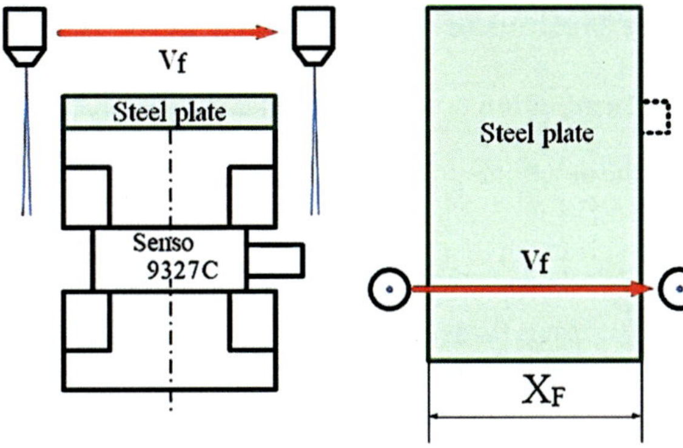

Fig. 1. Measurements setup.

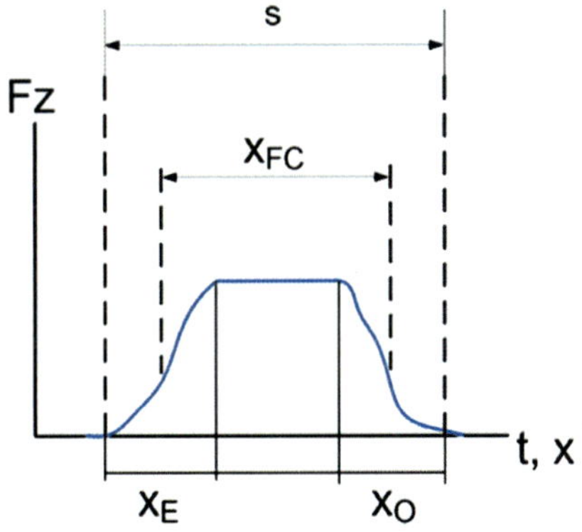

Fig. 2. The principles of jet force measurements.

According to Fig. 1 and Fig. 2 the jet diameter can be presented as a multiplication of the traverse velocity and the time which the jet requires to enter or to leave the plate which is written in:

$$D_{WJ} = S - X_{FC} = X_E = X_O \tag{1}$$

Due to the complexity of jet behaviour the following parameters are changed for the purpose of understanding the range of process stability and conditions of profile measurement. The stand-off distance between the waterjet nozzle and the sensor plate was varied in the interval between 40 mm and 80 mm. The feed velocity of the jet V_f was 100/200/400/800/1000 mm/min. The water pressure was selected in a range from 100 MPa to 300 MPa with a step of 100 MPa.

Because the shape of the jet cross-section may deviate from perfect circularity a single measurement does not provide sufficient information to determine the diameter. Eight separate measurements of the diameter of the jet cross-section are made, in steps of 45° according to the direction of the feed rate vector, and each measurement consists of entrance and output values so as to provide more accurate results.

3 Measurement Method of the Force Signals

3.1 Correlated Determination of the Jet Diameter

According to [3] for the determination of the jet diameter the "Three Lines Method" can be used, as it is shown in Fig. 3. The algorithm linearizes of the slope of the response curve from the force measurement as third line. Then two horizontal lines give the extrapolation of the plateaus representing constant values before and after the full cross-section of the waterjet impacts the plate. The "Three Lines Method" is able to determine the jet diameter using the projection dc of the third line segment on the x-axis. The line Fn is parallel to the x-axis and sometimes coincides with it. The second line Fn2 is fitted to the mean value of force after the jet enters the plate. The third line Fn3 is found by two points A1 and A2 which are located on the line. Pointes A1 and A2 are defined in the force signal curve.

The method is used to locate the entrance point (the point at which the jet first impacts the sensor plate) and the exit point. Knowing the duration of the full impact time and the feed rate of the machine, it is possible to calculate the full diameter (Eq. 1) of the jet with core, fog, and droplet zones.

It is also possible to exclude the fog zone from the force signal curve by high-pass filtering, since it is present as noise. Separating the droplet zone can be much more complicated, as the droplets sometimes have sufficient energy to exceed the threshold beyond which they actually remove material from the surface of the steel plate. A threshold can be determined experimentally, below which the jet lacks of pressure to remove material from the surface.

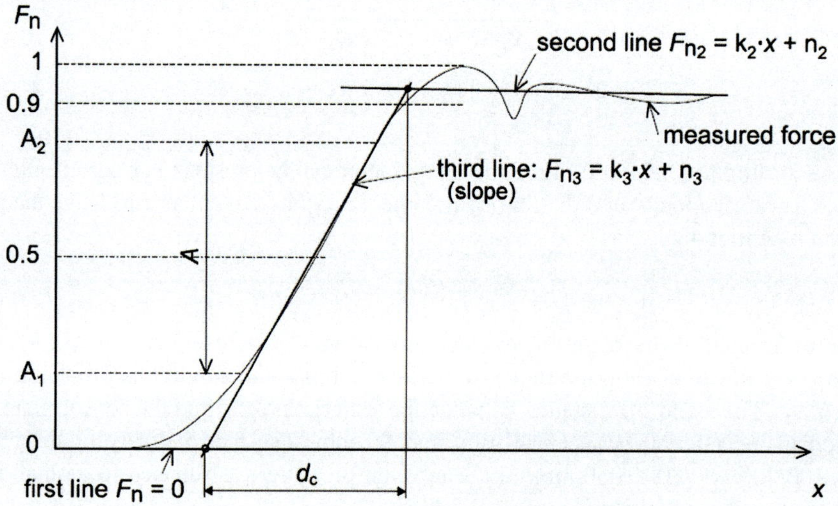

Fig. 3. The Three Lines Method, according to [3].

3.2 Evaluation of the Jet's Pressure

The idea of pressure measurements is shown in Fig. 4. The measurement is provided to define the window dt where dF(t) will increase significantly. The precision and stability of this method depends on the dt interval. It is possible to determine the jet's force intensity using the intensity method, but the "Three Lines Method" shows greater reliability.

It is possible to determine the intensity of jet profile by using the intervals in Fig. 3 and dividing the dF(t) function by dt. This method will be used for the visualization of the jet power distribution in the next chapter. If in the next interval the average of dF(t) is much higher (in current work it is 30%) than in the previous interval and in the last intervals the average value of dF(t) is increasing, then the force signal is significantly increasing and the diameter of the jet can be detected from the beginning to the end of the force signal increase.

This type of signal is activated when the sensor plate is far enough from nozzle, that means that the core zone is narrow and the surrounding droplet zone therefore increases in size.

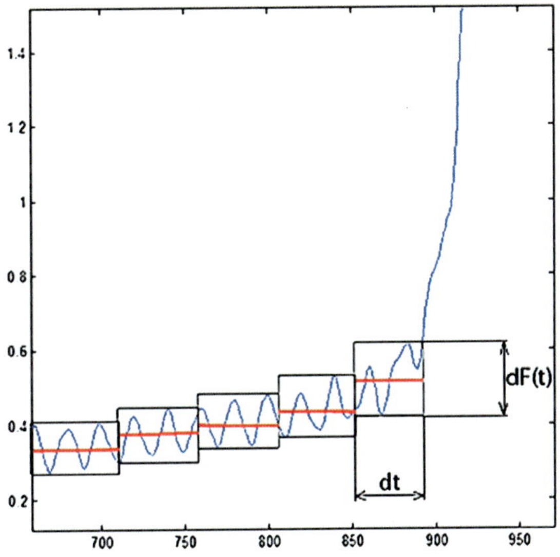

Fig. 4. Jet entry point search graph.

The circular shape of the jet is detected in experiments, and then the passage of the jet through the sensor plate is shown in Fig. 5.

The jet pressure distribution can be found by the force signal transformation as

$$F(x) = \int_{x}^{R} \int_{-Y(x)}^{Y(x)} P(\xi, \eta) d\xi d\eta \tag{2}$$

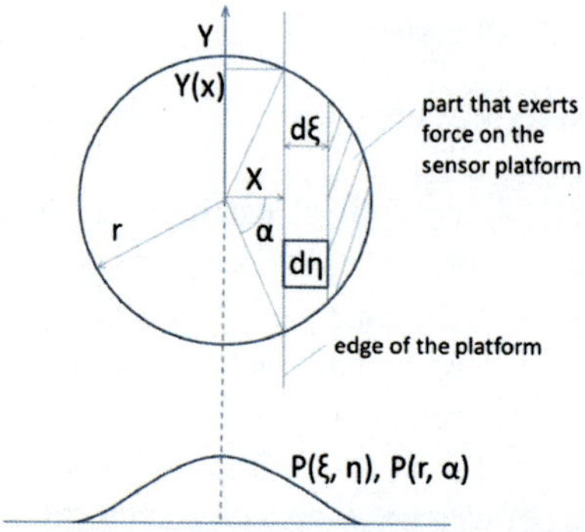

Fig. 5. Jet intensity transformation from the force sensor signal.

or in the polar coordinates

$$F(\mathrm{x}) = \int_{x}^{R} P(r) \cdot 2r\mu(\mathrm{r})\mathrm{dr} \tag{3}$$

The graph in Fig. 6 shows the intensity profile. From the graph can be seen that the main part of force growth is in the small period of time, in other words the fog zone is larger than droplet and core zones. The intensity profile is found as interval on force graph by "Three lines method" [1] between entrance and exit points within the transformation in Eq. 3.

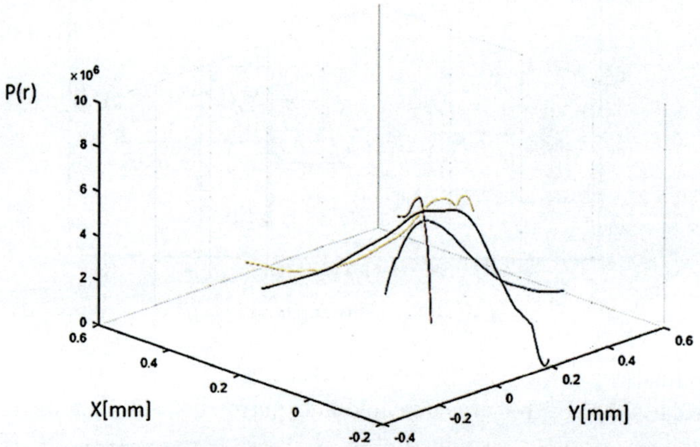

Fig. 6. Jet intensity graph.

3.3 Evaluation of the Ablation Threshold Determined by Force Signals

The ablation threshold is shown in Fig. 7. The idea of ablation threshold is based on the physical properties of the machined material. Two zones are shown on Fig. 8. The core zone and a part of the droplet zone of the jet have enough intensity to remove material from workpiece surface, whereas the other part of droplet zone is too weak to produce any significant impact on the surface. The ablating threshold can therefore only be defined by experiment.

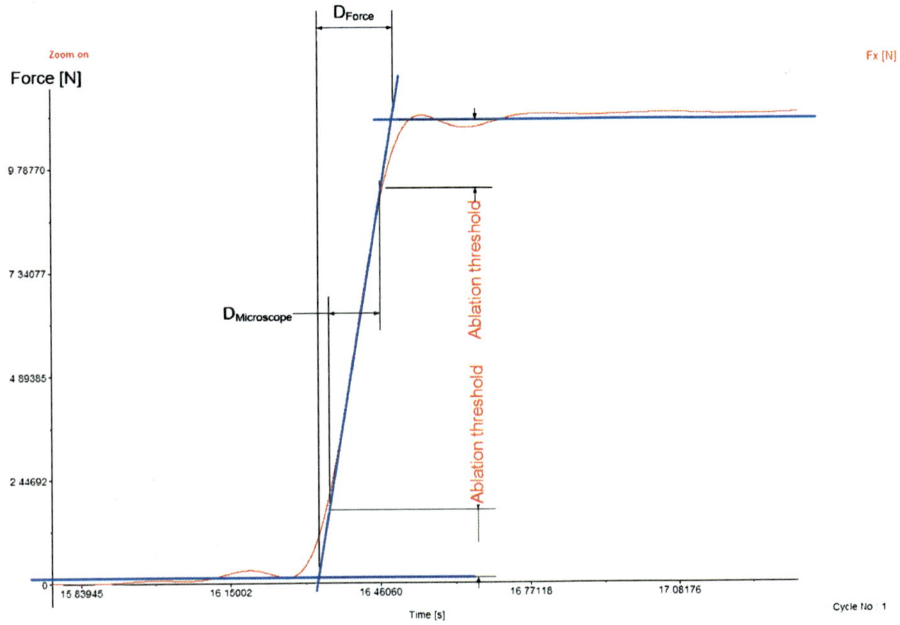

Fig. 7. Ablation threshold applied by the "Three Lines Method".

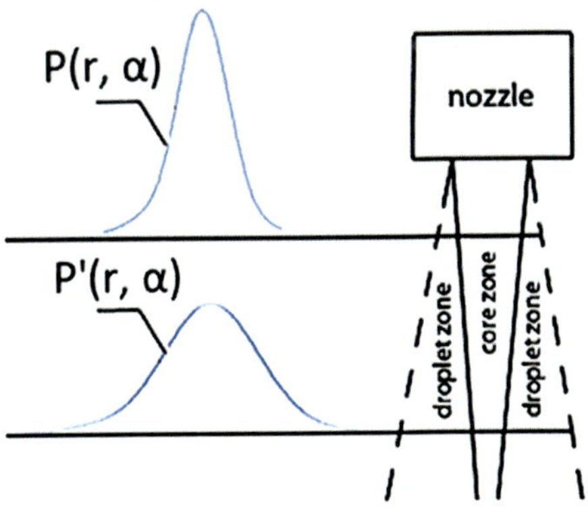

Fig. 8. Effect of stand-off distance on waterjet characteristics

Conventionally the jet is separated out 3 zones:

- Fog zone (outer ring zone): In this zone the intensity is too low to leave any geometrical traces on the surface of Titanium as observed by confocal microscope and magnification of 20X.
- Droplet zone: This middle ring zone is made up of drops detached from the main jet. The droplets have sufficient intensity and damage the sample around the zone of direct processing.
- Core zone: the zone of waterjet processing, where the waterjet has enough intensity to substantially remove material. The ablating threshold can be found from the comparison between force signal and impact profile by observing via microscope. The threshold values should be added on the top and the bottom of the force signal, because of the jet energy distribution.

4 Force and Microscope Measurements and Evaluation

4.1 Verification of the Waterjet Characterization Method by Microscope

Microscope measurements are applied to compare them with force signal measurements. The measurements were made by using confocal microscope Leica DCM-3D and machined material was titanium. The profile of the jet path is presented in Fig. 9. 1.4 mm of the waterjet path is measured on the material; the average profile of the path is in Fig. 9.

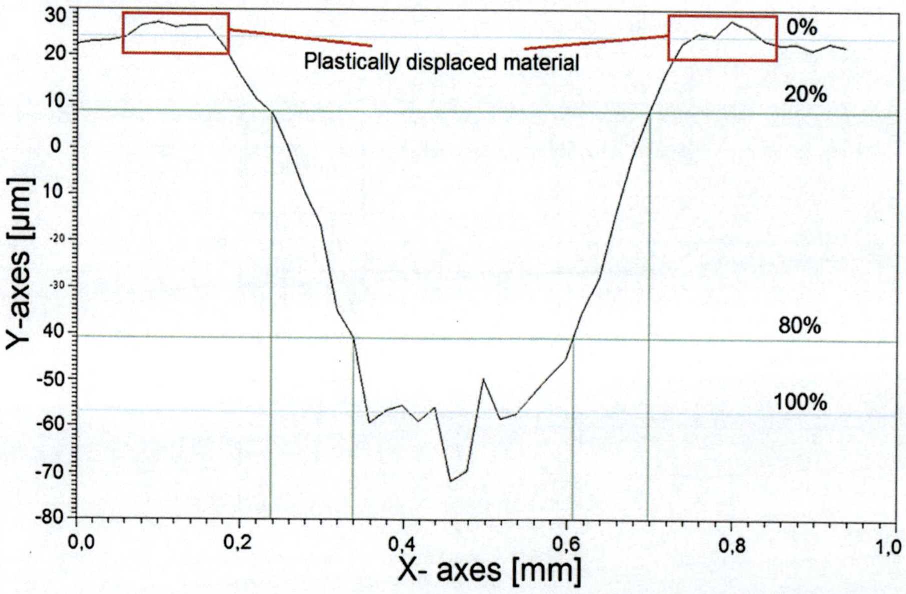

Fig. 9. Microscope measurements of the average path of waterjet on the titanium workpiece.

Because of complex geometry of the profile, the diameter of the jet was calculated in Eq. 3 as a mean value of width in 20% and 80% of depth. Values 20% and 80% were found through experiment and employed to avoid uncertainty of diameter on the top and on the bottom. The threshold in microscope measurements is used to separate the droplet zone of the jet from the core zone and avoid uncertainty of the profile the bottom and in the plastically displaced area. The same values are used for ablation threshold in force measurement.

$$D_{mean} = \frac{D_{20\%} + D_{80\%}}{2} \tag{4}$$

The microscope measurements are made in respect to the force measurements and to provide the reliability of results.

4.2 Jet Shape Determination and Evaluation of Results

The numerical information about microscope and force measurement comparison are presented in Tables 1 and 2. By using the value of standard deviation σ the precision of two methods is shown. Microscope measurements were made in 1.4 mm of jet path and the average value of it (as it is described in Sect. 4.1) is presented in Table 1 and performed using the same plate as for the force measurements. All experiments are made in four directions to get a complete intensity profile and detect the jet beam's deviation from rotational symmetry.

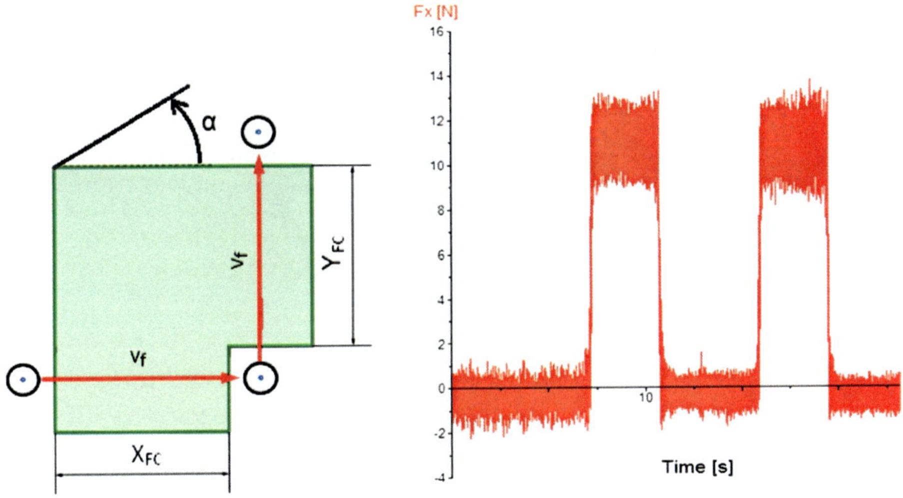

Fig. 10. The jet path and graph of force against time.

To show a 3-dimensional shape of the waterjet, the method explained in chapter 3.2 is applied to measurements, done in four different directions, i.e. 0°, 45°, 90° and 135°. The signal and experiment set up are shown in Fig. 10. Angles of 90° and 135° are reached by the direction Y of the jet, α is angle of the plate. It is applied to reach directions 45°, 135°.

Table 1. Microscope measurements.

α	Dx [um]	б x [um]	α	Dy [um]	б y [um]
α = 0°	1406.9	50.646	α = 0°	1512.3	81.934
α = 450°	1472.2	89.266	α = 45°	1462.5	13.370

Table 2. Force measurements.

α	Dx [um]	б x [um]	α	Dy [um]	б y [um]
α = 0°	1369.08	6.2401	α = 0°	1334.27	27.301
α = 45°	1357.30	25.743	α = 45°	1329.19	47.881

Results from Tables 1 and 2 are represented as a circularity of jet with standard deviation in Fig. 11. Big standard deviation in microscope measurements happens because of non-uniformity of jet during measurements. Instable impact from droplet zone energy can be the reason of such non-uniformity.

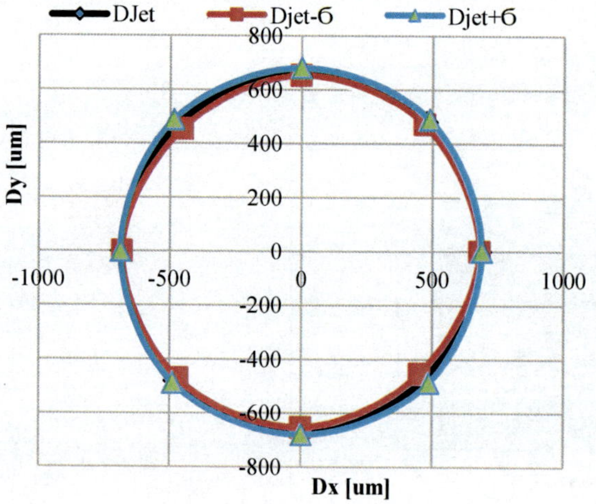

Fig. 11. Jet circularity.

5 Discussion of the Results

5.1 Force and Microscope Methods in Comparison

After force measurements and geometrical footprint identification it is possible to compare results the results are shown in Fig. 12. The microscope measurements were applied

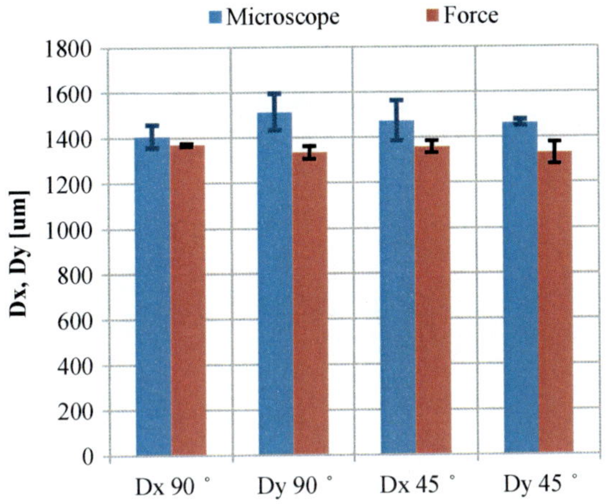

Fig. 12. Force measurements and geometrical footprint identification comparison.

after force measurements on the same plate. The machining is always performed in the different place to avoid the edge blunting.

Figure 12 shows the diameter defined by method of force measurements and method of microscope observations. From the diagram it can be seen that the value, measured by microscope observations is always higher than the diameter measured by force. This phenomenon was explained before and suggests the effect from stochastic nature of droplet zone force distribution.

5.2 Force and Microscope Methods in Comparison

The graphs in Fig. 13 are demonstrating the effect of velocity to the jet entrance time and diameter measurement. The measured diameter is increase with increase of traverse velosity and the jet entrance time remains the same after 300 mm/min.

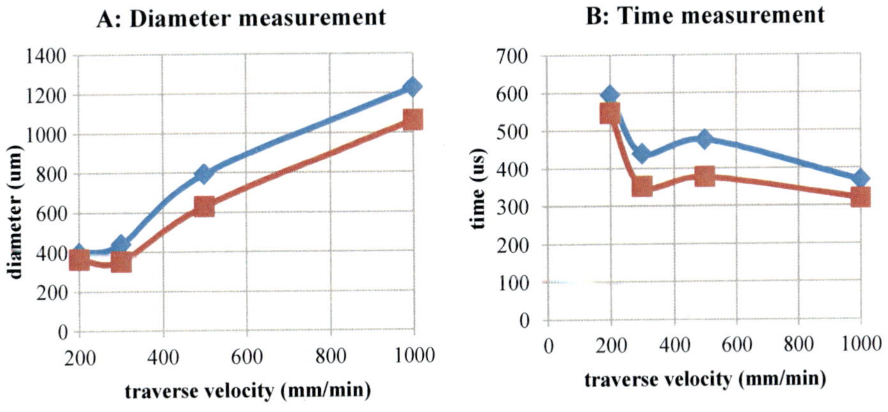

Fig. 13. Effect of traverse velocity to the jet diameter (A) and jet entrance time (B) detected by force sensor. The blue colour curve is X-direction and the red colour curve is Y-direction.

The probable cause of waterjet diameter extension using the high velocity measurements is the inertia of droplets around the waterjet. This phenomenon which is shown in Fig. 14 has a big influence on the process, especially at high traverse velocities (from 400 mm/min to 1000 mm/min).

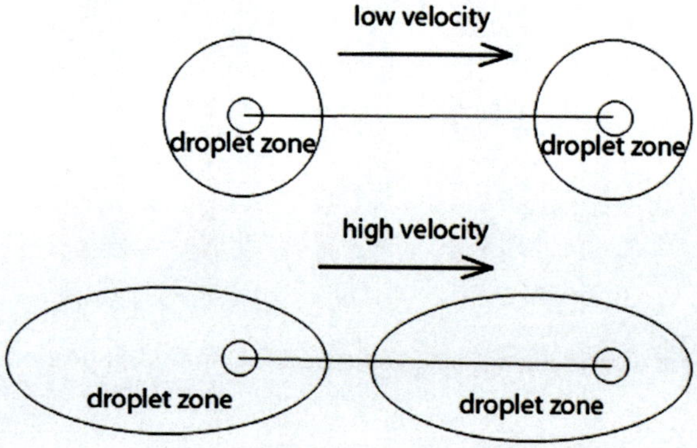

Fig. 14. Effect of traverse velocity.

6 Conclusions

The used force sensor generated very stable data compared to other e.g. optical measurement systems [7]. In case of using an optical systems the real jet shape can be hidden behind a fog zone. The used acoustic emission systems cannot provide any information about the jet structure and the related forces and is not useful for obtaining information about the expected profile of the processed surface.

The obtained results show the usability of the performed measurement process in general. Using a sensor system of this type allows to equip water jet machines with a quality control module which enables the operator to ensure steady jet shape conditions or to change the nozzle in case of any unwanted variance. The applicability under laboratory conditions was proven and this prototype system can now be used to collect data in order to aliment a process model for the prediction of manufacturing results. This prediction model needs to be capable of calculating overlap and feed rate accordingly.

Next step of applications of this sensor technology is the investigation of different nozzle geometries used at different jetting parameter combinations in order to collect empirical data input for mathematical process models. Also the identification of ablation thresholds of waterjet beams is possible with this kind of measuring platform.

Afterwards the obtained database will be the starting point for programming of a technology post processing unit for CAM systems.

References

1. Axinte, D., Kong, M.: An integrated monitoring method to supervise waterjet machining. CIRP Ann. **58**, 303–306 (2009)

2. Chillman, A., Ramulu, M., Hashish, M.: Waterjet and water-air jet surface processing of a titanium alloy: a parametric evaluation. J. Manuf. Sci. Eng. **132**, 011012 (2010)
3. Orbanić, H., Junkar, M., Bajsić, I., Lebar, A.: An instrument for measuring abrasive water jet diameter. Int. J. Mach. Tools Manuf **49**, 843–849 (2009)
4. Perianu, I., Ionescu, D., Verbitchi, V.: Method of measuring the abrasive-water jet diameter, for the cutting process control. J. Weld. Mater. Test. **1**, 6–10 (2017)
5. Prijatelj, M., Jerman, M., Orbanić, H., Sabotin, I., Valentinčič, J., Lebar, A.: Determining focusing nozzle wear by measuring AWJ diameter. Strojniški vestnik – J. Mech. Eng. **63**(10), 597–605 (2017)
6. Nedić, B., Baralić, J.: The wear of the focusing tube and the cut-surface quality. J. Tribol. Ind. **32**(2), 38–43 (2010)
7. Folkes, J., Li, Z.J.: Innovative method of waterjet jet diameter measurement for process and quality control. In: Proceedings of the 20th International Conference on Water Jetting, 20–22 October, Graz, Austria (2010)
8. Jerman, M., Lebar, A., Junkar, M.: Measurement and modelling of the water jet shape. In: Proceedings of the MIT & SLIM 2011 conference, 25–27 September, 2011, Fiesa, Slovenia (2011)

Effect of Ventilation to Abrasive Suspension Jet Under Submerged Condition

Guoyi Peng[1(✉)], Yukiteru Tamura[1], Yasuyuki Oguma[1], and Jiri Klich[1,2]

[1] College of Engineering, Nihon University, Koriyama, Fukushima, Japan
peng@mech.ce.nihon-u.ac.jp
[2] Institute of Geonics of the Czech Academy of Sciences, Ostrava, Czechia

Abstract. High-speed water jet shows a peculiar processing property in cutting submerged objects, but its processing ability decreases quickly with the standoff distance under submerged conditions. For improving the performance of submerged cutting by abrasive suspension jet (ASJ) a ventilation nozzle system has been developed but the flow structure of ventilated jet is still unclear. Focused on the effect of ventilation, high-speed camera observations of air ventilated submerged water jets are conducted in this work and the structure of air ventilated ASJ is identified by similarity analysis. Experiments show that air-coated suspension jet may be generated by increasing the ventilation flowrate to a certain extent so that ventilation air spreads out of the sheath in the form of large cavity covering the jet. The air-coated jet pulsates periodically and the dominant frequency of air-layer pulsation decreases with increasing of the ventilation flowrate. The Strouhal number of jet pulsation is dependent upon an altered cavitation number. The performance of air-ventilated ASJ is then examined by cutting tests with specimens of aluminum alloy under different ventilation conditions. The result demonstrates that the kerf depth of air-ventilated ASJ in submerged condition reaches to almost the same level of ASJ cutting in air. The cutting ability of ASJ in submerged condition is effectively enhanced by air ventilation.

Keywords: Abrasive suspension jet · Submerged jet · Air ventilation

1 Introduction

High-speed water jets, where pressured water or fluid mixture is jetted from a small nozzle at high speed, have been developed and widely applied to many fields of industry for its peculiar processing property [1]. As a means of specially machining for various materials, abrasive water jets have attracted much attention. Compared to the conventional abrasive injection jet (AIJ), abrasive suspension jet (ASJ hereafter), which is formed by propelling a premixed abrasive suspension through a nozzle, has a much greater material removal capacity and is easy to be applied to cut underwater structures such as disposing of nuclear power plant [1, 2]. However, the processing ability of ASJs decreases quickly with the increase of standoff distance, especially in the case of submerged working condition [2, 3]. In order to extend the effective standoff of ASJ under submerged condition

D. Klichová et al. (Eds.): Water Jet 2019, LNME, pp. 28–35, 2021.
https://doi.org/10.1007/978-3-030-53491-2_3

a new sheathed nozzle unit with ventilation, where a pipe-like sheath with a ventilation hole is set to the exit of a jet nozzle, has been developed and its effectiveness has been confirmed by experiments [3]. However, the flow structure of submerged abrasive water jet issuing from the sheathed nozzle is still unclear for its complexity.

In this work, high-speed observation of air-ventilated submerged water jet is conducted and its unsteady property is investigated experimentally. However, in the case of ASJ, surrounding water of ASJ will become muddy immediately for the diffusion of abrasive particles and it is almost impossible to observe the motion of jet flows. Thus, high-speed camera observation is made on tap water jet being analogous to ASJ under the condition of flow dynamic similarity. The effect of ventilation on the behavior of submerged water jet is investigated by the method of high-speed camera shooting and image analysis. Then, the flow structure and unsteady property of ventilated ASJs is the reasoned from analogy to water jet under the same cavitation number denoting the dynamic similarity. Cutting tests by air ventilated ASJ are carried out with specimens of aluminum alloy under submerged condition, and the effect of ventilation is further validated.

2 Experiment Apparatus and Method

Figure 1 show the test rig consisted of an open type water tank for high-speed camera observation of air-ventilated water jets. Water jet is injected horizontally and the depth of water to the nozzle head is approximately set to 210 mm. During flow observation, the water surface is covered with a flat plate to prevent air being involved into through the water surface. The processing ability of submerged ASJ decreases dramatically compared to ASJ in air. Under the same working parameters, except jet surrounding conditions such as in-air and underwater, the only factor influencing the processing ability of ASJ should be the difference of jet surrounding condition. If the flow resistance in submerged condition could be reduced and ASJ can impinge on the working target at the same velocity as ASJ in air, the kerf depth of submerged ASJ may approach to that of ASJ in-air. In order to enhance the processing capability of air-ventilated ASJ, the flow behavior of air-bubbly layer around ASJ generated by ventilation need to be clarified. As known, visualization observation of submerged ASJ is difficult because of the muddiness. Thus, high-speed observation is conducted on tap water jets at similarity condition of fluid dynamics. The spreading and breaking of jet is mainly dependent on the suction pressure in the sheath. Therefore, a nozzle head with relative long sheath is adopted for water jet, which has the same air suction property as a sheathed nozzle used for ASJ.

A conical convergent nozzle followed by a straight passage, called nozzle A10 [3], is used in the experiments. The diameter d and the length of the nozzle straight passage are 1.0 mm and 9.6 mm, respectively. A pipe-like sheath with ventilation hole is equipped to the nozzle. The diameter of the sheath is taken to be $d_{sheath} = 3$ mm, but the length L_{sheath} is adjusted to be different for water jet nozzle and ASJ nozzle to obtain the same air suction property. The pressure within the sheath, P_{sheath}, is measured by a pressure hole located at 7 mm from the sheath exit. The standoff distances measured from the nozzle exit and the sheath exit are denoted X and X', respectively.

Figure 2 shows the scheme of experiment device for submerged cutting test of air-ventilated abrasive suspension jet. The pressured slurry is generated by an in-house ASJ system, which consists of a high-pressure pump, abrasive storage tank of approximately 10 L in capacity and abrasive mixture unit. The maximum working pressure of the system is 35 MPa. The detail of the ASJ system may refer to reference [3]. Abrasive material used in the cutting tests is garnet whose grain size is sorted to 125–250 μm, and approximately 75 w% of abrasive particles are in the range of 150 to 212 μm. The ASJ nozzle assembly is attached to the arm of a six axes robot (Motoman HP20) via an extension shaft. The jet impinges perpendicular to an aluminum test specimen (JIS A5032P) that is fixed under the water surface with the water tank. The depth to the nozzle exit is set approximately 100 mm a/o 700 mm. The abrasive concentration of suspension generated by the ASJ system is measured to be 23 w%. Air is supplied to the sheath via a plastic tube with an inner diameter of 6 mm. The air flowrate is adjusted by ball valve and measured by mass air flow meter.

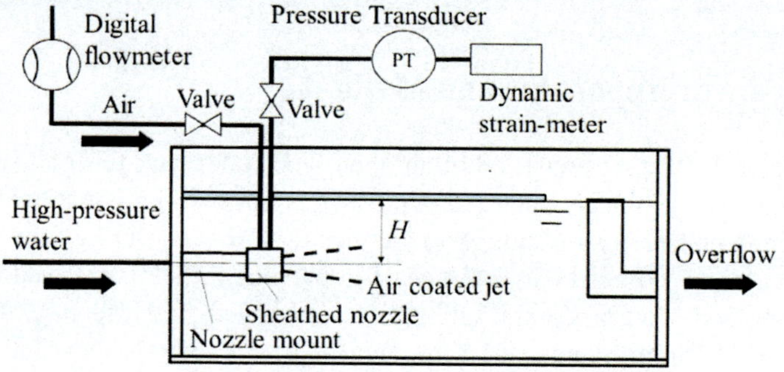

Fig. 1. Experiment device for high-speed observation of air-ventilated water jet.

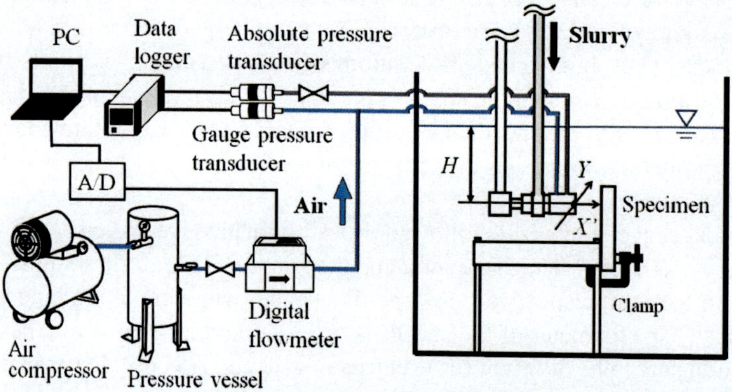

Fig. 2. Experiment setup of air ventilated abrasive suspension jet.

3 Experiment Results and Discussion

Different to the case of water jet in air, cavitation is an importance phenomenon for the case of submerged jet. Static pressure in the nozzle throat shown in Fig. 2 decreases greatly and the value of cavitation number defined by the lowest pressure may reaches to 0.001 and intensive cavitation will be caused. For the present ventilation nozzle assembly, static pressure in the sheath varies with ventilation air flowrate and the flow behavior of ventilated jet such as splitting and breaking of liquid jet is dependent on the suction pressure [4, 5] in the sheath. As a dimensionless index to demonstrate the fluid dynamic performance of the sheath, an altered cavitation number σ_c is defined by pressure in the sheath as follows [6].

$$\sigma_c = \frac{p_d - p_{sheath}}{P_i - p_d} \quad (1)$$

where P_i denotes the jet injection pressure, p_d does the surrounding pressure near the sheath exit, and p_{sheath} the pressure in the sheath measured form the pressure hole.

Figure 3 shows the variation of pressure in the sheath measured in the case of ASJ under different ventilation flowrates by solid symbol ▲ when the depth of nozzle head $H = 100$ mm. The jetting pressure of ASJ is set to 30 MPa. The length of the sheath pipe $L_{sheath}/d = 16$. It is demonstrated that p_{sheath} increases with the ventilation flowrate. In order to determine a proper operating condition of ventilation the flow structure of air-ventilated ASJ needs to be further clarified. However, visualization observation of submerged ASJ is difficult because of the muddiness. Thus, visualization observation is conducted on air ventilated water jet (WJ) under a fluid dynamic similarity condition instead of ASJ. The structure of WJ nozzle head is similar to that for ASJ but the length of sheath is extended to $L_{sheath}/d = 46$. Experiments are conducted under the same jetting pressure and water depth. The variation of pressure within the sheath for the case of ventilated WJ is shown by hollow symbol ○ in Fig. 3. As demonstrated in the figure the air suction property of both the sheaths for ASJ and WJ are almost the same. The behavior of ventilated ASJ should be in similarity to that of ventilated WJ under the same operating condition.

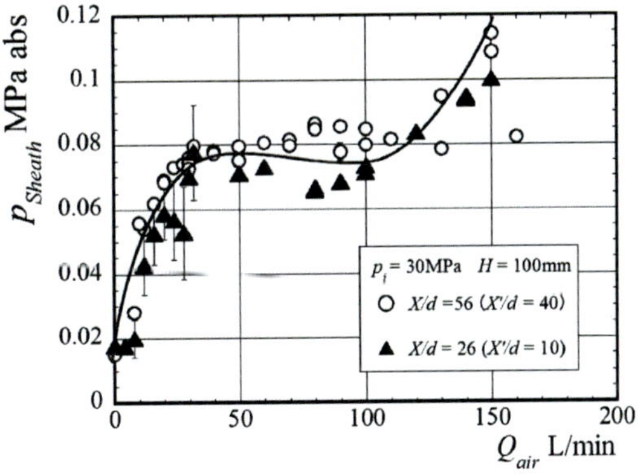

Fig. 3. Variations of sheath pressure with the ventilation air flow rate.

High-speed observation of air ventilated water jets under submerged condition are carried out by using a transparent water tank. Water jet is issued horizontally and the depth of the nozzle head is approximately 210 mm. During flow observation, the surface of water is covered by a plate in order to prevent air from flowing in from the water surface. Figure 4 shows a series of instantaneous images of ventilated water jet taken by high-speed camera when the jetting pressure P_i is 10 MPa and the ventilation air flowrate Q_{air} is 36 L/min. Here jet flows are illuminated from the opposite side of the camera lens and the air ventilated submerged jet appears to be a black band contains many bubbles. From these photographs, we may see a narrow band in relatively light grey color near the nozzle axis, which demonstrates a central liquid jet coated by air cavities. High-speed camera observation reveals that relative large cavities covering the liquid jet are formed when the ventilation flowrate is increased to a certain extent and a liquid jet coated by air layer, which is called as air-coated jet, is generated. The air layer coating the liquid jet shrinks and sheds downstream periodically. The length of continuous air layer and the frequency of continuous air layer oscillation vary with the ventilation flow rate. That is to say, a surrounding condition similar to that in air may be generated for water jet under submerged condition by ventilation.

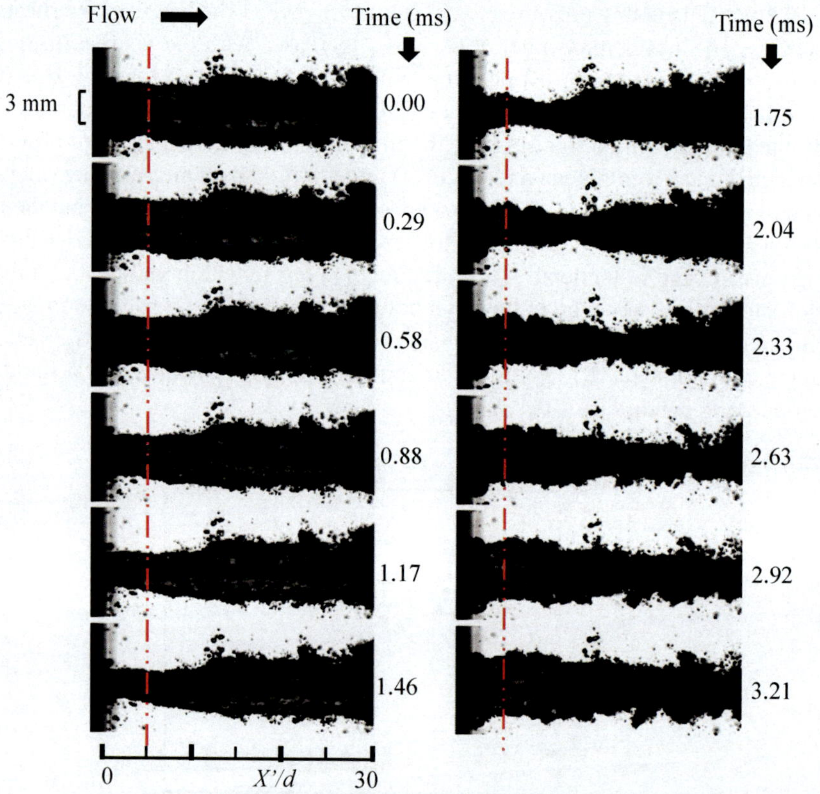

Fig. 4. Instantaneous photographs of air-ventilated water jet in time sequence.

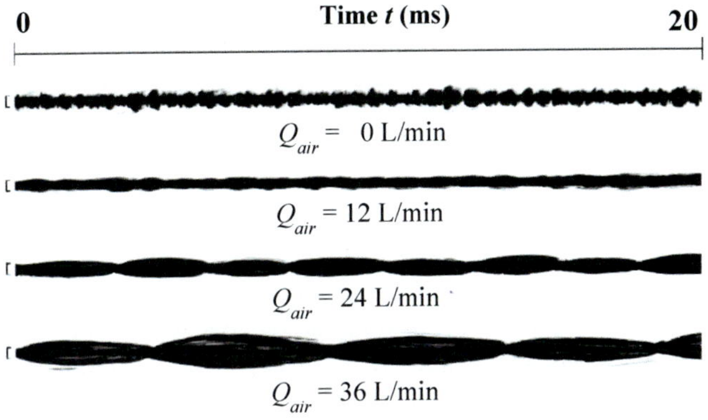

Fig. 5. Temporal variations of air-ventilated jet width at X'/d = 5 (Pi = 10 MPa, Qair = 36 L/min).

From these photograph we may see a band area in relatively light grey color demonstrating liquid jet at the central of the jet flows. In order to evaluate qualitatively the characteristic of the air-bubbly layer pulsation, image analysis of these photographs is carried out. As shown in Fig. 4 by the red dashed line, sampling images of one pixel width in the axial direction at $X'/d = 5$ were taken out from high-speed video camera photographs, and then they are arranged in time series. Figure 5 shows one pixel sampling images arranged in time order under different ventilation flowrate. Thus the vertical width of the black band corresponds to the instantaneous diameter of the air-bubbly layer at $X'/d = 5$. According to Fig. 5 we understand that the air-bubbly layer covering the jet pulsates periodically and its time period increases with the ventilation flowrate.

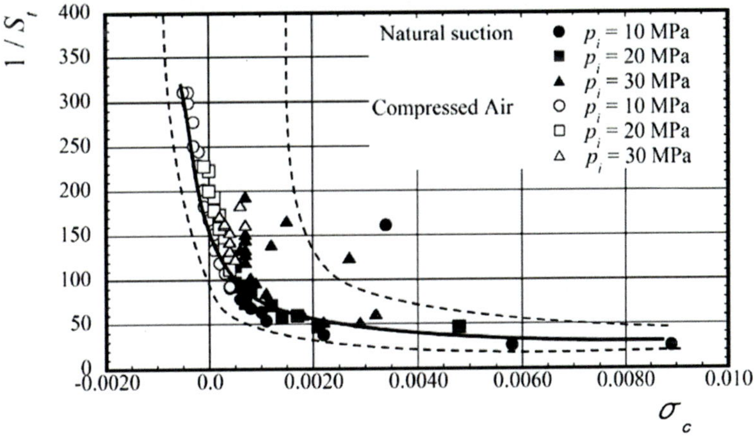

Fig. 6. Relationship between $1/St$ and σ_c.

The average grey level of these images is evaluated and the temporal variation of the grey level is further treated by FFT (Fast Fourier Transform). Then, the dominant frequency f of the air-bubbly layer pulsation is investigated by FFT analysis. As a dimensionless index of unsteady periodicity a Strouhal number St is defined by the dominant frequency f.

$$St = \frac{f \cdot d_{Sheath}}{V_{th}} \tag{2}$$

where V_{th} denotes the theoretical jet velocity calculated by neglecting all hydraulic losses. The relation between the Strouhal Number St and the altered cavitation number σ_c is then investigated. Figure 6 shows the result evaluated for different air ventilation flowrate under different injection pressures. The abscissa denotes the cavitation number σ_c, and the ordinate does the inverse of Strouhal Number $1/St$. As shown in the figure, all the values of the Strouhal Number under different working conditions locate within the region between the dotted lines and vary along nearly the solid line. In other words, the Strouhal Number is primarily dependent upon the pseudo cavitation number.

Figure 7 shows the variation of kerf depth of submerged ASJ under different air ventilation flowrate, where the jetting pressure P_i is 30 MPa and the water depth H at the nozzle exit is 100 mm. The non-dimensional standoff distance X/d is set to two different cases of 26 and 56, respectively. The nozzle traverse speed is 2 mm/s. The symbols ○ and ● denote the results when $X/d = 26$ ($X'/d = 10$ from the sheath exit). In comparison, the symbols △ and ▲ does results when $X/d = 56$ ($X'/d = 40$). The solid symbols denote results obtained under natural suction and hollow ones do the results obtained by ventilation with compressed air. Under the same working condition the kerf depth is examined to be $h_{in\text{-}air1} \cong 20.9$ mm averagely when $X/d = 26$,and $h_{in\text{-}air2} \cong 16.5$ mm when $X/d = 56$. It demonstrates that the largest kerf depth of submerged cutting at when $X/d = 26$ almost reaches to 100% of the cutting depth in air under an optimum ventilation condition. As shown in the figure, the kerf depth at both the different standoff

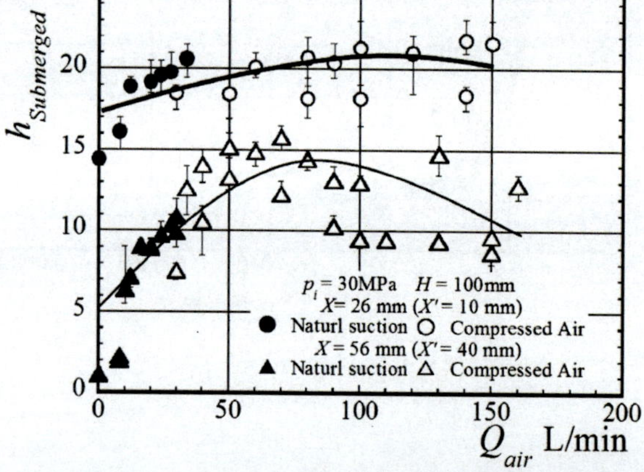

Fig. 7. Kerf depth of abrasive suspension jet varying with the ventilation airflow rate.

positions increases with the ventilation flow rate. However, as the ventilation flow rate gets over a certain value the kerf depth declines reversely.

4 Conclusions

The flow structure of air-ventilated ASJ in submerged condition has been investigated by high-speed camera observation of water jet under flow dynamic similarity condition and the effect of ventilation on the cutting ability of ASJs is evaluated through submerged cutting tests with specimens of aluminum alloy. The main conclusions are summarized as follows.

1. Air-coated suspension jet may be generated by increasing the ventilation flowrate to a certain extent so that ventilation air spreads out of the sheath in the form of large cavity covering the jet.
2. The continuous air layer surrounding the jet pulsates periodically, and the dominant frequency of air-layer pulsation decreases with increasing of the ventilation flowrate. The Strouhal number of air-layer pulsation is dependent upon the altered cavitation number, and the reverse of *St* is arranged to be a non-linear function of σ_c under different working conditions.
3. The kerf depth of ventilated ASJ increases with the ventilation flow rate in a certain range. The kerf depth reaches to almost the same level to that of ASJ in air when $Q_{air} = 100$ L/min while $X/d = 26$, and the cutting ability of air-ventilated ASJ in submerged condition is effectively enhanced.

References

1. Summers, D.-A.: Water Jetting Technology. CRC Press, New York (2003)
2. Hung, T.N., Louis, H., Peter, D., Senne, T.: Submerged application of abrasive water suspension jets. In: Water Jetting, pp. 205–216. BHR Group (2004)
3. Shimizu, S., Nishikata, H., Peng, G., Oguma, Y.: Submerged cutting characteristics of abrasive suspension jet (in Japanese). Tran. JSME (B) **81**(831), 15–00361 (2015)
4. Peng, G., et al.: Periodic behavior of cavitation cloud shedding in submerged water jets issuing from a sheathed pipe nozzle. J. Flow Control, Measur. Vis. **6**(1), 15–26 (2018)
5. Peng, G., Yang, C., Oguma, Y., Shimizu, S.: Numerical analysis of cavitation cloud shedding in a submerged water. J. Hydrodynamics (B) **28**(6), 986–993 (2016)
6. Shimizu, S., Peng, G., Oguma, Y.: Air coated abrasive suspension jets under submerged condition. Mod. Mach. Sci. J. **3**, 2214–2217 (2018)

Regular Papers

Systematic Change of Abrasive Size Distribution

Thomas Bergs[1], Manuel Schüler[1(✉)], Tim Herrig[1], Jan Fernolendt[2], and Marco Linde[2]

[1] Fraunhofer Institute for Production Technology IPT, Aachen, Germany
manuel.schueler@ipt.fraunhofer.de

[2] ANT Applied New Technologies AG, Lübeck, Germany

Abstract. The continuous suspension (ConSus) technology is a new coming approach that promises performance advantages over industrially used abrasive waterjet (AWJ) systems in production scenarios. However, it currently lacks practical experience and process parametrization is typically still empirical of either using a 2-phase abrasive suspension jet (ASJ) or a 3-phase injection AWJ tool. A practical comparison of respective AWJ tool characteristics of prior work piece contact is needed for fundamental parametrization.

An experimental approach was used to compare the respective AWJ systems. The abrasive material was compared before and after going through either an injection AWJ head or a ConSus ASJ system. Typical abrasive materials and mesh sizes for AWJ machining were systematically investigated. This paper shows that current AWJ cutting systems yield different results when using equal abrasive material. ConSus shows almost no significant effect on abrasive size distribution while commonly used injector systems shifts a tremendous portion of the original abrasive to smaller grain sizes. Therefore, the results enhance process understanding and revision of existing process models for future applications.

Keywords: Suspension · Abrasive · Recycling

1 Introduction

The future demand for high-performance materials is increasing significantly throughout all knowledge-based European industrial sectors. Abrasive waterjet (AWJ) machining has proven to be one of the most flexible non-conventional production techniques for cutting high-performance or hard-to-machine materials [1]. In addition to AWJ cutting, the potential of waterjet milling for difficult-to-machine materials has long been acknowledged. Since the first investigation of its economic and technological characteristics in 1987 by Hashish [2], waterjet milling has been considered as an alternative to rough milling in the field of hard-to-machine materials. However, waterjet machining of complex geometries has not been widely used in industry due to the difficulty to predict the results of the machining process. Existing models were developed empirically and therefore consist of unknown constants or were limited in parametrization [3–5]. Faced with a great variety of waterjet industrial applications, a deep understanding of its physical phenomena is highly demanded [6, 7].

D. Klichová et al. (Eds.): Water Jet 2019, LNME, pp. 39–48, 2021.
https://doi.org/10.1007/978-3-030-53491-2_4

Common AWJ systems widely used in industry are injector waterjets in which the abrasive medium is entrained into a high-velocity plain waterjet. Systemically, the AWJ injection head can be treated as a supplementary of a plain waterjet configuration (Fig. 1 - left). The velocity of the waterjet develops a local vacuum, which sucks an airstream containing abrasive particles into a mixing chamber by injection principle. Alternatively, abrasive suspension or slurry jet (ASJ) systems were used in which abrasive is added continuously to the pressurized water flow before being squeezed through the nozzle (Fig. 1 - right). Both AWJ variants can be subdivided according to the type and location of the abrasive acceleration and its phase composition:

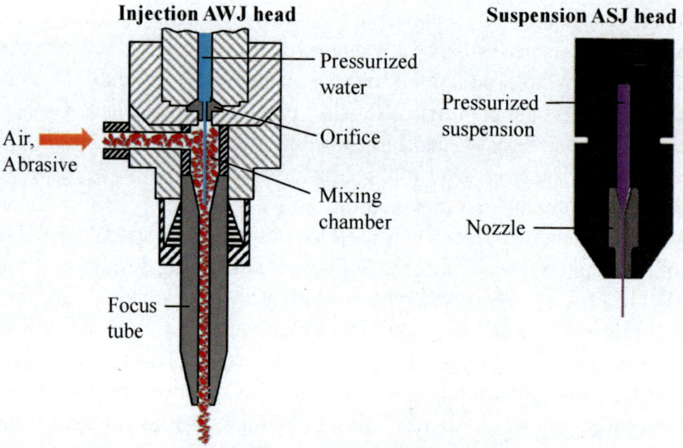

Fig. 1. Schematic principle of injection AWJ (left) and suspension ASJ (right) system heads.

- Suspension ASJ (two phase – water, abrasive)
- Injection AWJ (three phase – water, abrasive, air)

The load and the mechanical interference of the abrasive material affects the physical interaction of the high-velocity jet. In contrast to the primary plain waterjet, the injection AWJ system sucks air into the mixing chamber. A post-mixing process occurs that creates a chaotic jet composition of involved phases of water, air and supplementary solids. Prior work piece contact, initial mechanical stresses and wall contacts induce wear and alter cutting performance. Therefore, increased system pressure, accompanying with higher wear of the orifice and other high-pressure components, must be taken into account to compensate the drawbacks in cutting performance. This is different for ASJ systems, where a premixed suspension of water and abrasives is accelerated through an orifice with the absence of air, systematically similar to the plain waterjet setup. The velocity of the jet remains unchanged with different abrasive load ratios [8]. Therefore, the ASJ technique is another process variant for AWJ systems that promises an increased cutting performance at lower water pressure. Former systems with a discontinued ASJ technique were commonly used for dismantling or mobile applications of limited use, due to the required process interruption to refill the suspension unit [9]. With the availability of a novel continuous suspension (ConSus) mixing unit, the potential of ASJ technique for industrial cutting applications was renewed. From its physical characterization and its performance with smaller jet diameter, the ASJ promises to be superior in some circumstances over its injection counterpart. Despite a higher constructional expense of pressurized parts, the ASJ is advantageous in terms of lower system pressure and the flexibility of either dry or wet abrasive feed. However, a drawback is the lack of experience in production scenarios.

The most important factor in the AWJ material removal of high-strength materials is the erosion of the material by kinetic energy transfer of the abrasive particles in interaction with the work piece surface. At given jet velocities, the abrasive size distribution directly affects the kinetic energy of the solid particles. Early erosion models were developed in the 1960s, prior to the inception of AWJ technology. These erosion models have formed the basis of many models employed in AWJ, which describe the rate of material removal rather than the spatial distribution [10]. The common erosion models of abrasive particle interaction with the work piece are based on the works of Finnie [11] and Bitter [12, 13]. Recent work already showed that AWJ injector systems shift abrasive sizes to smaller values due to particle break up while accelerating down the focus tube [14–16]. In particular when machining high-strength or brittle hard materials, the material removal phenomena of the abrasive properties are predominant for material removal characteristics. Consequently, a systematical change of the abrasive size distribution within the AWJ tool is fundamental for enhanced process understanding and for further improvement of existing process models.

This study investigates the abrasive size distribution between injection and suspension AWJ systems. Different standard abrasive materials were compared prior and after passing the AWJ system. The solids were caught exiting the AWJ head prior work piece contact. Finally, a comparison was made to its primary state.

2 Procedure

The abrasive material was sieved and its original delivery state prior was used as a reference. Table 1 gives an overview of the AWJ parameters used for experiments. Nine test sieves covering a range from 25 μm up to 500 μm were chosen for categorical analysis (Table 1). An amplitude of 0.75 mm was set for a duration of five minutes at circular motion of the sieve cascade. Every content of the sieves was weighted and its relative percentage over total mass collected was calculated. A mobile catching device was designed and installed at the exit of the ASJ machine head (Fig. 2 – left). Table 1 gives also an overview of the AWJ parameters used for experiments. Because the injection AWJ contains a huge amount of air to spill the water column out of the catcher rapidly, the active time was restricted in practical manner. To keep the abrasive concentration inside the catcher at a recent level, the AWJ duration was set around 30 s. Supplementary, the injection AWJ was shot into a water basin at inclined angle for analysis (Fig. 2 – right). After the catcher was loaded, a minimum of 20 min was allowed for sedimentation. Both the spillway and the collected material inside the catching container were used for analysis. The solids were dried with the help of a heat gun aiming at the container wall from the outside to avoid thermal damage or a blow away. In analogy to the primary analysis, the collected abrasive was sieved and weighted again. A comparison of abrasive disintegration after going through the AWJ head was made by its percentage shift of the categorical size.

Table 1. Process parameters, set-up and test sieve sizes used for AWJ experiments.

Process parameter	Injection AWJ	Suspension ASJ
Orifice-/Focus tube diameter	0.25 mm/0.76 mm	0.55 mm
Water pressure p	360 MPa, 420 MPa	100 MPa, 150 MPa
Abrasive mass flow $\dot{m}_A$ (all mesh sizes)	400 g/min 500 g/min –	400 g/min 500 g/min 800 g/min
Test sieve grid size [μm]	500, 250, 150, 100, 90, 80, 50, 25	

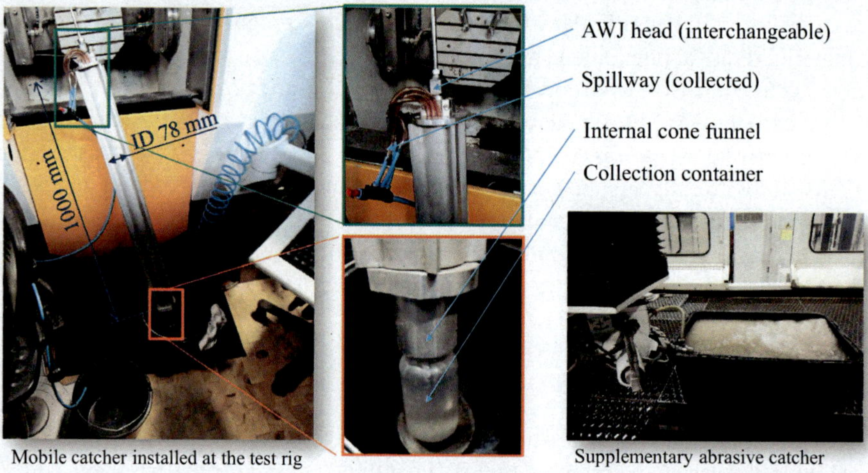

Fig. 2. Catcher used to collect abrasive after passing the AWJ head.

3 Equipment

The experiments of injection AWJ were carried out using an H.G. Ridder 5-axis waterjet machine type HWE-P2030, driven by a UHDE 45 KW intensifier pump capable of up to 600 MPa water pressure. Suspension experiments were conducted with a novel continuous suspension system type ConSus 1500 capable of 150 MPa maximum system pressure. This unit is exclusively manufactured by ANT-Applied New Technologies and it was installed at a 5-axis Satisloh G3 machine tool as a test rig at Fraunhofer IPT. A horizontal sieve shaker AS200 by Retsch was used for grain size analysis. A sieve cascade of nine test sieves allows a categorical size exploration. For all experiments, garnet abrasives supplied by Kuhmichel Recycling GmbH was used in two standard grain sizes of #80 mesh and #120 mesh respectively. Finally, an optical 3D micro-coordinate measurement system Alicona type InfiniteFocus G5 was used to analyze the achieved surface quality of the cutting samples.

4 Experimental Results

The particles masses obtained from the injection AWJ revealed an intense particle disintegration. When using #80 mesh garnet abrasive, the harsh acceleration conditions lowers the size category of greater than 250 µm from 50% prior passing the AWJ head to remaining 5% within this category afterwards (Fig. 3). In other words, a huge percentage of the biggest abrasive particles has shifted to smaller sizes. Here, it should be mentioned that particles smaller than 90 µm could not be re-used economically for cutting applications due to the amount of dust and foreign material to be removed. The waste portion has grown by roughly 30% without work piece contact – a deep drawback for resource optimization or recycling matters. Changing to #120 mesh abrasive of the same type, only a 5% level of the 250 µm size category remains from its original state. The outstanding size shift is now located around the 150 µm category. Hence, the waste category smaller than 90 µm has grown by 40% respectively. A comparison of both abrasive mesh sizes reveals that #80 mesh original abrasive has been lowered by the injection AWJ system to a size distribution smaller than #120 original state (Fig. 4).

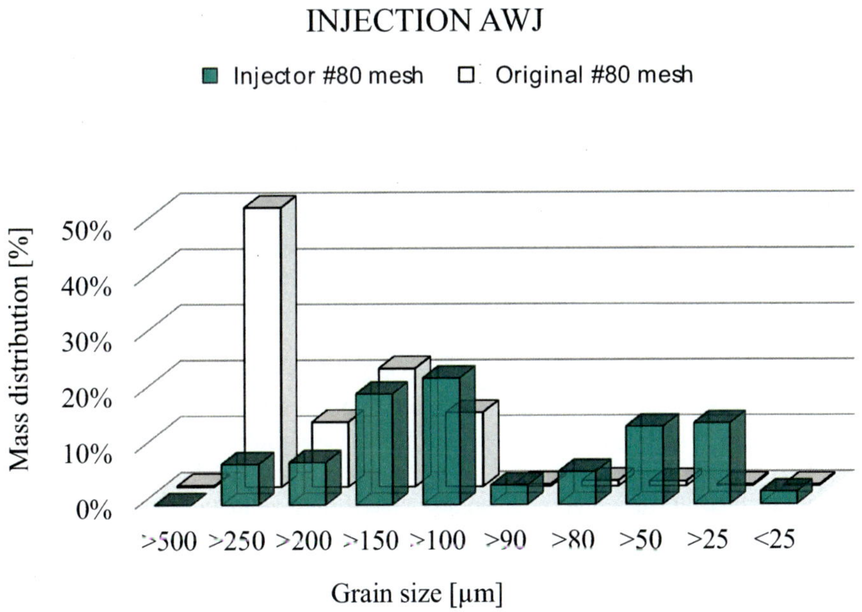

Fig. 3. Injection AWJ particle shift of #80 mesh garnet ($p = 420$ MPa, $\dot{m}_A = 400$ g/min).

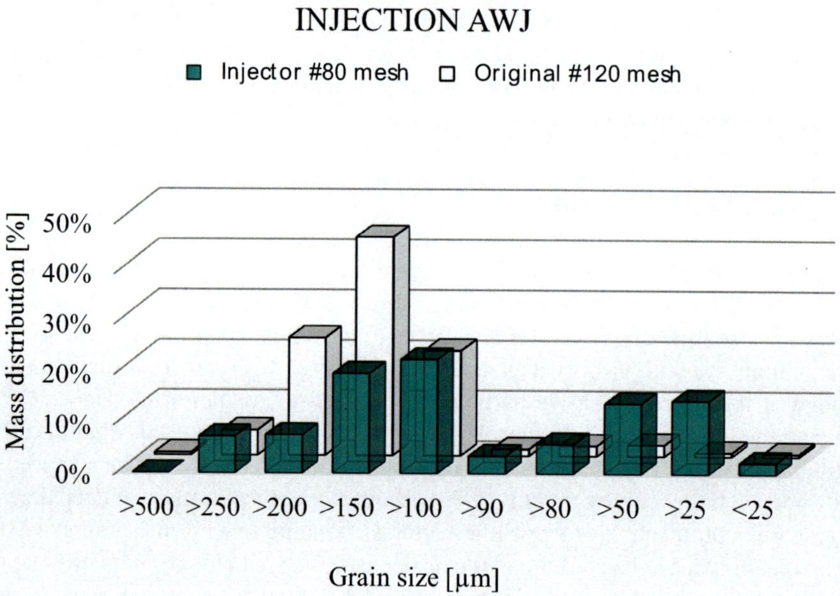

Fig. 4. Injection AWJ particle comparison of #80 mesh garnet ($p = 420$ MPa, $\dot{m}_A = 400$ g/min).

Contrary, the abrasive size distribution of #120 mesh garnet passing the ASJ head reacts almost unchanged to its original state. In particular, at medium 100 MPa system pressure, which is quite comparable in cutting performance in most conditions to an injection AWJ configuration of 400 MPa, the original abrasive size distribution remains constant prior and afterwards (Fig. 5). When rising the cutting performance of ConSus to maximum system pressure $p = 150$ MPa and doubled abrasive mass flow $\dot{m}_A = 800$ g/min, the obtained results slightly change. Obviously, the probability of internal collisions of the ASJ has risen even though the velocity profile of the abrasive particles remains constant at higher abrasive loads [14]. Here, a relative size shift of less than 10% was evaluated. During the experimental evaluation, a natural scattering of the abrasive sieving around 5% was observed. Noticeably, the growth of the waste category was also within this limit, meaning that this non re-usable abrasive sizes were within a natural scattering deviation. For a better comparison, the individual percentage of the systematic abrasive size distribution were related to its original state. The systematic response of both injection AWJ at moderate conditions and suspension ASJ at maximum performance for #120 mesh abrasive is illustrated in Fig. 6. It should be noted that at these conditions, the suspensions cutting rate is usually up to two or three times higher than for injection AWJ.

After analysis of the systematic abrasive size shift, a validation of the findings was done by real cutting tests. An analysis of the surface quality machined was intended to link the shift in the abrasive sizes. Here, a nickel-based alloy 718 was used for experiments, which acts very yielding and is conventionally hard to machine. The cutting edge surface was analyzed at the exit side of the AWJ leaving the work piece where the striations are typically apparent. All measurements were performed 1 mm off the outer edge to prevent irregularities. The average surface characteristics of a 7 mm long image

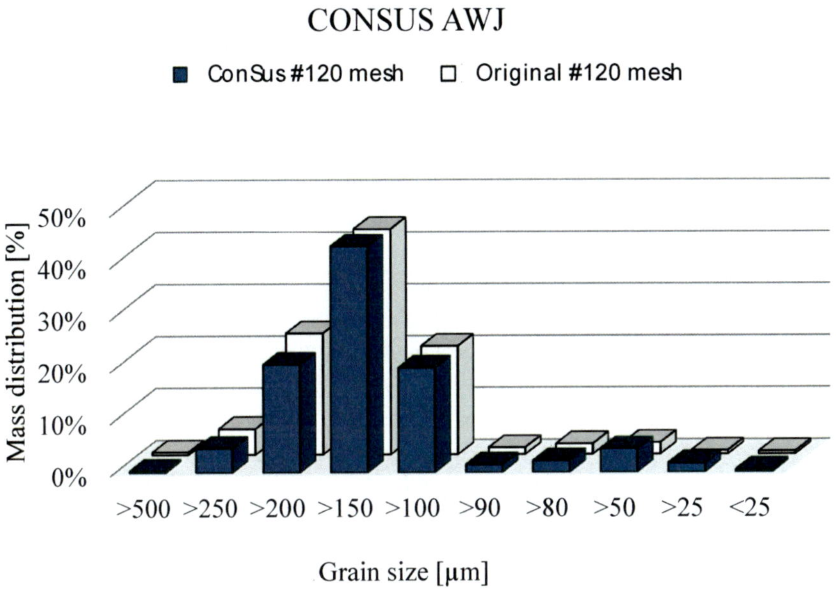

Fig. 5. Suspension ASJ particle shift of #80 mesh garnet ($p = 100$ MPa, $\dot{m}_A = 400$ g/min).

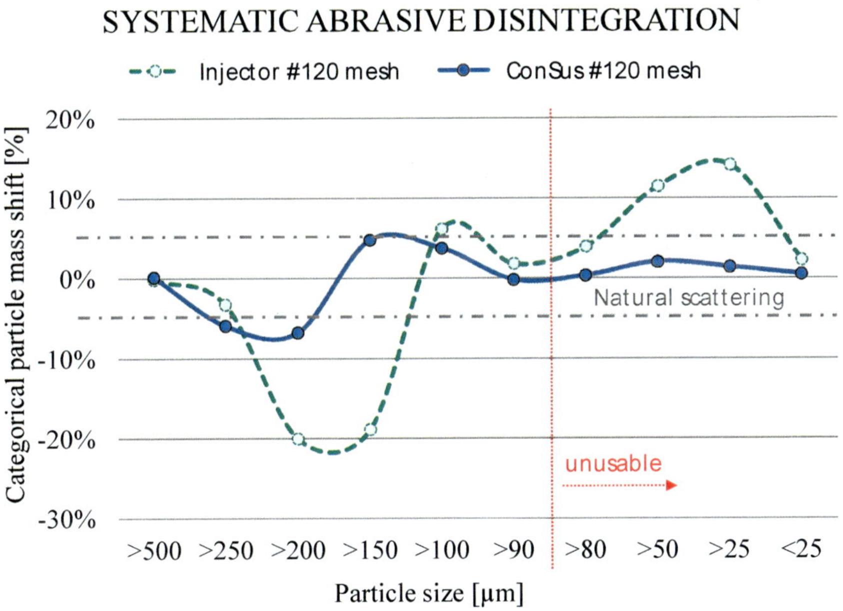

Fig. 6. System response: abrasive size shift of moderate injection AWJ ($p = 420$ MPa, $\dot{m}_A = 400$ g/min) compared to high-performance suspension ASJ ($p = 150$ MPa, $\dot{m}_A = 800$ g/min).

field was considered for analysis. Figure 7 shows a plot of results. In analogy to the previous findings, the injection AWJ of using #80 mesh abrasive enters with very good surface quality values that rapidly rises to worse values. Here, it should be mentioned that preliminary experiments revealed that the cutting performance of injection AWJ when using #120 mesh was reduced by about 25%. Controversy, the matching ConSus AWJ curve of #80 mesh enters beyond that Sa values but increases quite slower. As already revealed in the previous experiments, the cutting results of injection AWJ that was fed with #80 mesh is close to the surface quality of using ConSus with #120 mesh. It should be noted that higher feed rates could be achieved here. This proofs the findings of shifted particle size. Particle disintegration leads to a larger number of cutting edges despite their individual kinetic energy is lowered. Hence, smaller particles are restricted in cutting performance. In other words, smaller particles may produce smoother surfaces at very low feed rates in AWJ cutting applications or when cutting depth is not required. Finally, the cutting test was supplemented by comparing the results to ConSus AWJ fed with comparable smaller #120 mesh abrasive. Here, the surface values at low feed rate were in pair with the injection AWJ results of #80 mesh originally. The lower gradient of the ConSus curve could be a result of delimited abrasive size distribution and optimized erosion interaction.

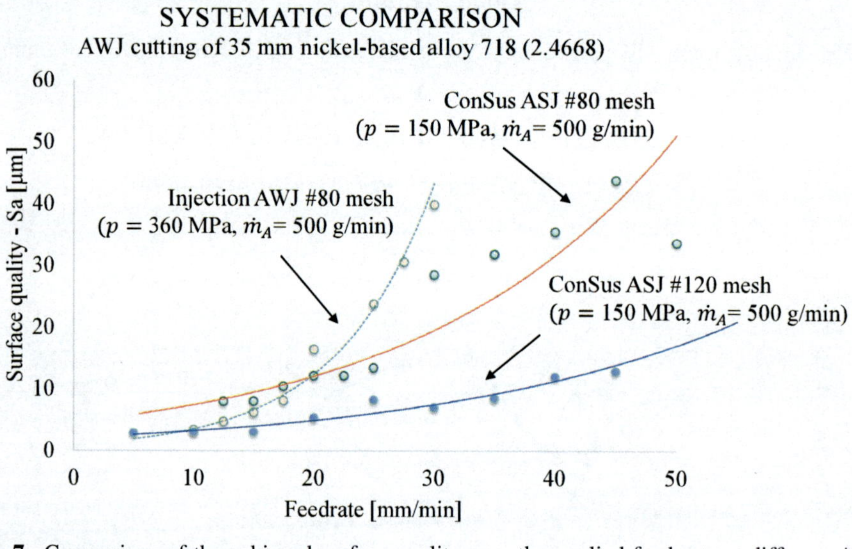

Fig. 7. Comparison of the achieved surface quality over the applied feed rate at different AWJ configuration.

5 Summary and Outlook

In this paper, the systematic size distribution of common AWJ systems has been documented. A 3-phase injection AWJ was compared with a 2-phase continious suspension ASJ tool by their characteristics on the particle disintegration prior work piece contact. An appropriate catching device was designed and installed at dedicated AWJ systems. The abrasive passing the AWJ head was collected and sieved. A comparison was made by the categorical sizes of a test sieve cascade to its original state. Finally, the influence on the machining result was accounted for the surface quality at real cutting tests.

This study revealed that injection AWJ systems intensively shifts abrasive size distribution to smaller values. In contrast, the same abrasive sizes when using suspension ASJ remains nearly unchanged within natural scattering. For #80 mesh garnet, the majority of the 250 μm size category was shifted to smaller values while passing an injection AWJ head. Transferable, the finer #120 mesh lost intensive amounts around the corresponding 150 μm category. Up to 40% of the original abrasive particles were shifted to values smaller than 90 μm, typically considered as waste and not applicable for recycling in cutting applications. This could be a strong drawback for injection AWJ systems in production scenarios where resource efficient approaches are demanding. For better comparison of the systematic abrasive disintegration, a system response curve was introduced, which indicates the relative size shift of the abrasive material to its original state prior to passing the AWJ system. The findings of the abrasive disintegration analysis were validated at cutting samples that revealed the optimization potential of dedicated AWJ systems for future applications. However, systematic particle disintegration may vary for natural abrasives due to physical characteristics and chemical composition. High strength materials that could withstand harsh mechanical loads could be an alternative for special applications.

In the future, both the systematic and the parametric influence of the AWJ tool could help to optimize the eroded surface in terms of surface integrity or performance of the machined parts. In addition, alternative abrasive materials should be tested to withstand harsh conditions of injection AWJ systems better or to have improved recycling capabilities. Existing models should be revised to cover the systematic response of abrasive material. Future work could help to further understand and optimize the AWJ process for high-demanding applications and CAM development.

Acknowledgements. Part of the work was funded by the Deutsche Forschungsgemeinschaft (DFG, German Research Foundation) – Project No. 223500200 – TRR 136, in collaboration with ANT Applied New Technologies AG.

References

1. Axinte, D.A., Karpuschewski, B., Kong, M.C., Beaucamp, A.T., et al.: High energy fluid jet machining (HEFJet-mach): from scientific and technological advances to niche industrial applications. CIRP Ann. Manuf. Technol. **63**(2), 751–771 (2014)
2. Hashish, M.: Milling with abrasive waterjets: a preliminary investigation. In: Proceedings of the 4th US Waterjet Conference (1987)
3. Kong, M.C., Axinte, D., Voice, W.: Challenges in using waterjet machining of NiTi shape memory alloys: an analysis of controlled-depth milling. J. Mater. Process. Technol. **211**(6), 959–971 (2011)
4. Momber, A., Kovacevic, R.: Principles of Abrasive Waterjet Machining. Springer, London (1998)
5. Klocke, F., Schreiner, T., Schüler, M., Zeis, M.: Material removal simulation for abrasive waterjet milling. Procedia CIRP **68**, 541–546 (2018)

6. Uhlmann, E., Männel, C.: Modelling of abrasive waterjet cutting with controlled depth for near-net-shape fabrication. In: 52nd CIRP Conference on Manufacturing Systems (2019)
7. Brinksmeier, E., Meyer, D., Heinzel, C., Lübben, T., et al.: Process signatures - the missing link to predict surface integrity in machining. Procedia CIRP **71**, 3–10 (2018)
8. Foldyna, J., Zelenak, M.: The Measurement of the velocitiy of individual particles in the ASJ downstream the Nozzle. Technical report, Institute of Geoneics of the ASCR, OSTRAVA (2014)
9. Putz, M., Dix, M., Morczinek, F., Dittrich, M.: Suspension technology for abrasive waterjet (AWJ) cutting of ceramics. Procedia CIRP **77**, 367–370 (2018)
10. Hoogstrate A.M.: Towards high-definition abrasive waterjet cutting: a model based approach to plan small-batch cutting operations of advanced materials by high-pressure abrasive waterjets. Dissertation, Technology University Delft (2000)
11. Finnie, I.: Erosion of surfaces by solid particles. Wear **3**(2), 87–103 (1960)
12. Bitter, J.G.: A study of erosion phenomena part I. Wear **6**(1), 5–21 (1963)
13. Bitter, J.G.: A study of erosion phenomena: Part II. Wear **6**(3), 169–190 (1963)
14. Fowler, G., Shipway, P.H., Pashby, I.R.: A technical note on grit embedment following abrasive water-jet milling of a titanium alloy. J. Mater. Process. Technol. **159**(3), 356–368 (2005)
15. Perec, A.: Disintegration and recycling possibility of selected abrasives for water jet cutting. DYNA **84**(203), 249–256 (2017)
16. Zaremba, D.; Heese, P.; Bauer, M.; Maier, H. J.; Hassel, T.: Particle disintegration in the abrasive water injection jet. In: WJTA-IMCA Conference and Expo, New Orleans, Louisiana (2015)

Effect of Abrasive Water Jet Machining on the Geometry of Shapes in Selected Tool Steels

František Botko[1(✉)], Petr Hlaváček[2], Dominika Lehocká[1,2], Vladimír Foldyna[2], Michal Hatala[1], and Vladimir Simkulet[1]

[1] Faculty of Manufacturing Technologies with a seat in Prešov, Technical University of Kosice, Prešov, Slovakia
frantisek.botko@tuke.sk
[2] Institute of Geonics of the Czech Academy of Sciences, Ostrava, Czechia

Abstract. The presented article is focused on the abrasive water jet machining of tool steel material. Selected tool steels Uddeholm Dievar and Böhler K110 were machined and final pockets were evaluated based on the side angle of the pocket. Sequentially were created pockets with increasing depth from 1 to 5 mm for both materials. Evaluation of pocket angle is crucial factor for determination of machining allowance for finishing cuts. Also metallography analysis for both materials was created for observation of removal process and for detection of cracks produced during abrasive water jet machining. In terms of corrosion of presented tool steels can be stated that due to higher number of alloying elements is corrosion on machined surface acceptable for further processing of created shapes. Obtained results can extend application possibilities of AWJ milling for rough machining of tool steels, especially for thin-walled components with reduction of induced residual stresses and deformations of shapes.

Keywords: Abrasive water jet · Milling · Depth · Side angle

1 Introduction

Abrasive water jet technology (AWJ) [1, 2] is used in various branches of the engineering industry for cutting and disintegrating materials of various properties [3]. AWJ is characterized by the use of abrasive particles [4, 5] to enhance the cutting or machining effect. The main advantages of the technology are dust-free process [6] and absence of heat-affected zone formation. The general disadvantages are the possibility of corrosion on some metals and limited depth of penetration (caused by jet lag). By appropriately combining the properties of the abrasive water jet with the material properties, AWJ technology can be used in various specific operations that are difficult to machine by conventional methods [7–9]. Nowadays intensive research is underway on the use of AWJ as a substitute for some turning [10, 11], milling [12] and drilling [13, 14] operations. The reimbursement concerns areas in which conventional machining [15] is associated with

D. Klichová et al. (Eds.): Water Jet 2019, LNME, pp. 49–55, 2021.
https://doi.org/10.1007/978-3-030-53491-2_5

technological production problems or health risks. Machining using AWJ is associated with lower manufacturing accuracy and the need for finishing operations.

The presented article is focused on the possibility of using water jet technology in roughing operations of shape milling of tool steel. AWJ milling is most often used in machining of difficult-to-machine materials where the use of conventional milling is expensive and technologically demanding. The negative thermal effects of conventional milling can be avoided, by using AWJ "cold cut". Metals milling using AWJ was investigated by several authors. AWJ milled materials can be different, such as carbon foam [16], borosilicate glass [17], carbon/epoxy laminates in aircraft industry [18, 19], composites [20], titanium [21, 22], aluminum 2024 T3 [23], Inconel [24] and comparison of AWJ milling on different steel grades (stainless steel, mild steel, die steel and mold steel) was discussed by Gupta et al. [25]. Possibilities of machining (milling) tool steel by abrasive water jet were discussed by Pal and Tandon [26].

2 Experimental Procedure and Measurement

Materials used in experimental testing were tool steels Uddeholm Dievar and Böhler K110 in the soft annealed condition. Condition of Uddeholm Diever steel (Fig. 1) was soft annealed with an average hardness 171 HV1 and Böhler K110 steel (Fig. 2) was soft annealed with an average hardness 256 HV1.

The experimental investigation was performed in the laboratory at the Institute of Geonics of the Czech Academy of Sciences in Ostrava. Experimental milling was realized on a technological device for the machining of flat materials 2D X-Y cutting table PTV WJ2020-2Z-1xPJ. The abrasive was Australian garnet (MESH 80) and abrasive mass flow was 100 g.min^{-1}. The working water pressure was generated by PTV 75–60 pump and the value of working pressure was 120 MPa. The stand-off distance of nozzle from experimental material was 4 mm. The diameter of the nozzle orifice was 0.33 mm and focusing tube diameter was 1.02 mm. The traverse speed was 500 mm.min^{-1}.

The experimental procedure consists of creating 5 pockets for both materials. Pockets were machined with an increasing number of passes, starting with one pass and the final pocket was created by 5 passes. The depth of cut for one pass was approximately 1 mm per pass with a linear strategy. Exact depths for each pocket are shoved on graphical dependence (Fig. 4).

The main focus of the experiment was to determine the side angle of each pocket. This value is important for the determination of suitability AWJ milling for roughing shapes with thin walls with a small depth of cut for finishing operations.

AWJ milling due to a different mechanism of material removal is more suited for creating thin-wall components compared to conventional milling.

Measurement of the side angle of pockets (Fig. 3) was performed using a digital microscope Keyence VHX-5100. Cyan lines represent profile of the pockets. Measured values of pockets depth and side angle for both experimental materials are shown in graphical dependences (Fig. 4).

Fig. 1. Uddeholm Dievar – pockets created using AWJ milling.

Fig. 2. Böhler K110 – pockets created using AWJ milling.

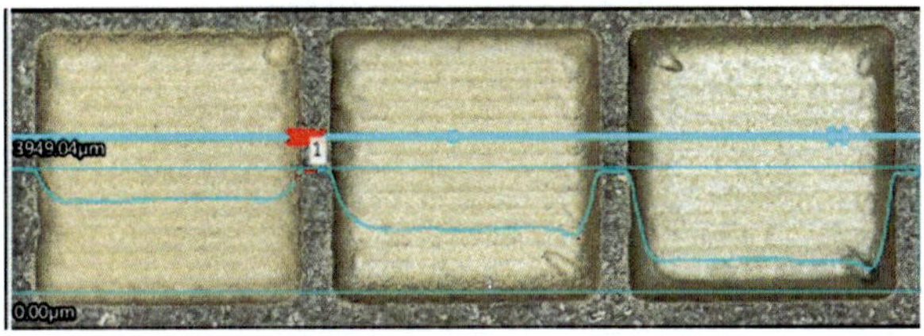

Fig. 3. Measurement of depth of pocket and side angle.

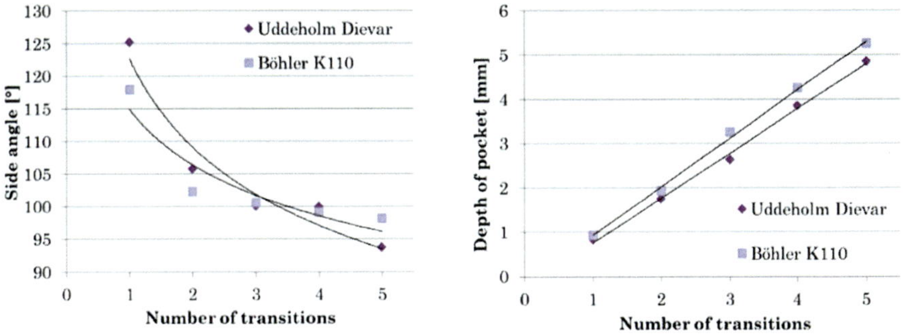

Fig. 4. Graphical dependences number of transitions on the depth of pocket and side angle.

3 Results

Following Fig. 5 represents the closest view on machined bottoms of the pockets created in both experimental materials. It can be observed that for material Uddeholm Dievar is the area of the bottom slightly bigger compared with Böhler K110 which corresponds with bigger side angle of the pocket in Böhler K110. Pockets in the presented experiment investigation are exhibiting tracks of corrosion on the surface, which is caused by longer exposition to the water (visible in Fig. 5.). The last (deepest) pocket of each material shows no track of corrosion, which is caused by removing samples from the machine desk and drying by pressure air. For avoiding corrosion on machined parts will be for the future experiments final passes on each pocket performed rearmost and immediately dried.

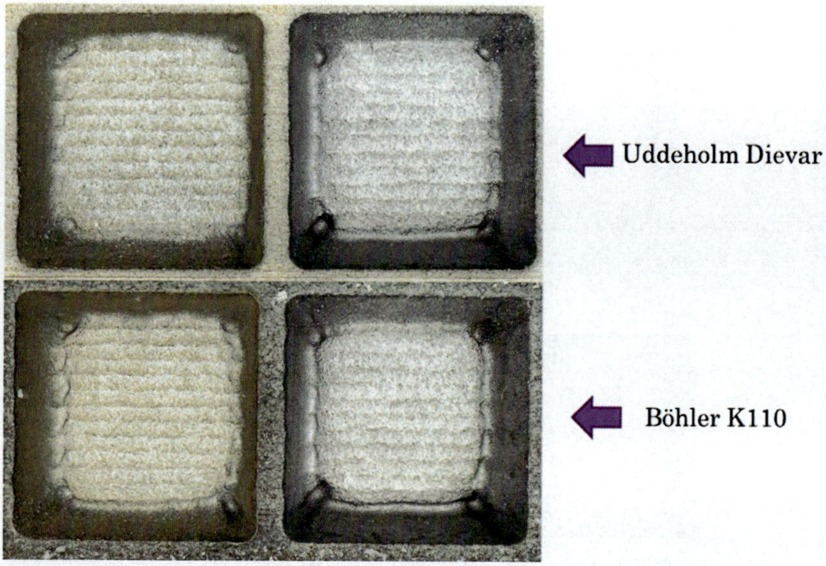

Fig. 5. Final pockets created in Uddeholm Diever and Böhler K110.

Figure 6 shows the microstructure of the observed Uddeholm Diever steel and Böhler K110 ground on 500, 1200, 2000 and 4000 SiC sandpaper, polished using a 1 μm diamond suspension and etched in 4% Nital. A more uniform loss of material removal by AWJM technology is observed for Böhler K110.

The results of the measurement were as follows:

- Depth of cut for one transition was approximately 1 mm per transition
- Slightly higher depth of cut was observed on steel K110
- Side angle is decreasing with increasing depth of the pocket
- After the first transition was higher angle observed on material Uddeholm Dievar compared to Böhler K110 steel (125, 19° to 117, 95°)
- Deepest pocket (approx. 5 mm 5 transitions) shows higher side angle for Böhler K110 compared to Uddeholm Dievar (98, 15° to 93,72)
- Subsurface cracks after the mechanical action were observed
- Material Böhler K110, which also exhibited higher hardness which suggests lower toughness, these subsurface cracks are more pronounced.
- After AWJ milling there is a necessity for finishing operation using conventional milling or EDM

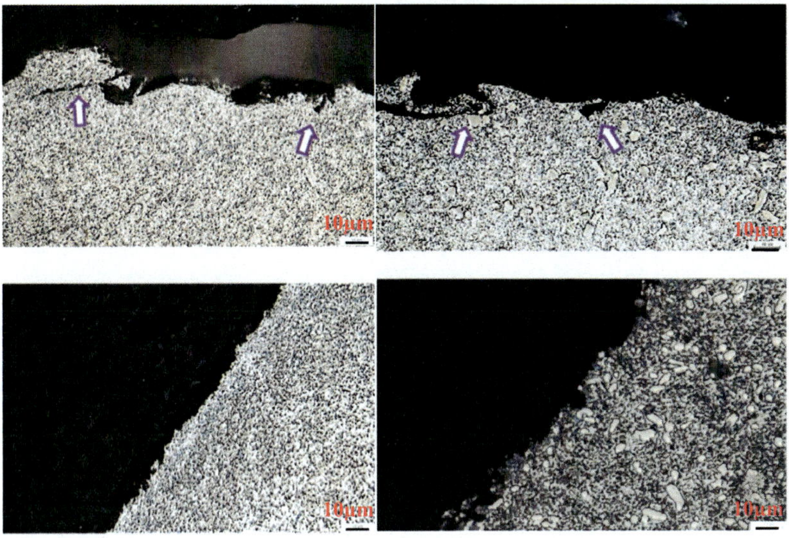

Fig. 6. The microstructure of the observed Uddeholm Diever (left) and Böhler K110 (right).

4 Conclusion

Engineering practice is constantly looking for new ways of approach to different machining processes. Using AWJ milling technology is one of the opportunities to machine a variety of shapes. The advantage of AWJ milling technology is that the main direction of the force is perpendicular to the material base and this gives an opportunity to create thin-walled shapes in the different materials. As is evident from the obtained results, the side angle of the pockets is on the depth circa 5 mm are appropriate for final machining process. Also, metallography shows promising results from the perspective of cracks created during machining. There were observed cracks only near surface which do not span to deeper layers and can be easily removed in subsequent final cuts. Based on obtained results can be stated, that AWJ machining of blind shapes shows promising perspective for subsequent research and optimization of process for practical applications.

Acknowledgements. This work was supported by the Slovak Research and Development Agency under the contract No. APVV-15-0696. This work was supported by projects: VEGA 1/0492/16 - Research of the Possibilities of Elimination Deformations of the Thin Components with the Use of High Speed Machining, project No. FV 30233 and TAČR TH04020218 - Development of special abrasives for machining of hard-to-machine materials by AWJ technology including prototype of Smart recycling line based on Industry 4.0 principles. The experiments were conducted with the support of the Institute of Clean Technologies for Mining and Utilization of Raw Materials for Energy Use - Sustainability Program, reg. no. LO1406 financed by the Ministry of Education, Youth, and Sports of the Czech Republic, and supported by a project for the long- term conceptual development of the research institution RVO: 68145535. This publication is the result of the Project implementation: University Science Park TECHNICOM for innovative applications with the support of knowledge technologies - Phase II, ITMS2014+: 313011D232, supported by the European Regional Development Fund.

References

1. Kinik, D., Ganovska, B., Hloch, S., Monka, P., Monkova, K., Hutyrova, Z.: On-line monitoring of technological process of material abrasive water jet cutting. Tehnicki Vjesnik Tech. Gaz. **22**(2), 351–357 (2015). https://doi.org/10.17559/TV-20130904111939
2. Molitoris, M., Pitel, J., Hosovsky, A., Tothova, M., Zidek, K.: A review of research on water jet with slurry injection. In: International Conference on Manufacturing Engineering and Materials, ICMEM 2016, vol. 149, 333–339 (2016). https://doi.org/10.1016/j.proeng.2016.06.675
3. Liu, X.C., Liang, Z.W., Wen, G.L., Yuan, X.F.: Waterjet machining and research developments: a review. Int. J. Adv. Manuf. Technol. **102**(5–8), 1257–1335 (2019). https://doi.org/10.1007/s00170-018-3094-3
4. Foldyna, J., Zelenak, M., Klich, J., Hlavacek, P., Sitek, L., Riha, Z.: The measurement of the velocity of abrasive particles at the suction part of the cutting head. Tehnicki Vjesnik Tech. Gaz. **22**(6), 1441–1446 (2015)
5. Zelenak, M., Foldyna, J., Riha, Z.: The measurement of abrasive particles velocities in the process of abrasive water jet generation. In: XIX. The Application of Experimental and Numerical Methods in Fluid Mechanics and Energetics 2014, pp. 276–280. Amer Inst Physics, New York (2014). https://doi.org/10.1063/1.4892749
6. Perec, A.: Environmental aspects of abrasive water jet cutting. Rocznik Ochrona Srodowiska **20**, 258–274 (2018)
7. Krolczyk, G., Legutko, S., Stoic, A.: Influence of cutting parameters and conditions onto surface hardness of duplex stainless steel after turning process. Tehnicki Vjesnik Tech. Gaz. **20**(6), 1077–1080 (2013)
8. Fecova, V., Michalik, P., Zajac, J., Hanisko, E., Zajac, J.: Coherencies of temperature and surface roughness by milling procedure. In: Precision Machining VII, vol. 581, pp. 348–353 (2014). https://doi.org/10.4028/www.scientific.net/KEM.581.348
9. Fecova, V., Michalik, P., Zajac, J., Mihok, J., Berthoty, J.: The chip in the up and down milling process. In: Material Science and Engineering Technology II, vol. 856, pp. 379–383 (2014). https://doi.org/10.4028/www.scientific.net/AMR.856.379
10. Nag, A., Scucka, J., Hlavacek, P., Klichova, D., Srivastava, A.K., Hloch, S., Dixit, A.R., Foldyna, J., Zelenak, M.: Hybrid aluminium matrix composite AWJ turning using olivine and Barton garnet. Int. J. Adv. Manuf. Technol. **94**(5–8), 2293–2300 (2018). https://doi.org/10.1007/s00170-017-1036-0
11. Hutyrova, Z., Scucka, J., Hloch, S., Hlavacek, P., Zelenak, M.: Turning of wood plastic composites by water jet and abrasive water jet. Int. J. Adv. Manuf. Technol. **84**(5–8), 1615–1623 (2016). https://doi.org/10.1007/s00170-015-7831-6
12. Klocke, F., Schreiner, T., Schuler, M., Zeis, M.: Material removal simulation for abrasive water jet milling. Procedia CIRP **68**, 541–546 (2018). https://doi.org/10.1016/j.procir.2017.12.110
13. Zhao, J., Zhang, G.C., Xu, Y.J., Lin, A.G., Zhao, J.Y., Yang, D.Y.: Enhancing rate of penetration in a tight formation with high pressure-water jet (HPWJ) via a downhole pressurized drilling tool. J. Petrol. Sci. Eng. **174**, 1194–1207 (2019). https://doi.org/10.1016/j.petrol.2018.11.042
14. Deepak, D., Pai, K.A.: Study on abrasive water jet drilling for graphite filled glass/epoxy laminates. J. Mech. Eng. Sci. **13**(2), 5126–5136 (2019). https://doi.org/10.15282/jmes.13.2.2019.24.0421
15. Maruda, R.W., Feldshtein, E., Legutko, S., Krolczyk, G.M.: Analysis of contact phenomena and heat exchange in the cutting zone under minimum quantity cooling lubrication conditions. Arab. J. Sci. Eng. **41**(2), 661–668 (2016). https://doi.org/10.1007/s13369-015-1726-6

16. Pahuja, R., Ramulu, M.: Study of surface topography in abrasive water jet machining of carbon foam and morphological characterization using discrete wavelet transform. J. Mater. Process. Technol. **273**, 116249 (2019). https://doi.org/10.1016/j.jmatprotec.2019.05.030
17. Haghbin, N., Ahmadzadeh, F., Papini, M.: Masked micro-channel machining in aluminum alloy and borosilicate glass using abrasive water jet micro-machining. J. Manuf. Process. **35**, 307–316 (2018). https://doi.org/10.1016/j.jmapro.2018.08.017
18. Hejjaji, A., Zitoune, R., Toubal, L., Crouzeix, L., Collombet, F.: Influence of controlled depth abrasive water jet milling on the fatigue behavior of carbon/epoxy composites. Compos. Part A Appl. Sci. Manuf. **121**, 397–410 (2019). https://doi.org/10.1016/j.compositesa.2019.03.045
19. Hejjaji, A., Zitoune, R., Crouzeix, L., Le Roux, S., Collombet, F.: Surface and machining induced damage characterization of abrasive water jet milled car-bon/epoxy composite specimens and their impact on tensile behavior. Wear **376**, 1356–1364 (2017). https://doi.org/10.1016/j.wear.2017.02.024
20. Srinivasu, D.S., Axinte, D.A.: Mask-less pocket milling of composites by abrasive waterjets: an experimental investigation. J. Manuf. Sci. Eng. Trans. ASME. **136**(4), 0410052014 (2014). https://doi.org/10.1115/1.4027181
21. Kanthababu, M., Ram, M.R., Emannuel, P.N., Gokul, R., Rammohan, R.: Experimental investigations on pocket milling of titanium alloy using abrasive water jet machining. FME Trans. **44**(2), 133–138 (2016). https://doi.org/10.5937/fmet1602133K
22. Popan, I.A., Balc, N., Popan, A.I., Carean, A., Ceclan, V.A.: Analysis on the abrasive water-jet milling process on titanium alloys. In: International Conference on Production Research - Regional Conference Africa, Europe and the Middle East (ICPR-AEM 2016) and 4th International Conference on Quality and Innovation in Engineering and Management (QIEM 2016), pp. 407–411. Technical University Cluj-Napoca, Cluj-Napoca (2016).
23. Cenac, F., Zitoune, R., Collombet, F., Deleris, M.: Abrasive water-jet milling of aeronautic aluminum 2024–T3. Proc. Inst. Mech. Eng. Part L J. Mater. Des. Appl. **229**(1), 29–372015 (2015). https://doi.org/10.1177/1464420713499288
24. Escobar-Palafox, G.A., Gault, R.S., Ridgway, K.: Characterisation of abrasive water-jet process for pocket milling in Inconel 718. In: Fifth Cirp Conference on High Performance Cutting, pp. 404–408. Elsevier Science BV, Amsterdam (2012). https://doi.org/10.1016/j.procir.2012.04.072
25. Gupta, T.V.K., Ramkumar, J., Tandon, P., Vyas, N.S.: A study on deviations of the jet with traverse speeds on different materials in pocket milling using abrasive water jet machining process. In: Advanced Materials Design and Mechanics II, pp. 402–405 (2013). https://doi.org/10.4028/www.scientific.net/AMM.372.402
26. Pal, V.K., Tandon, P.: Identification of the role of machinability and milling depth on machining time in controlled depth milling using abrasive water jet. Int. J. Adv. Manuf. Technol. **66**(5–8), 877–881 (2013). https://doi.org/10.1007/s00170-012-4373-z

Erosion of Titanium and Aluminium Alloys Using Pulsating Water Jet: Effect of Standoff Distance

Dominik Čuha[1], Akash Nag[2], Alice Chlupová[3], and Sergej Hloch[1](✉)

[1] Faculty of Manufacturing Technologies TUKE with a seat in Prešov, Prešov, Slovakia
sergej.hloch@tuke.sk
[2] Indian Institute of Technology (Indian School of Mines), Dhanbad, Dhanbad, Jharkhand, India
[3] Institute of Physics of Materials of the Czech Academy of Sciences, Brno, Czechia

Abstract. The disintegration of the material by PWJ technology depends upon many technological parameters of the equipment and particular material properties. This paper deals with the erosive effects of PWJ on the polished surfaces of metallic materials such as titanium (Ti6Al4V) and aluminium (ENAW1050AH24) alloy. The evaluation and comparison of the extent of disintegration are done on the basis of the erosion depth achieved. The nozzle movement over the material followed an ascending equidistant trajectory from $z = 15$ to 41 mm with a vertical step height of 2 mm. PWJ transitions over the material were performed with a sonotrode frequency $f = 20.19$ kHz, liquid pressure $p = 40$ MPa and a feed rate $v = 0.25$ mm.s^{-1}. The depth value of the grooves formed was then measured using a MicroProf FRT contactless measuring device and evaluated in the SPIP 6.7.8 software. For aluminium sample, the deepest average depth value is $h = 1583.2$ μm ($z = 29$ mm), while the titanium material has eroded to a much smaller depth, $h = 283.2$ μm ($z = 31$ mm). By comparing the achieved hydrodynamic effects of PWJ on titanium and aluminium, significant differences in the depth of erosion and its course can be observed due to the different mechanical properties of the materials.

Keywords: Pulsating water jet · Titanium · Aluminium · Polished surface · Erosion

1 Introduction

In the technical practice, there are several modifications of the water jet technology. Depending on the technological process, high-speed fluid flow enables erosion of different materials [1–3], hydro abrasive disintegration of rotating workpieces [4–11], drilling, cleaning and surface treatment [12–14] of materials. In the majority of processes, continuous (WJ) or abrasive (AWJ) water jet is commonly used. The development of the technology has modified the continuous stream into discontinuous, resulting in the formation of pulsating water jet (PWJ). Ultrasonically generated pulsations allow efficient

D. Klichová et al. (Eds.): Water Jet 2019, LNME, pp. 56–66, 2021.
https://doi.org/10.1007/978-3-030-53491-2_6

material disintegration using relatively low fluid pressures compared to continuous flow [15]. As a result of the gradual process of formation of individual pulses after the liquid exit from the nozzle, the technology has a wide range of applications in terms of the nozzle distance above the material. From this point of view, the PWJ liquid column can be divided into different efficiency zones - initial, transient and final [16]. The material present in the initial zone may be subjected to a process of strengthening its surface and subsurface layers (peening) [17, 18]. Measurable material erosion is the result of the transition zone effect [19–21] and the material roughening represents the final (droplet) PWJ efficiency zone. Other PWJ parameters that play an important role in these processes include fluid pressure (p), frequency (f), amplitude (A) and power (P) of the ultrasonic excitation, nozzle diameter (d), feed rate (v) and acoustic chamber length (*Lc*) [22]. However, the effect of the PWJ also depends on the target material used. The mechanical properties and initial surface roughness of the workpiece material are the most important characteristics. In the literature, it is possible to search for several studies carried out to investigate the erosive effect of PWJ on various kinds of materials. The authors studied the disintegration ability of PWJ on samples of copper and brass alloys [23, 24], which were subsequently evaluated in terms of the resulting morphology and anisotropy of the material surface. The findings confirmed the dependence of eroded surface on PWJ technological parameters - feed rate, fluid pressure, number of impacts, number of passes. Similar experimental tests were performed on a rolled aluminium sample to investigate the erosive effect of PWJ in terms of different values of ultrasonic excitation amplitude, standoff distance and fluid pressure [25]. The results of the tests showed that with increasing pressure, the PWJ efficiency transition zone is shifted further away from the material surface. The passage of PWJs at various selected standoff distances confirmed the existence of individual zones of efficacy. At the same time, it has been shown that at a higher amplitude of the ultrasonic excitation results in a larger material removal. In order to observe measurable erosion during the experiments, on-line measurement of acoustic emission was recorded during disintegration of aluminium alloy [26]. Comparison of PWJ efficiency on the resulting topography and microstructure of the material surface was carried out while disintegrating aluminium alloy and stainless steel [27]. Determination of the PWJ disintegration effect was also performed on various treated surfaces of aluminium samples [28]. In order to achieve different initial surface roughness, the samples were subjected to fine and coarse milling, planning and rolling processes. The experiment was performed at the selected constant standoff distance and varying feed rate values. The conclusions of the experiment supported the fact that the initial surface roughness plays an important role in material disintegration, but at the same time, it is necessary to consider a possible strengthening process for specific types of surface treatments. Stainless steel material has also been exposed to PWJ, investigating the microstructure and microhardness of the surface [27], the maximum disintegration capacity of the PWJ [29], and the impact of the number of impacts on the erosion stages of the material [30]. In view of the above studies, the absence of PWJ´s effect on polished material surfaces over the entire range of PWJ disintegration capability is observed. To address this research gap, the aim of this study is to describe the disintegration ability of PWJ on polished titanium and aluminium alloy. In order to clearly observe the progressive erosive effect of PWJ within its transition zone of efficiency [16], an equidistantly increasing standoff distance is chosen as the nozzle trajectory above the material. The objective of the evaluation is to achieve deep erosion of samples and their mutual comparison. The study would serve as a basic study for further observations associated with the disintegration of the polished samples.

2 Experimental Setup

Titanium (Ti6Al4V) and aluminium (ENAW1050AH24) alloy samples were selected as the workpiece material in the experiment. The chemical composition and basic mechanical properties of the materials are shown in Table 1. The surface roughness of the individual samples was lowered by the fine grinding process prior to the experiment.

Table 1. Chemical and mechanical properties of Ti6Al4V and EN AW-1050A.

Chemical Composition Ti6Al4V [%]				
Ti	Al	V	O	Fe
90	6	4	<0.2	< 0.25
Mechanical Properties Ti6Al4V				
Tensile Strength R_m [MPa]	Yield Strength $R_{p\ 0.2}$ [MPa]	Elongation at Fracture A [%]	Hardness H [HRB]	
950	880	14	334	

Chemical Composition EN AW-1050A [%]					
Al	Fe	Si	Mn	Zn	Cu
99.5	<0.4	<0.25	< 0.05	< 0.05	< 0.05
Mechanical Properties EN AW-1050A					
Tensile Strength R_m [MPa]	Yield Strength $R_{p\ 0.2}$ [MPa]	Elongation at Fracture A [%]	Hardness H [HRB]		
110	84	6.8	34		

The technological equipment (Fig. 1) consisted of a 2D XY table type PTV WJ2020 - 2Z - 1xPJ adapted for material disintegration (Fig. 1a), hydraulic high-pressure pump Hammelmann HDP 253 with a maximum operating pressure of 160 MPa and a maximum liquid flow of 67 l.min^{-1} (Fig. 1b), a ABB IRB 6640-180 // 2.55 manipulation robot with a nozzle head and a StoneAge circular nozzle mounted in the nozzle head (Fig. 1a, c). The generation of pulses in the liquid was provided by a piezoelectric transducer actuating a sonotrode with a nominal oscillation frequency of 20 kHz. The power source for the piezoelectric transducer was the Ecoson WJ-UG 630-40 ultrasonic generator with a maximum output power of 800 W (Fig. 1d). The surface of the individual samples was disintegrated at the selected technological parameters shown in Table 2. The grooves formed were scanned using a MicroProf FRT optical profilometer (Fig. 1e). The obtained data were then exported to SPIP 6.7.8, in which depth measurements of individual grooves created by PWJ were performed.

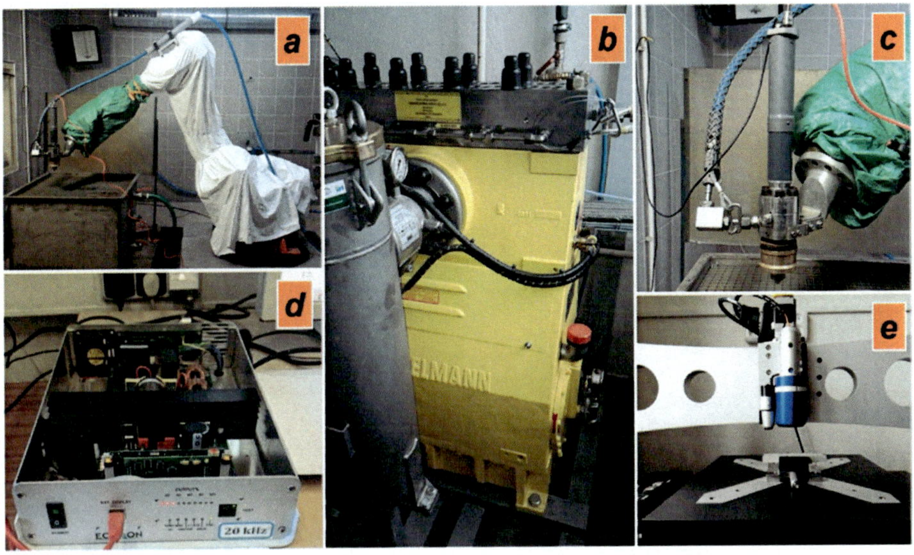

Fig. 1. Equipment used in the experiment: a - ABB handling robot with 2D XY table, b - Hammelmann hydraulic pump, c - PWJ nozzle head, d - Ecoson ultrasonic generator, e - MicroProf FRT optical profilometer.

Table 2. Selected parameters for the experiment.

Technological parameters		
Variable	Standoff distance	$z = 15–41$ mm
Constant	Liquid pressure	$p = 40$ MPa
	Sonotrode frequency	$f = 20.19$ kHz
	Sonotrode amplitude	$A = 3$ µm
	Sonotrode power	$P = 250–260$ W
	Acoustic chamber length	$L_c = 12$ mm
	Nozzle diameter	$d = 0.6$ mm
	Feed rate	$v = 0.25$ mm.s^{-1}
Theoretical	Liquid velocity	$v_W = 254.81$ m.s^{-1}
	Liquid flow rate	$Q_W = 4.32$ l.min^{-1}

Before starting the experiment, a program for equidistant vertical movement of the robot arm and handling the robot was written.

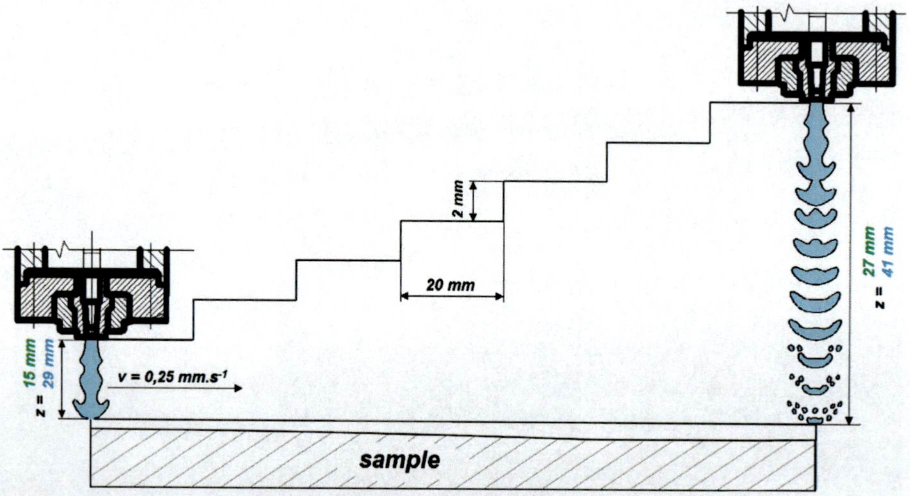

Fig. 2. Experimental scheme.

For the starting standoff distance above the material surface, the value $z = 15$ mm was determined (Fig. 2). This standoff distance value was determined from the knowledge obtained in previous test experiments, that the effects of PWJ at given system parameters do not have the desired disintegration capability at a smaller distance. The subsequent nozzle trajectory was performed by a linear movement of 20 mm in the parallel direction above the material. At the end of this section, the distance from the material was increased by standoff distance of the nozzle in the vertical axis by 2 mm. The whole process of horizontal and vertical movement was repeated. The described equidistant trajectory of the nozzle ended after the last horizontal movement at a height of $z = 27$ mm above the sample surface. Previously tested results have shown that the PWJ disintegration effect continues at a higher distance above the material. However, the length of the sample did not allow further material disintegration. For this reason, the given equidistant passage of the nozzle was repeated next to the first groove formed with a subsequent initial distance $z = 29$ mm above the material (Fig. 2). At the end of the sample, the nozzle was at a height of $z = 41$ mm above the material. The total standoff distance in the experiment was represented by a range from $z = 15$ mm to $z = 41$ mm above the material with a 2 mm difference between the individual horizontal movements. The horizontal movement of the nozzle with respect to the stationary sample was carried out at a constant feed rate $v = 0.25$ mm.s^{-1}. Both polished samples were disintegrated in the manner described. The liquid pressure in the system was kept constant at $p =$ 40 MPa throughout the experiment. This value represents the pressure energy of the liquid needed to achieve the desired PWJ disintegration effects on the polished samples used. Using the ultrasonic generator display, the oscillation frequency of the sonotrode was tuned to $f = 20.19$ kHz with respect to its maximum efficiency. Simultaneously, the power values of the sonotrode P [W] were read and recorded. The length of the acoustic chamber ($L_c = 12$ mm) was set according to the previous tests with respect to the greatest disintegration capability of the PWJ in combination with the nozzle diameter used ($d =$ 0.6 mm) and the fluid pressure ($p = 40$ MPa). At the end of the experiment, the grooved samples were cleaned and dried. Subsequently, the MicroProf FRT optical profilometer was used to scan the grooves formed (Fig. 1e). Later, the obtained data were exported to SPIP 6.7.8, where the depths of the individual grooves were measured and analysed. For each standoff distance of the nozzle from the material, 5 depth measurements of the corresponding groove section were made. The graph plotted is based on the calculation of arithmetic means of groove depth values and also shows the standard deviation of the depth from its mean value.

3 Results and Discussion

The deep disintegration of polished titanium and aluminium samples by a pulsating water jet generates a different course over a given standoff distance (z = 15–41 mm) (Fig. 3). Due to the different mechanical properties of the materials used, elastic-plastic deformation occurs at different standoff distances of the nozzle from the surface of the samples.

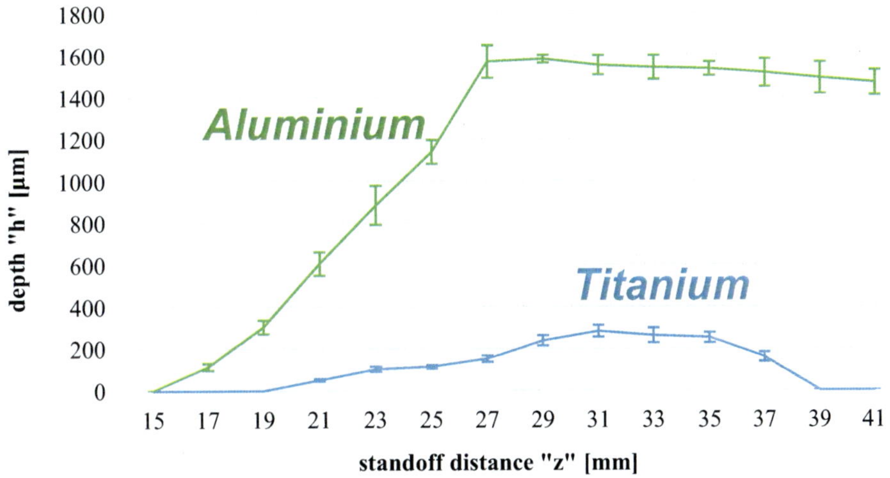

Fig. 3. Comparison of the achieved disintegration depth of titanium and aluminium alloy.

Figure 3 shows the disintegration depth patterns of the titanium and aluminium samples. The blue plot, it can be concluded from the graph that measurable depth erosion does not occur till standoff distance z = 19 mm. At the same time, it can be concluded that under the given experimental conditions up to this limit, the material is within the initial PWJ efficacy zone [16]. The impact of the pulsating water jet for achieving deep disintegration is not sufficient due to the initial stage of formation of individual fluid pulses [25]. The material is exposed to pulsating water jet which so far only causes elastic deformation. In the context of the erosion curve of the material, the incubation stage can be considered for all imaginary elementary points within the material section for standoff distance varying from z = 15 mm to z = 19 mm [31]. Each imaginary point of the material of the given section was exposed to the impact of the PWJ, which was not able to exceed the incubation time limit with the applied technological parameters. Therefore, there is no measurable plastic deformation of the material and no measurable depth erosion. A change in the PWJ disintegration effect can be observed at a nozzle distance $z = 21$ mm above the sample surface. The depth of erosion in this section takes an average depth value of $h = 50$ μm. The cause of material removal can be considered as an advanced stage of fluid pulse development. The material is located at the interface of the initial and transition zone of the PWJ efficiency. The impact pressure and the subsequent lateral fluid flow from the impact point have sufficient energy to cause plastic deformation of the polished surface. The individual imaginary elementary points of the material are exposed to this energy at the impact of the PWJ, which causes the transition from the incubation period to the area of measurable erosion on their erosion curves [31]. The maximum mean disintegration depth reading of the titanium sample over the entire z [mm] range is $h = 283.2$ μm at $z = 31$ mm. The given standoff distance from the material represents the greatest erosive effect of PWJ due to completely formed pulses of liquid. The material is located at the most optimal distance of transition PWJ efficacy zone. By further increasing the distance of the nozzle from the material, the erosive effect of PWJ is reduced. The last disintegration depth ($h = 160.4$ μm) can be observed at a distance $z = 37$ mm above the surface. The given standoff distance from the material represents the termination of the transition zone effect and the onset of the final PWJ zone. From the graph it is also possible to observe a sudden change in the erosive effect of PWJ as it moves from distance $z = 37$ mm to distance $z = 39$ mm. From the value of $h = 160.4$ μm at a standoff distance of $z = 37$ mm to the value of $h = 0$ μm at $z = 39$ mm, the PWJ disintegration abruptly weakens. By comparing the onset (from $z = 19$ mm to $z = 21$ mm) and the completion (from $z = 37$ mm to $z = 39$ mm) of the erosive effect of the PWJ, a different steep course can be observed. It follows that under the given experimental conditions the transition between the initial and transition zones is smoother. On the contrary, there is a more significant imaginary boundary between the transition zone and the final zone. It can be concluded that at a smaller distance, the change in the nozzle standoff distance plays a smaller role compared to the same height change over an overall longer distance. The reason is the aerodynamic resistance from the environment to the kinetic energy of the jet. At a higher standoff distance from the material, further increasing standoff distance results in a more noticeable difference in PWJ effect due to the ever-increasing liquid cone and its contact with the surrounding environment. The contact surface of the cone with the environment with increasing standoff distance increases many times. In this way, the jet breaks up into tiny droplets and loses its disintegration ability. The polished surface of the sample is also likely to play a role in the situation. Due to the smaller surface irregularities forming an obstacle to the lateral jet, greater impact energy is required to achieve disintegration. Consequently, the changes in disintegration will be more pronounced.

The aluminium sample curve (Fig. 3) shows that no disintegration depth was measured up to standoff distance $z = 15$ mm from the material. The material is in the initial efficiency zone of the PWJ fluid column. For the imaginary elementary points of the material in this section, the incubation period of the erosion curve can be considered. For a given material and technological parameters, the transition zone of PWJ efficiency starts at a value of $z = 17$ mm ($h = 114.34$ μm). Disintegration begins with the formation of microcracks and subsequent measurable depth erosion. The disintegration effect of PWJ increases with increasing standoff distance and reaches its maximum at a distance $z = 29$ mm above the surface ($h = 1583.2$ μm). This value is preceded by a slightly smaller depth value $h = 1570.8$ μm at a distance $z = 27$ mm. Up to this value, the depth disintegration curve increases steadily. With a further increase in standoff distance by 2 mm to $z = 29$ mm, the increase in PWJ effect is minimal. For this reason, a slight difference in erosion can be neglected and both $z = 27$ mm and $z = 29$ mm can be regarded as the standoff distance with the highest disintegration effect in the full range $z = 5$ to 41 mm. The sudden change in the groove depth increasing trend at $z = 27$ mm reflects the achievement of the optimal standoff distance with the fully developed shape of the liquid pulses. At the same time, a rapid change may be the result of a polished sample surface that highlights changes in PWJ effects. Further increasing the standoff distance leads to a decrease in the disintegration depth. The material reaches beyond the optimum distance and moves to the lower limit of the transition zone. However, the decreasing course is relatively smooth in nature. At the highest distance from the material ($z = 41$ mm), the PWJ disintegration capacity is represented by a depth $h =$ 1470.2 μm. From the graph, it can be concluded that the erosive effect of PWJ continues with the further increase of the standoff distance.

These materials have different mechanical properties (Table 1), which lead to considerable differences in the resulting disintegration depth pattern (Fig. 3). The first measurable disintegration recorded in the case of the titanium sample is shifted by 4 mm compared to the aluminium sample in terms of the standoff distance of the nozzle from the material (Ti - $z = 21$ mm, Al - $z = 17$ mm). At the same time, the initial erosion of the titanium sample results in a smaller groove depth (Ti - $h = 50.22$ μm, Al - $h =$ 114.34 μm). From this, it can be concluded that the incubation time of the erosion curve for the titanium is longer as compared to aluminium. The material is able to withstand the effects of PWJ for a longer period of time and at the same time undergo less erosion after crossing this limit. The PWJ transition zone also differs for both materials and has a different impact on them. In the case of an aluminium sample, the development of deep disintegration takes a much steeper course compared to titanium. As the standoff distance increases, the erosion of aluminium becomes more significant with increasing material removal. The PWJ achieves maximum disintegration depth at a lower standoff distance ($z = 27$, (29) mm - Al, $z = 31$ mm - Ti). The optimum nozzle distance for the greatest PWJ effect (within the equidistant trajectory used) for both materials does not show a significant difference. However, the groove depth is more than 5.5 times higher for the aluminium sample ($h = 1583.2$ μm - Al, $h = 283.2$ μm - Ti). This is due to much lower resistance to compressive and shear stresses induced by the PWJ impact for aluminium compared to titanium at used technological parameters. The course of PWJ erosion decreases considerably for both the sample. The groove formed in aluminium tends to continue beyond the standoff distance more than $z = 41$ mm. The depth value at this distance is $h = 1470.2$ μm. However, the groove formed in titanium ceases to exhibit a measurable disintegration depth at a standoff distance $z = 39$ mm. From the above, it is apparent that while the titanium sample at $z = 39$ mm is in the final PWJ efficiency zone, for the aluminium sample the given distance still represents the effective transition zone of the liquid column. Therefore, it is particularly important to define under which conditions and on what material the experiment is carried out in the context of determining the individual zones of PWJ efficiency. In order to display the range of values obtained when measuring the individual groove depth sections, standard deviations were plotted (Fig. 3). When comparing the experimental samples, a smaller depth range for the titanium material is visible. This may be due to the higher resistance of titanium to the effects of PWJ resulting from its mechanical properties. Due to the greater strength compared to the aluminium sample, the titanium material erodes in smaller quantities. Those results in smaller variations in depth values within the individual sections of the groove formed. Within both samples, a relatively approximate standard deviation value can be observed over the entire course of disintegration. The polished surface of the samples is considered to be the reason, which in the experiment represents constant conditions in terms of surface roughness.

4 Conclusion

The objective of the present study was to determine the effect of standoff distance ($z =$ 15 to 41 mm) on disintegration depth (h) of the polished sample of Aluminium alloy and Ti6Al4V. Following conclusions were drawn from the results of the experiment:

1. Various efficacy zone such as initial, transition and final (droplet) zone of the water column and its formation with varying standoff distance for different materials were determined.
2. The transition and droplet formation stage were formed at standoff distance $z =$ 15 mm, (unknown value) mm for aluminium alloy and $z = 19$ mm, 39 mm for titanium alloy.
3. The deepest disintegration depth achieved throughout the variation of standoff distance was 1583.2 μm for aluminium alloy as compared to 283.2 μm for titanium alloy. This difference in the depth was attributed to the physical properties of both the materials.
4. Disintegration depth curve for aluminium alloy showed that erosion of the material would continue with further increase in the standoff distance ($z > 41$ mm).

Acknowledgement. This study was supported by the Slovak Research and Development Agency under Contract No. APVV-17-0490.

References

1. Hloch, S., Nag, A., Pude, F., Foldyna, J., Zeleňák, M.: On-line measurement and monitoring of pulsating saline and water jet disintegration of bone cement with frequency 20 kHz. Measurement **147**, 106828 (2019). https://doi.org/10.1016/j.measurement.2019.07.056
2. Lehocká, D., Klich, J., Botko, F., Foldyna, J., Hloch, S., Kepič, J., Kovaľ, K., Krejči, L., Storkan, Z.: Pulsating water jet erosion effect on a brass flat solid surface. Int. J. Adv. Manuf. Technol. **97**, 1099–1112 (2018). https://doi.org/10.1007/s00170-018-1882-4

3. Raj, P., Hloch, S., Tripathi, R., Srivastava, M., Nag, A., Klichová, D., Klich, J., Hromasová, M., Muller, M., Miloslav, L., Chattopadhyaya, S., Adamcik, P.: Investigation of sandstone erosion by continuous and pulsed water jets. J. Manuf. Process. **42**, 121–130 (2019). https://doi.org/10.1016/j.jmapro.2019.04.035
4. Nag, A., Srivastava, A.K., Dixit, A.R., Chattopadhyaya, S., Mandal, A., Klichová, D., Hlaváček, P., Zeleňák, M., Hloch, S.: Influence of abrasive water jet turning parameters on variation of diameter of hybrid metal matrix composite. In: Singh, M., Kushvah, B., Seth, G., Prakash, J. (eds.) Applications of Fluid Dynamics, pp. 495–504. Springer, Singapore (2018)
5. Hloch, S., Hlaváček, J., Vasilko, K., Cárach, J., Samardžić, I., Kozak, D., Hlavatý, I., Ščučka, J.J., Klich, J., Klichová, D.: Abrasive waterjet (AWJ) titanium tangential turning evaluation. Metalurgija **53**, 537–540 (2014)
6. Hlaváček, P., Cárach, J., Hloch, S., Vasilko, K., Klichová, D., Klich, J., Lehocká, D.: Sandstone turning by abrasive waterjet. Rock Mech. Rock Eng. **48**, 2489–2493 (2015). https://doi.org/10.1007/s00603-015-0719-9
7. Cárach, J., Hloch, S., Hlaváček, P., Ščučka, J., Martinec, P., Petrů, J., Zlámal, T., Zeleňák, M., Monka, P., Lehocká, D.: Tangential turning of Incoloy alloy 925 using abrasive water jet technology. Int. J. Adv. Manuf. Technol. **82**, 1747–1752 (2016). https://doi.org/10.1007/s00170-015-7489-0
8. Hutyrová, Z., Ščučka, J., Hloch, S., Hlaváček, P., Zeleňák, M.: Turning of wood plastic composites by water jet and abrasive water jet. Int. J. Adv. Manuf. Technol. **84**, 1615–1623 (2016). https://doi.org/10.1007/s00170-015-7831-6
9. Cárach, J., Hloch, S., Petrů, J., Nag, A., Gombár, M., Hromasová, M.: Hydroabrasive disintegration of rotating Monel K-500 workpiece. Int. J. Adv. Manuf. Technol. **96**, 981–1001 (2018). https://doi.org/10.1007/s00170-018-1653-2
10. Srivastava, A.K., Nag, A., Dixit, A.R., Tiwari, S., Scucka, J., Zelenak, M., Hloch, S., Hlavacek, P.: Surface integrity in tangential turning of hybrid MMC A359/B4C/Al2O3by abrasive waterjet. J. Manuf. Process. **28**, 11–20 (2017). https://doi.org/10.1016/j.jmapro.2017.05.017
11. Nag, A., Ščučka, J., Hlavacek, P., Klichová, D., Srivastava, A.K., Hloch, S., Dixit, A.R., Foldyna, J., Zelenak, M.: Hybrid aluminium matrix composite AWJ turning using olivine and Barton garnet. Int. J. Adv. Manuf. Technol. **94**, 2293–2300 (2018). https://doi.org/10.1007/s00170-017-1036-0
12. Srivastava, M., Tripathi, R., Hloch, S., Chattopadhyaya, S., Dixit, A.R.: Potential of using water jet peening as a surface treatment process for welded joints. Procedia Eng. **149**, 472–480 (2016). https://doi.org/10.1016/j.proeng.2016.06.694
13. Srivastava, M., Hloch, S., Gubeljak, N., Milkovic, M., Chattopadhyaya, S., Klich, J.: Surface integrity and residual stress analysis of pulsed water jet peened stainless steel surfaces. Measurement **143**, 81–92 (2019). https://doi.org/10.1016/j.measurement.2019.04.082
14. Hloch, S., Srivastava, M., Krolczyk, J.B., Chattopadhyaya, S., Lehocká, D., Simkulet, V., Krolczyk, G.M.: Strengthening effect after disintegration of stainless steel using pulsating water jet. Teh. Vjesn. **25**, 1075–1079 (2018). https://doi.org/10.17559/TV-20170327134630
15. Foldyna, J., Sitek, L., Švehla, B., Švehla, S.: Utilization of ultrasound to enhance high-speed water jet effects. Ultrason. Sonochem. **11** (2004). https://doi.org/10.1016/j.ultsonch.2004.01.008
16. Chillman, A., Ramulu, M., Hashish, M.: A general overview of waterjet surface treatment modeling. In: American WJTA Conference and Expo (2009)
17. Srivastava, M., Hloch, S., Krejci, L., Chattopadhyaya, S., Dixit, A.R., Foldyna, J.: Residual stress and surface properties of stainless steel welded joints induced by ultrasonic pulsed water jet peening. Meas. J. Int. Meas. Confed. **127**, 453–462 (2018). https://doi.org/10.1016/j.measurement.2018.06.012

18. Srivastava, M., Hloch, S., Tripathi, R., Kozak, D., Chattopadhyaya, S., Dixit, A.R., Foldyna, J., Hvizdos, P., Fides, M., Adamcik, P.: Ultrasonically generated pulsed water jet peening of austenitic stainless-steel surfaces. J. Manuf. Process. **32** (2018). https://doi.org/10.1016/j.jmapro.2018.03.016
19. Hloch, S., Foldyna, J., Pude, F., Kľoc, J., Zeleňák, M., Hvizdoš, P., Monka, P., Smolko, I., Ščučka, J., Kozak, D.: Experimental in-vitro bone cements disintegration with ultrasonic pulsating water jet for revision arthroplasty. Tech. Gaz. **22**, 1609–1615 (2015). https://doi.org/10.17559/TV-20150822145550
20. Hloch, S., Foldyna, J., Sitek, L., Zeleňák, M., Hlaváček, P., Hvizdoš, P., Kľoc, J., Monka, P., Monková, K., Kozak, D., Magurová, D.: Disintegration of bone cement by continuous and pulsating water jet. Tech. Gaz. **20**, 593–598 (2013). https://hrcak.srce.hr/106686
21. Nag, A., Hloch, S., Dixit, A.R., Cuha, D.: Investigation on pulsating liquid jet with physiological saline on aluminium surface. In: Hloch, S., Klichová, D., Krolczyk, G., Chattopadhyaya, S., Ruppenthalová, L. (eds.) Advances in Manufacturing Engineering and Materials, pp. 63–71. Springer, Cham (2019). https://doi.org/10.1007/978-3-319-99353-9_8
22. Nag, A., Hloch, S., Čuha, D., Dixit, A.R., Tozan, H., Petrů, J., Hromasová, M., Müller, M.: Acoustic chamber length performance analysis in ultrasonic pulsating water jet erosion of ductile material. J. Manuf. Process. **47**, 347–356 (2019). https://doi.org/10.1016/j.jmapro.2019.10.008
23. Lehocka, D., Klich, J., Foldyna, J., Hloch, S., Krolczyk, J.B., Carach, J., Krolczyk, G.M.: Copper alloys disintegration using pulsating water jet. Measurement **82**, 375–383 (2016). https://doi.org/10.1016/j.measurement.2016.01.014
24. Lehocká, D., Klichová, D., Foldyna, J., Hloch, S., Hvizdoš, P., Fides, M., Botko, F.: Comparison of the influence of acoustically enhanced pulsating water jet on selected surface integrity characteristics of CW004A copper and CW614N brass. Measurement **110**, 230–238 (2017). https://doi.org/10.1016/j.measurement.2017.07.005
25. Foldyna, J., Sitek, L., Ščučka, J., Martinec, P., Valíček, J., Páleníková, K.: Effects of pulsating water jet impact on aluminium surface. J. Mater. Process. Technol. **209**, 6174–6180 (2009). https://doi.org/10.1016/j.jmatprotec.2009.06.004
26. Hloch, S., Adamčík, P., Nag, A., Srivastava, M., Čuha, D., Müller, M., Hromasová, M., Klich, J.: Hydrodynamic ductile erosion of aluminium by a pulsed water jet moving in an inclined trajectory. Wear **428–429**, 178–192 (2019). https://doi.org/10.1016/j.wear.2019.03.015
27. Lehocka, D., Klich, J., Botko, F., Simkulet, V., Foldyna, J., Krejci, L., Storkan, Z., Kepic, J., Hatala, M.: Comparison of ultrasonically enhanced pulsating water jet erosion efficiency on mechanical surface treatment on the surface of aluminum alloy and stainless steel. Int. J. Adv. Manuf. Technol., 1–10 (2019). https://doi.org/10.1007/s00170-019-03680-8
28. Klich, J., Klichova, D., Foldyna, V., Hlavacek, P., Foldyna, J.: Influence of variously modified surface of aluminium alloy on the effect of pulsating water jet. Stroj. Vestn./J. Mech. Eng. **63**, 577–582 (2017). https://doi.org/10.5545/sv-jme.2017.4356
29. Lehocka, D., Klich, J., Pitel, J., Krejci, L., Storkan, Z., Duplakova, D., Schindlerova, V., Sajdlerova, I.: Analysis of the pulsating water jet maximum erosive effect on stainless steel. In: Hloch, S., Klichová, D., Krolczyk, G., Chattopadhyaya, S., Ruppenthalová, L. (eds.) Advances in Manufacturing II, pp. 233–241. Springer, Cham (2019). https://doi.org/10.1007/978-3-030-16943-5_2
30. Foldyna, J., Klich, J., Hlavacek, P., Zelenak, M., Scucka, J.: Erosion of metals by pulsating water jet. Tech. Gaz. **19**, 381–386 (2012). https://hrcak.srce.hr/83919
31. Thomas, G.P., Brunton, J.H.: Drop impingement erosion of metals. Proc. R. Soc. A Math. Phys. Eng. Sci. (1970). https://doi.org/10.1098/rspa.1970.0022

Water Jet as a Promising Tool to Disperse Carbon Nanotubes in Water Solution

Vladimir Foldyna[1,2](✉), Josef Foldyna[1], and Michal Zelenak[1]

[1] Institute of Geonics of the Czech Academy of Sciences, Ostrava, Czechia
Vladimir.foldyna@ugn.cas.cz
[2] Nanotechnology Centre, VSB-Technical University of Ostrava, Ostrava, Czechia

Abstract. This paper is focused on obtaining a homogenous dispersion of carbon nanotubes in water solution treated by the pulsating and continuous water jet technology. Our experimental setup consists of Kärcher HD 6/18 plus pump, ABB robot IRB 6640, pulsating jet generator and Hammelmann nozzle with a diameter $d = 0.3$ mm. Carbon nanotubes were in the concentration of 0.375 wt%. Three types of samples were prepared. Carbon nanotubes were mixed with no surfactant, half concentration of surfactant and with a standard concentration of surfactant, which was optimized in previous experiments (1.5 wt%). Based on our previous experimental results, we used a naphthalene based superplasticizer. A spectrum of pressures of 1 MPa, 5 MPa, 10 MPa and 15 MPa was used to induce both cavitation and non-cavitation phenomena. Samples were kept in idle state to investigate if they were subject to re-agglomeration. Consequently, all Carbon nanotubes solutions were measured on the Malvern nanosizer.

Keywords: Continuous/pulsating water jet · Carbon nanotubes · Dispersion

1 Introduction

Carbon nanotubes (CNTs) can be described as hexagonal nets created from rolled-up graphene into cylindrical nanoscale tubes. They are either with open ends or ended with fullerene hemispheres. CNTs can be divided into two main groups: single-walled carbon nanotubes (SWCNTs) and multi-walled carbon nanotubes (MWCNTs). SWCNTs are in the form of a single rolled-up sheet of graphene (usually prepared by the process of laser ablation) and MWCNTs consist of two or more graphene layers. MWCNTs are cheaper to prepare (usually made by a method of chemical vapor deposition), so they are used more frequently than SWCNTs [1]. MWCNTs have many applications in industry (as electronic materials, in medicine and chemistry) [2] due to their exceptional chemical and physical properties (mechanical strength, flexibility and corrosion resistance, electrical conductivity) [3]. These properties make them excellent reinforcing materials for various types of nanocomposites. CNTs are commonly used as additives to reinforce composite materials, such as concrete polymethyl methacrylate, epoxy, alumina and petroleum pitch composites.

D. Klichová et al. (Eds.): Water Jet 2019, LNME, pp. 67–72, 2021.
https://doi.org/10.1007/978-3-030-53491-2_7

However, their applications are limited due to the effective inter-tube van der Waals interactions, which leads to clusterization, poor dispersibility, and re-agglomeration in various solvents [4]. In recent years, many scientific papers on the topic of CNTs dispersion have been published. Standard methods, i.e. ultrasonication or mechanical mixing, combined with various surfactants such as sodium dodecyl sulfate, sodium deoxycholate, gum Arabic, cetyltrimethylammonium bromide, and others, are often used for obtaining a homogenous and stable dispersion [5].

2 Experimental

Experiments oriented at the dispersion of carbon nanotubes in water by pulsating (PWJ) and continuous (CWJ) water jet treatment were performed in the Waterjet laboratory of the Institute of Geonics.

2.1 Laboratory Equipment and Testing Conditions

The experimental assembly for the water jet treatment evaluation consisted of a pulsating water jet head (with a 20 kHz ultrasonic generator) installed on the arm of an ABB IRB 6640 robot and a low-pressure Kärcher HD 6/18 plus pump. A spectrum of pressures of 1 MPa, 5 MPa, 10 MPa and 15 MPa was used to induce both cavitation and non-cavitation phenomena. The commercially available Hammelmann nozzle with a diameter $d =$ 0.3 mm was used in the experiment to generate both the PWJ and the CWJ. A proprietary method of the pulsating liquid jet formation based on the generation of acoustic waves by the action of an acoustic transducer on the pressure liquid and their transmission via a pressure system to the nozzle (so-called acoustic generator of pressure pulsations) was used to create the PWJ. The acoustic generator of pressure pulsations used in experiments formed pressure pulsations at the frequency of about 20 kHz (exact frequency depends on the actual geometrical configuration of the generator), the amplitude of vibrations of the acoustic transducer was set to 7 μm. More details on the method of pulsating water jet generation can be found in appropriate literature, see e.g. [6, 7]. The experimental setup is visualized in Fig. 1.

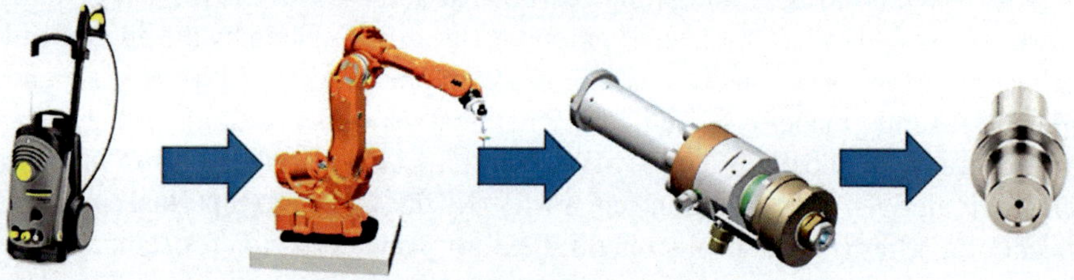

Fig. 1. Experimental setup: Kärcher HD 7/18 C plus pump [8], ABB robot IRB 6640 [9], pulsating jet generator and Hammelmann nozzle of 0.3 mm.

2.2 Preparation of Testing Samples

Multi-walled carbon nanotubes from Yurui (Shanghai) Chemical Co., Ltd. were used for the preparation of a CNT/concrete composite in the presented work. They were prepared by chemical vapor deposition with a purity of more than 95%. Their properties are summarized in Table 1.

Table 1. Properties of MWCNTs.

Properties	
Internal diameter	Approx. 5 nm to 12 nm
Outer diameter	Approx. 30 nm to 50 nm
Length	10 μm to 20 μm
Bulk density	0.22 g·cm^{-3}
True density	2.1 g·cm^{-3}

A defined volume of CNTs was stirred in 30 l barrel. Three types of samples were prepared: Solution A – without any surfactant (a naphthalene based superplasticizer), solution B with half amount of the surfactant (0.75 wt%) and solution C with a standard amount of the surfactant (1.5 wt%). This amount was optimized based on previous experiments in our laboratory, a specially designed sonotrode rosette vessel was used to enable homogenous mixing of nanomaterials. A more precise description of this method was published in scientific papers [10, 11]. The above-mentioned spectrum of pressures was selected based on the numerical simulation method showing that the cavitation process for pulsating water jet occurs at a pressure of 8 MPa. This means that solutions treated with the pulsating water jet at pressures of 1 and 5 MPa were exposed both to cavitation effects and pressure pulsations. Solutions treated with the pulsating water jet at pressures of 10 and 15 MPa were only exposed to pressure pulsations. Secondly, the solution was treated by the continuous water jet. In both cases, treated solutions were captured in a vessel at a distance of approximately 30 cm. Solution stability of all types of samples was observed for more than a month to determine if they re-agglomerate and sediment.

2.3 Particle Size Measurements

In solutions (dispersions) with no observable instability or re-agglomerations, particle size and particle size distribution were analyzed by a specialized automated Zetasizer® instrument (Malvern, Netherlands). Malvern Zetasizer software v.7.01 was used to evaluate and statistically process the obtained results.

3 Results and Discussion

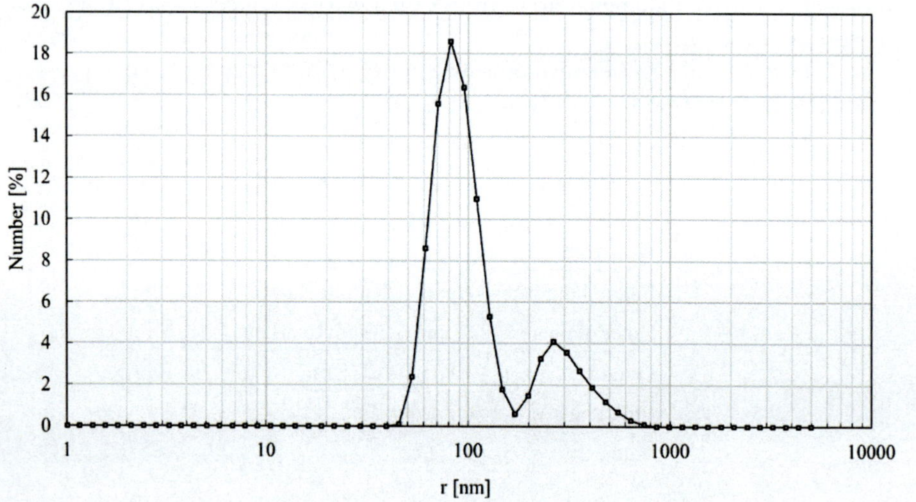

Fig. 2. Example of the particle size distribution of solution C using a CWJ at a pressure of 15 MPa.

A typical shape of the particle size distribution is illustrated in Fig. 2. Most of the realized experiments showed a typical bimodal distribution of particles. Small-size (nano)particles having the median size usually under or around 100 nm and big-size (nano)particles having the median size over 200 nm were present. In experiments with solution A at a pressure of 5 MPa (with and without pulsation) and with solution C at pressures of 10 and 15 MPa (with pulsation), only one peak of the particle size distribution occurred. In these experiments, the one peak corresponded to small- or big-size particles in relation to the median size value (small-size $\leq$ 150 nm; big-size > 150 nm) (Table 2).

Table 2. Results of solution A dispersed by pulsating and continuous water jet.

Method	PWJ		CWJ	
Pressure [MPa]	Size [nm]		Size [nm]	
1	42.5	302.4	85.9	269.6
5	–	349.7	106.4	257.8
10	78.2	273.2	73.6	256.2
15	73.9	416.5	74.2	295.4

Table 3. Results of solution B dispersed by pulsating and continuous water jet.

Method	PWJ		CWJ	
Pressure [MPa]	Size [nm]		Size [nm]	
1	93.3	255.1	–	–
5	106.7	–	112.6	–
10	117.2	–	70.8	240.9
15	80	261.4	129.3	–

Table 4. Results of solution C dispersed by pulsating and continuous water jet.

Method	PWJ		CWJ	
Pressure [MPa]	Size [nm]		Size [nm]	
1	–	–	–	–
5	117.4	414.3	–	216.7
10	–	404	94.7	419.1
15		200	88.4	316.2

Generally, the method of water jet treatment using a continuous and pulsating water jet at various pressures can be applied for homogenous dispersion of carbon nanotubes in water solution. However, the obtained results show that the pressure of 1 MPa is too low and thus unsuitable for solutions C dispersion by PWJ and CWJ (Table 4), and solution B dispersion by CWJ (Table 3). The results indicate that the cavitation process does not have a significant effect on the dispersion of CNTs. Other pressures (i.e. pressures of 5, 10 and 15 MPa) are suitable to disperse CNTs in tested solutions. Most of the samples had a much higher peak of the small-size particle distribution, excluding the following 4 samples: solution A dispersed by PWJ at a pressure of 1 MPa and by CWJ at 5 MPa; solution C dispersed by PWJ at pressures of 10 and 15 MPa.

4 Conclusions

Method of water jet treatment can be used for homogenous dispersion of carbon nanotubes in water solution at various pressures and using a continuous or pulsating water jet. Both methods of pulsating and continuous water jet application showed similar results of the dispersion of all tested solutions. It was proven that the cavitation process did not have a significant effect on the dispersion of CNTs. The pressure of 1 MPa was too low to disperse CNTs. Contrary to the authors' expectations, better results were achieved for solutions A and B. Therefore, it is preferable to use lower surfactant concentrations or even solution without any surfactant for the tested method of CNTs dispersion by water jets.

The presented research should further continue with:

- Measurement of the Zeta potential of the best samples
- Preparation of CNTs/composite materials based on the best samples of dispersed solutions in cooperation with Brno University of Technology
- Realization of high-resolution transmission electron microscopy images of the above-mentioned composites at the Institute of Physics of Material of the Czech Academy of Sciences in Brno
- Comparison of composites prepared by means of the water jet technology with samples previously made using the ultrasonication (flexural and compressive strength, the effect of the pulsating and continuous water jet on the surface).

References

1. Kang, S., Seo, J., Park, S.: The characteristics of CNT/cement composites with acid-treated MWCNTs. Adv. Mater. Sci. Eng. **2015**, 1–9 (2015). https://doi.org/10.1155/2015/308725
2. Grobert, N.: Carbon nanotubes – becoming clean. Mater. Today **10**(1–2), 28–35 (2007). https://doi.org/10.1016/S1369-7021(06)71789-8
3. Popov, V.N.: Carbon nanotubes: properties and application. Mater. Sci. Eng.: R: Reports **43**(3), 61–102 (2004). https://doi.org/10.1016/j.mser.2003.10.001
4. Jiang, L., Gao, L., Sun, J.: Production of aqueous colloidal dispersions of carbon nanotubes. J. Colloid Interface Sci. **260**(1), 89–94 (2003). https://doi.org/10.1016/s0021-9797(02)00176-5. ISSN 00219797
5. Luo, J., Duan, Z., Li, H.: The influence of surfactants on the processing of multi-walled carbon nanotubes in reinforced cement matrix composites. Physica status solidi (a) (2009). https://doi.org/10.1002/pssa.200824310. ISSN 18626300
6. Foldyna, J., Svehla, B.: Method of generation of pressure pulsations and apparatus form implementation of this method. US patent no. 7,934,666, USPTO, Aleksandria
7. Foldyna, J., Sitek, L., Scucka, J., Martinec, P., Valicek, J., Palenikova, K.: Effects of pulsating water jet impact on aluminum surface. J. Mater. Process. Technol. **209**(20), 6174–6180 (2009). ISSN 0924-0136
8. High pressure HD 7/18 C Plus. In: Karcher Data sheet PDF: https://s1.kaercher-media.com/documents/datasheets/machines/cs_CZ/1.151-614.0_PI_cs_CZ.pdf
9. IRB 6640. A strong robot for numerous applications. In: ABB, Data Sheet PDF
10. Foldyna, V., Foldyna, J., Klichova, D., Klich, J., Hlavacek, P., Bodnarova, L., Jarolim, T., Kutlakova, K.M.: Effects of continuous and pulsating water jet on CNT/CONCRETE COMPOSITE. J. Mech. Eng. **63**(10), 583–589 (2017). ISSN 0039-2480
11. Foldyna, V., Foldyna, J., Klichova, D.: Effects of water jets on CNTs/concrete composite. MM Sci. J. **2018**(01), 2229–2233 https://doi.org/10.17973/mmsj.2018_03_201775. ISSN 18031269

Influence of Concrete Age on Resistance to Fast-Flowing Liquids

Petr Hlaváček[1(✉)], Libor Sitek[1], Lenka Bodnárová[2], Rudolf Hela[2], Kamil Souček[1], Vendula Zajícová[1], and Vladimír Foldyna[1]

[1] Institute of Geonics of the Czech Academy of Sciences, Ostrava, Czechia
petr.hlavacek@ugn.cas.cz
[2] Faculty of Civil Engineering, Brno University of Technology, Brno, Czechia

Abstract. It is generally known that the properties of concrete, particularly the strength, vary depending on the concrete age due to the chemical-mechanical processes occurring in the structure of concrete during ageing (maturing). The issue often is the *time n*eeded for *concrete* to reach its desired safety before a concrete structure made of fresh concrete can be partially or completely loaded. In structures coming into contact with flowing liquids, the resistance of surface layers to erosion caused by abrasion or cavitation is also important, in addition to the strength. Therefore the analysis of effects of the concrete age from 7 days to 1 year on the concrete resistance to fast-flowing liquids is the subject of this paper. A method of accelerated hydraulic-mechanical simulation of the surface layer loading by means of a pure high-speed water jet and a water jet with abrasive particles was used in order to investigate the concrete erosion resistance. The influence of the concrete age on the resistance to fast-flowing liquids has been proven. After a year, the concrete resistance increased by about 36% in the case of the pure water jet, and 28% in the case of the abrasive water jet.

Keywords: Concrete aging · Resistance · Fast-flowing liquid · Abrasive water jet

1 Introduction

Concrete surfaces affected by flowing liquids for a long time are exposed to a combination of several degradation processes. The gradual erosion of surface layers is caused not only by abrasion but also by the effects of chemical and bacterial impacts. In addition, cavitation occurs at higher fluid flow rates [1]. Fortunately, the aforementioned degradation mechanism for concrete structures is the relatively slow disintegration rate of the surface layers. However, this is an obstacle for quick and reliable prediction of the behavior of concrete structures. Visible deterioration can occur within some months, more often after several years or decades. Due to the slow process of degradation, it is thus difficult to verify the resistance of a particular concrete mixture in real conditions.

High-speed water jets represent a promising [2] and relatively popular [3, 4] method for simulating the flow of contaminated liquids that contain high or low amounts of free

D. Klichová et al. (Eds.): Water Jet 2019, LNME, pp. 73–80, 2021.
https://doi.org/10.1007/978-3-030-53491-2_8

hard particles of different sizes. The interaction of the water jet with the concrete surface causes the same degradation mechanisms that occur in real concrete structures. Molding abrasive particles in the high-speed water jet may further improve the disintegration process. The erosion is rapid and real erosion processes can be thus faithfully simulated with an appropriate combination of water jet parameters and test settings [5].

2 Effects of Concrete Age

Concrete strength changes over time. The strength development of concrete depends on many factors; in particular, the type of cement used, concrete density, water coefficient, type of additives and admixtures used and finally, climatic conditions (especially the ambient temperature). The variability of the progression of concrete strength increase is described, for example, in the standard [7]. According to the strength coefficient, the increase in strength is classified from very slow (ratio of the average compressive strength after 2 days to the average compressive strength after 28 days is less than 0.15) to rapid (ratio greater or equal to 0.5). The concrete strength increases even after 28 days from the concrete production. In terms of the concrete composition, the increase in concrete strength over a longer period depends mainly on the type of admixture used (i.e. slag, fly ash). It is generally expected that higher concrete strength leads to better resistance of the concrete to negative effects of flowing water. Due to the complexity and extent of hydraulic structures, the duration of their exposure to the flowing water is longer than the process of concrete curing in 28 days.

Our research is focused on investigating the influence of concrete age on its resistance to the abrasive erosion using a simulation method based on the application of high-speed water jets (for example [2]).

3 Method, Material and Equipment

The method is based on directing a high-speed water jet perpendicular on the surface of a concrete specimen and moving it along a predetermined zig-zag trajectory in order to affect a larger area of concrete [6]. One impact cycle consists of two mutually perpendicular trajectories, creating a pocket in the surface layers of concrete. One or two disintegration cycles were made in the same place in the concrete specimen in order to deepen the pocket and thus to increase the amount of the concrete removed. Two contactless methods were used for evaluation of the removed volume V_e: a portable 3D scanner (GO!SCAN SPARK, Creaform, Quebec, Canada) (see Fig. 1) and a digital microscope (VHX 5000, Keyence, Osaka, Japan). Results obtained with the 3D scanner showed approximately 5% higher values of the removed volume. This is because the method enables the measurement of cavities that are "invisible" for the digital microscope. Therefore, only the values of removed material obtained by 3D scanning method were used for further analysis.

The volumetric erosion rate V_r was applied as a measure for the concrete resistance to fast-flowing liquids

$$V_r = V_e \cdot t_e^{-1} \quad (1)$$

where V_e is the removed volume and t_e is the exposure time.

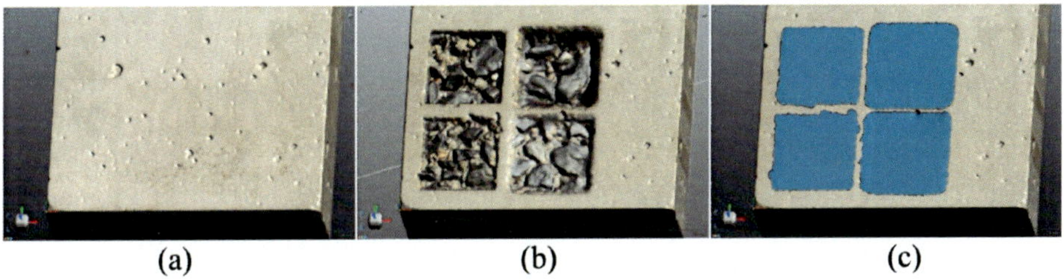

Fig. 1. Concrete specimen (a) before the test, (b) after water jet erosion, (c) removed volume evaluated based on 3D scanning.

Inner structure of concrete was analysed by means of an X-ray CT. A micro X-ray CT scanner XT H 225 ST by Nikon Metrology was used. One 3D CT model was constructed from 3141 2D radiographic projections of a test specimen obtained during a full-angle rotation. The exposure time of one X-ray projection was 16 s. Edge size of individual cubic voxels in the 3D CT volume is approximately 17 μm. VG Studio Max v. 2.2 software was used for the CT data analysis. The test specimens were scanned prior to treatment with a pure water jet or a water jet with abrasive particles. After erosion tests, the same concrete specimens were re-scanned in order to evaluate the effects of erosion on integrity of the inner concrete structure. The analyzed volume was located in the central part of the scanned test specimens to eliminate the effects of external boundary conditions.

Concrete C30/37 [7] specimens in the shape of a cube with dimensions 150 × 150 × 150 mm were prepared for testing. High-speed water jet erosion tests were performed on these specimens after 7, 28, 60 and 365 days from manufacturing. The following raw materials were used for the concrete mixture preparation: Portland cement CEM I 42.5 R, four fractions of aggregate (sand 0/4 mm, coarse crushed aggregates fraction 4/8, 8/16, 11/22 mm) and ground blast furnace slag. Water from a water supply system was used. A poly(carboxylate ether)-based superplasticizer was dosed in the amount of 1% by cement weight. The consistency of fresh concrete was determined using the concrete slump test [8]. The slump was 110 mm – S3.

The following properties of concrete specimens were determined: density of the hardened concrete [9], compressive strength [10] and depth of penetration of the water under pressure [11]. Properties of the concrete after 28 days of ageing are given in Table 1.

Table 1. Compressive strength, density of the hardened concrete and depth of penetration of the water under pressure (after 28 days).

Concrete property	Value
Compressive strength	40.2 MPa
Density of the hardened concrete	2380 kg.m^{-3}
Depth of penetration of the water under pressure	Max. 14 mm

High-pressure water was supplied to a cutting head by a PTV75-60 high-pressure pump with two pressure intensifiers capable to deliver up to 7.8 l·min^{-1} of water at operating pressure from 40 up to 415 MPa (max power 75 kW). An X-Y cutting table PTV WJ2020-2Z-1xPJ – 2D with inclinable cutting head specially designed for cutting with water and abrasive water jets was used to traverse an abrasive water jet cutting head. A standard commercially available PTV Slice I cutting head equipped with a water nozzle diameter of 330 μm, focusing tube diameter of 1020 μm and focusing tube length of 76.2 mm was used. During experiments, the water jet parameters were set as follows: water pressure – 40 MPa (for abrasive water jet) and 80 MPa (for pure water jet), standoff distance (from the exit of the focusing tube to the top surface of the specimen) – 50 mm. The impact angle of the water jet was set perpendicular to the concrete surface. The traverse velocity of the water jet above the concrete surface was set at 500 mm·min^{-1}. In all experiments, the abrasive water jet consisted of a mixture of water and garnet abrasive grains having an average equivalent spherical diameter of

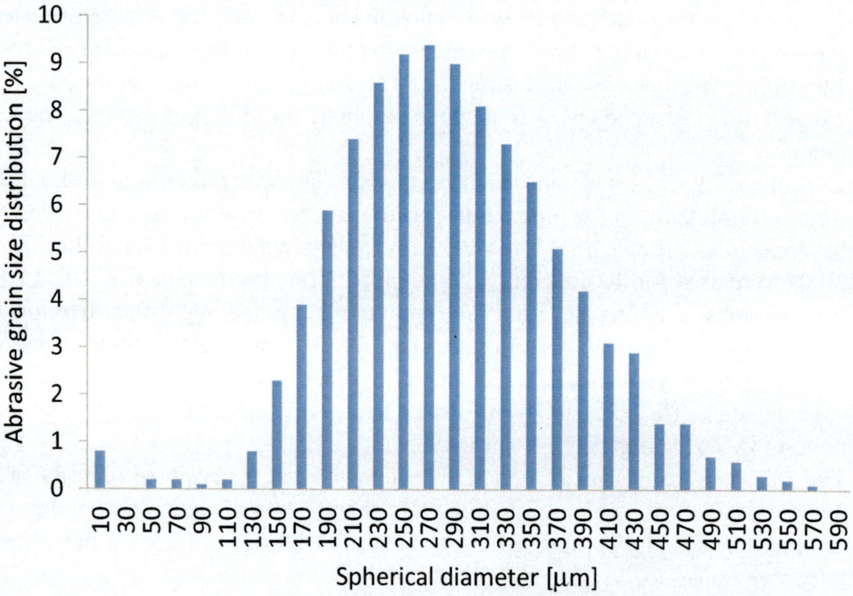

Fig. 2. Grain size distribution of the used abrasive material (Australian Garnet 80 MESH). The grain size distribution was measured using a Fritsch ANALYSETTE 22 NanoTec laser particle size analyzer.

288 μm (Australian Garnet 80 MESH produced by GMA in Australia). The abrasive particle size distribution measured by a laser particle size analyzer (ANALYSETTE 22 NanoTec, Fritsch) is shown in Fig. 2. The abrasive mass flow rate of 20 g·min^{-1} was used.

4 Results and Discussion

The results obtained so far have shown that the concrete age from 7 days to 1 year has an impact on the concrete resistance to fast-flowing liquids. Generally, due to the age, the concrete resistance to pure or abrasive water jets is slightly increased (see Figs. 3 and 4). Especially during the first two months when the compressive strength of concrete specimens increases [12], there is also a slight increase of concrete resistance to the perpendicular impact of pure/abrasive water jets. After a year, the concrete resistance increased by about 36% in the case of the pure water jet, and 28% in the case of the abrasive water jet. In case of the water jet penetration into greater depths of concrete caused by repeated loading of already disintegrated surface (erosion cycle repeated at the same place - see Figs. 3 and 4), the change of resistance depending on the concrete age is rather ambiguous. It actually appears that the concrete resistance to water jet penetration slightly decreases in deeper structures of concrete. Anyway, the increase of concrete resistance was clearly proven after one year of aging of concrete specimens using all testing methods (i.e. pure/abrasive water jet with one or two erosion cycles).

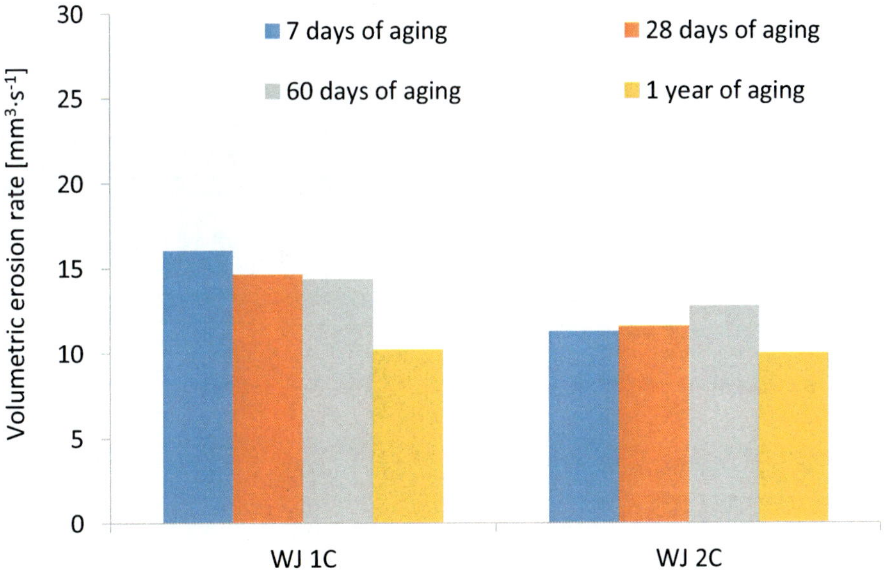

Fig. 3. Measurements of the volumetric erosion rate caused by a pure water jet at different concrete ages. Water pressure of 80 MPa was used. (WJ 1C – one cycle, pure water jet, WJ 2C - two cycles, pure water jet).

While pure water jet mainly removes the hardened cement paste and reveals the aggregate grains, the water jet with abrasive particles disintegrates in greater depths

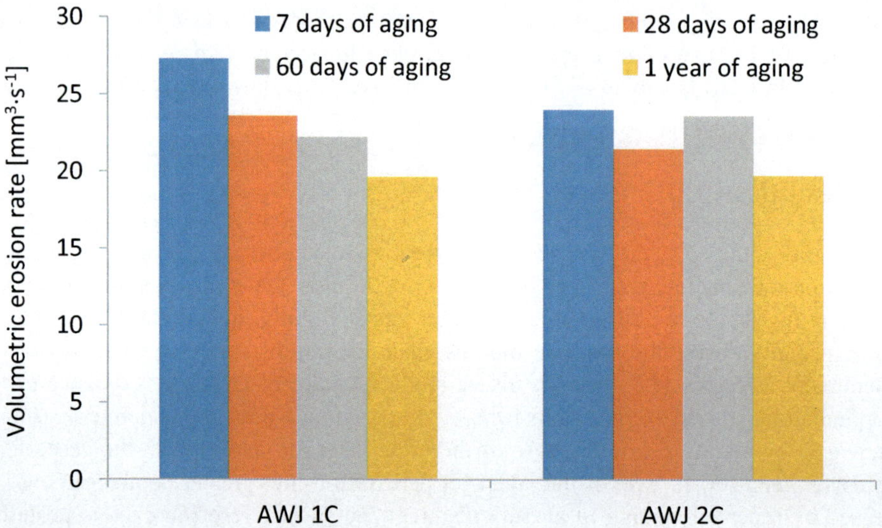

Fig. 4. Measurements of the volumetric erosion rate caused by an abrasive water jet at different concrete ages. Water pressure of 40 MPa and abrasive mass flow rate of 20 g·min^{-1} was used. (AWJ 1C - one cycle, abrasive water jet, AWJ 2C - two cycles, abrasive water jet).

and washes out entire aggregate grains, eventually amputates them. Finally, the abrasive water jet smoothens the entire surface.

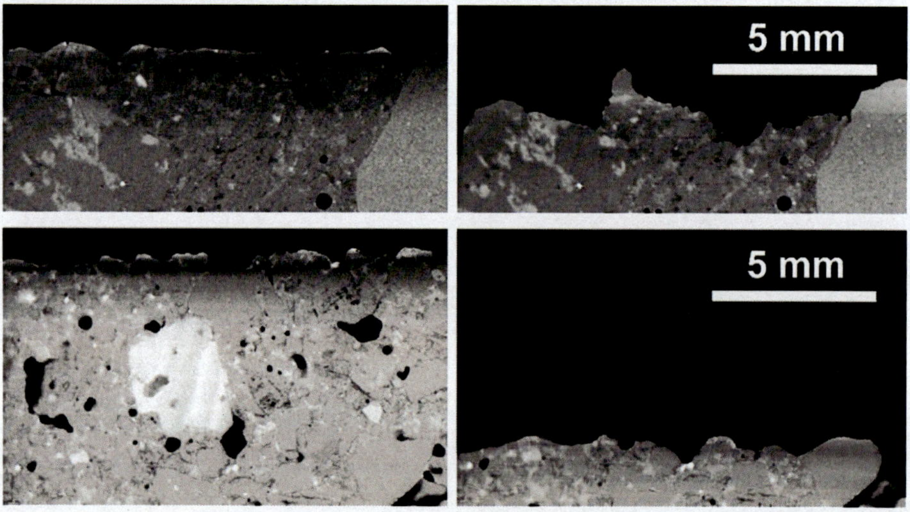

Fig. 5. X-ray CT analysis of subsurface concrete structures after the removal of surface layers by means of high-speed water jets (upper – pure water jet, bottom – water jet with abrasives, left – inner concrete structures before jet loading, right – the same structures after removal of concrete layers by relevant jet type).

Regardless of the age of concrete specimens, detailed inspection of subsurface structures confirmed that the high-speed water jets (without and with abrasives) used for simulations do not damage areas below the concrete surface. The loading by water jets does not cause initiation of new fractures or cracks. Any observed cracks and inhomogeneities in the structure of the concrete after the application of water jets were already present in the structure before their application (see Fig. 5). It corresponds to the actual erosion damage occurring in real hydraulic structures.

5 Conclusion

The concrete strength changes over time. The concrete strength development depends on many factors; in particular, the type of cement used, concrete density, water coefficient, type of additives and admixtures used and finally, climatic conditions (especially the ambient temperature).

The obtained results showed that the age of concrete (from 7 days to 1 year) affects the resistance of concrete to fast-flowing liquids, i.e. pure or with small abrasive particles. In general, the resistance of the tested C30/37 concrete was increased by several tens of percent during simulation of erosion effects by means of high-speed water jets, whether pure or with abrasive particles. Appearance of the resulting surface after the application of water jets reveals that the fast-flowing liquids primarily removes the hardened cement paste from concrete surfaces and reveals the aggregate grains. Larger amounts of abrasive particles in the water jet can cause washing out of whole aggregate grains or their amputation. The entire surface is then smoothened by particles in the water jet. This reduces the subsequent cavitation loading of the surface.

The simulation method using the high-speed water jets does not cause the initiation of secondary cracks under the newly created surfaces, which corresponds to the situation of real fluid-loaded structures (partly based on our previous research).

Acknowledgement. The authors are very thankful to the Czech Science Foundation which supported the research by the 18-25035S project and the Institute of Geonics of the Czech Academy of Sciences for the support from the institutional budget RVO 68145535.

References

1. Sitek, L., Hlaváček, P., Souček, K., Bodnárová, L., Foldyna, J., Zajícová, V., Berčáková, A., Foldyna, V.: X-Ray CT inspection of subsurface areas of concretes exposed to fast flowing liquids. New Trends Prod. Eng. **2**(1), 450–459 (2019)
2. Hlaváček, P., Sitek, L., Hela, R., Bodnárová, L.: Erosion test with high-speed water jet applied on surface of concrete treated with solution of modified lithium silicates. In: Hloch, S., Klichová, D., Krolczyk, G.M., Chattopadhyaya, S., Ruppenthalová, L. (eds.): Proceedings of the International Conference on Manufacturing Engineering and Materials, ICMEM 2018. Advances in Manufacturing Engineering and Materials, pp. 135–143. Springer Nature Switzerland AG (2019)
3. Hu, X.G., Momber, A.W., Yin, Y.G.: Hydro-abrasive erosion of steel-fibre reinforced hydraulic concrete. Wear **253**, 848–854 (2002)

4. Momber, A.W.: Effects of erodent flow energy and local exposure time on the erosion of cement-based composites at high-speed hydro-abrasive flow. Wear **378–379**, 145–154 (2017)
5. Sitek, L., Hlaváček, P., Bodnárová, L.: Use of high-speed water flows for accelerated mechanical modelling of erosive wear of concrete surfaces. In: MATEC Web of Conferences, Innovative Technologies in Engineering Production (ITEP 2018), Bojnice, Slovakia vol. 244, p. 02007 (2018)
6. Hlaváček, P., Sitek, L., Klichová, D., Bodnárová, L.: Effects of abrasives during accelerated simulation of mechanical corrosion of cement composites using abrasive water flow. In: Reiterman, P. (ed.) Acta Polytechnica CTU Proceedings. Special Concrete and Composites 201, vol. 22, pp. 31–37. Czech Technical University in Prague, Praha (2019)
7. EN 206+A1 concrete - specification, performance, production and conformity (2018)
8. EN 12350-2 testing fresh concrete - part 2: slump-test (2019)
9. EN 12390-7 testing hardened concrete - part 7: density of hardened concrete (2019)
10. EN 12390-3 testing hardened concrete - part 3: compressive strength of test specimens (2009)
11. EN 12390-8 testing hardened concrete - part 8: depth of penetration of water under pressure (2019)
12. Shetty, M.S.: Concrete Technology (Theory and Practice), 7th edn. S. Chand & Company Ltd., New Delhi (2006)

Optimal Abrasive Mass Flow Rate for Rock Erosion in AWJ Machining

Petr Jandačka[1] (✉), Jiří Ščučka[1], Petr Martinec[1], Miloslav Lupták[2], Ivan Janeček[1], S. M. Mahdi Niktabar[1], Michal Zeleňák[1], and Petr Hlaváček[1]

[1] Institute of Geonics of the Czech Academy of Sciences, Ostrava, Czechia
petr.jandacka@ugn.cas.cz

[2] Faculty of Materials, Metallurgy and Recycling, Technical University of Košice, Košice, Slovakia

Abstract. Erosion process of the Abrasive Water Jet (AWJ) machining on a rock material depends on various parameters. A brittle rock material is removed predominantly by the deformation wear. Abrasive grains can enhance the machining since their hardness and therefore material removal is better than work of pure water. However, to this day, determination of an optimal dosage of the abrasive matter entering into the mixing chamber was rather neglected. Based on the laws of conservations of momentum and energy combined with some empirical facts, it is possible to derive an equation for erosion of the material and subsequently another relation that provides estimation of the optimal mass flow rate of an abrasive matter. The main equation for erosion consists of six experimental coefficients, for an intensive AWJ only four, describing the effectivity of the water jet k_w and abrasive jet k_a impacting the material. It seems that the energy that is used for erosion is very small and the ideal dosage of the abrasive material is approximately $\dot{M}_a \approx \dot{m}_w \cdot (1 - 2k_w/k_a)$, where $\dot{M}_a$ and $\dot{m}_w$ are the abrasive and water mass flow rates. Such an equation could be used for setting of AWJ machines. The economical optimum (eo) for cutting is proposed via the relation $eo = E/E_{max} - R/R_{max}$, with the erosion E and its maximum value and the resistivity R and its maximum value.

Keywords: AWJ · Erosion · Rock material · Erosion model

Nomenclature (SI unit system)

a (m)	size of the cubic abrasive grain (length of the side)
ΔA (m^2)	change of the surface area of the material
b (m)	interatomic distance in a crystal
D (m)	nozzle diameter
D_f (m)	focusing tube diameter
e (Pa)	young modulus of the material
E (kg/s)	erosion of the material (removed mass of material per time)
E_{rel} (kg/kg)	relative erosion of the material (removed mass of the material/mass of abrasives)

D. Klichová et al. (Eds.): Water Jet 2019, LNME, pp. 81–90, 2021.
https://doi.org/10.1007/978-3-030-53491-2_9

$f(\alpha)$ (-) function of the impact angle α, for brittle rock materials Eq. $f(\alpha) = [1 + \sin(2\alpha + 90°)]/2$ is valid
h (m) depth of cut into the material
k_i (-) dimensionless constant less or equal to 1
$k_{01,02}$ (-) positive coefficients
k_a (-) dimensionless impact effectivity of the abrasive particles
k_w (-) dimensionless impact effectivity of the water jet
K (-) $K = a\rho_m/(6\gamma)$
K_1 (-) unknown dimensionless coefficient
m_a (kg) mass of the single abrasive grain
$\dot{m}_w$ (kg/s) mass flow rate of water exiting from the nozzle, general rule is $\dot{m}_w = dm_w/dt$
$\dot{m}_{w0}$ (kg/s) minimal mass flow rate of water exiting from the nozzle capable to start material erosion
$\dot{M}_a$ (kg/s) mass flow rate of abrasive grains exiting from the mixing chamber
$\dot{M}_{a0}$ (kg/s) minimal mass flow rate of abrasive grains capable to start material erosion
M_m (kg) mass of material grains
$\dot{M}_m$ (kg/s) mass flow rate of material grains
N (-) number of material grains
N (-) number of material grains per time unit
P (Pa) pressure of water in the nozzle
P_0 (Pa) minimal pressure of the water jet capable to remove material
q (-) quality parameter of the surface
$\dot{Q}$ (W) rate of heat production by the AWJ, dQ/dt
R (kg/kg) resistivity of the material (mass abrasive consumption/mass material wear)
r_m (Pa) strength of the material
t (s) time
u (m/s) speed of movement of nozzle above material
v_w (m/s) speed of the water jet exiting from the nozzle
v (m/s) speed of the AWJ exiting from the mixing chamber
ΔV (m^3) element of volume in the material
x (m) size of the cubic abrasive grain (length of the side)
γ (J/m^2) surface energy of the material
μ (-) discharge coefficient of the nozzle, $\mu \approx 0.6$
λ_i (m^{-1}) positive coefficient of damping of the erosion
ρ_a (kg/m^3) true density of abrasive grains
ρ_m (kg/m^3) true density of material grains

1 Introduction

Erosion is a process where an abrasive grain removes the material of a workpiece. The quantity is expressed in various units, typically mg/g, g/g, g, g/min, cm^3/g, etc. In the review work by Sodomka [1] results of material erosion by a stream of abrasive grains are presented. The main conclusions are:

1. Mass flow rate of material erosion reaches the minimal value when erosion is supposed to starts. The limit shifts toward zero with the increase in speed of grains.
2. General equation for erosion is $E = kv^n$, where k is a constant, v is the speed of grains and n is an exponent in the interval $2 < n < 2.5$ for ductile materials and >2.5 for brittle materials.
3. Impact angle, i.e. the angle between the speed vector of abrasive grains and the plane of the workpiece, significantly influences the erosion process. In ductile materials, extreme values of erosion can be reached. In case of the aluminium alloy, it is approximately at 20°, in case of the chromium steel at 30°. In case of the brittle glass, there exists a sinusoidal-like dependency with the maxima about 100°. In general, the best impact angle for brittle materials is the perpendicular angle.
4. Size of abrasive grains linearly improves the erosion of glass. However, in case of the organic glass, nylon and steel, the erosion reaches its maxima when using grains of 200–300 μm.
5. Hardness of abrasive grains improves the erosion according to the relation $E \sim H^{0.4}$, where H is the Vickers hardness. Erosion is expressed in relative mass units. If referred to [2] this effect was studied based on the correlation with the microhardness.
6. There is log-linear dependency between the material hardness and erosion
7. Finally, according to experience, it is easier to derive analytical equations for ductile materials than for brittle materials.

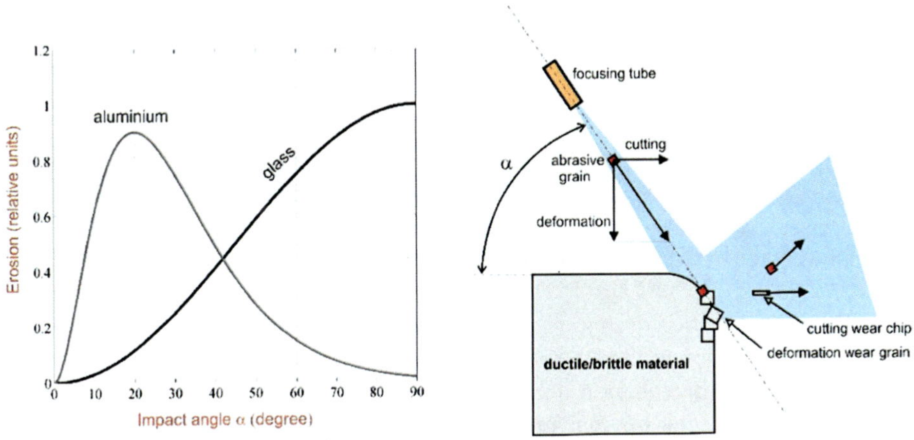

Fig. 1. Schematic presentation of the cutting and deformation wear principle. Ductile aluminium shows the ideal cutting wear in the abrasive impact testing with maximal erosion at about 20°. The glass as a brittle material exhibits almost harmonic dependence between the impact angle and erosion, with the maxima by approximately perpendicular impact [1].

General theory of the erosion is based on two types of material removal – cutting and deformation wear. The deformation wear is caused by a perpendicular component of the abrasive (AWJ) velocity vector; the cutting wear by a tangential component. The cutting wear depends on the yield strength of material. The deformation wear is more intensive in brittle materials and the cutting wear in ductile materials, as presented in Fig. 1. However, the combination of both of them is most common.

Erosion of a rock material by the Abrasive Water Jet (AWJ) technology is well described in the review works of [3, 4] It is well known that hard abrasive grains (typically garnets) significantly enhance the erosion [5]. The erosion quantity can be easily transformed to the parameter of the depth of cut. In the Kovacevic's article [4], the authors presented various relations for the erosion of ductile, brittle, concrete and other types of materials. Unfortunately, the optimal mass flow rate of the abrasive matter for rock materials, which is the subject of our interest, was not mentioned. Moreover, the model presented by Zeng and Kim [6] that focuses on rock-type materials is based on an experimental relation and subsequent regression. Despite the fact that this model is commonly used, it contains a lot of constants and exponents that have to be determined (calculated) before a material can be machined.

The above-mentioned model of Zeng and Kim shows a linear increase of the erosion effect at the increase in the abrasive mass flow rate. However, if we take into account the law of conservation of momentum and different impacts of water and the abrasive matter, the value of the erosion speed should show a decreasing trend and be close to zero at high abrasive mass flow rates. The reason is that the abrasive matter diminishes the kinetic energy of the water jet and, accordingly, of abrasive grains. In some doses of abrasives, the AWJ have not enough kinetic energy to overcome the surface energy of material grains and remove them from the block. In conclusion, the abrasive mass flow rate does not provide with complete information about the capability of the abrasive matter to erode a material. Further analysis is needed, since the optimal dosage of the abrasive material can significantly diminish the consumption, thus saving energy and making the AWJ technology more effective.

The goal of this article is to create a simple physical model for the erosion speed and derive the optimal abrasive mass flow rate that enables good estimations for an effective work with the abrasive matter.

2 Theoretical Modelling of the AWJ Erosion on Rocks

In the theoretical section of this article, two equations are derived. The first equation is the physical derivation of the erosion, more exactly the speed of erosion, completed by damping factors. In this equation, the size of abrasive particles is unimportant (irrelevant). From this equation, the optimal mass flow rate of abrasive particles is derived and presented as the second main equation. We assume in these derivations that material is composed of cubic mono-size grains having volume a^3 and abrasive particles are poly-size cubic grains with volume x^3. We consider the abrasive grains to be ideal firm grains; however, in reality they become slightly fragmented inside the mixing chamber and during the interaction with a material [5, 7, 8].

Two interaction processes occur during the AWJ machining. The first one takes place among water and abrasives in the mixing chamber. The second one exists among the AWJ and a material to be machined. During the interaction of abrasive grains with the water jet in the mixing chamber, the abrasive grains are accelerated to the velocity v to the detriment of the water jet. The original speed of water thus decreases from v_w to v.

The simplest form of the momentum conservation law (divided by time) in the nozzle direction is

$$\overbrace{\dot{m}_\mathrm{w} v_\mathrm{w}}^{\text{before interaction}} = \overbrace{\dot{m}_\mathrm{w} v + \dot{M}_\mathrm{a} v}^{\text{after interaction}} \quad (1)$$

It must be recognized that portion of water is not decelerated by abrasive grains due to no interaction with them in the mixing chamber. Moreover, this portion of water depends on $\dot{M}_\mathrm{a}$ and the shape of the mixing chamber, so the exact momentum law would be rather very complicated. From this point of view, the Eq. (1) is a simplification of the real process. The energy (power) conservation law for the interaction process in the mixing chamber is

$$\overbrace{\frac{1}{2}\dot{m}_\mathrm{w} v_\mathrm{w}^2}^{\text{before interaction}} = \overbrace{\frac{1}{2}\dot{m}_\mathrm{w} v^2 + \frac{1}{2}\dot{M}_\mathrm{a} v^2 + \dot{Q}}^{\text{after interaction}} \quad (2)$$

where the last quantity is the energy loss rate. If an equation for the erosion speed E is to be derived, the most appropriate is to use the power equation for the interaction of the AWJ with the material. This equation corresponds to the exact energy conservation law. However, some empirically-based assumptions must be included. The equation should be in the form

$$\frac{1}{2}\dot{m}_\mathrm{w} v^2 + \frac{1}{2}\dot{M}_\mathrm{a} v^2 = \gamma \cdot \Delta\dot{A} + k_\mathrm{w0}\frac{1}{2}\dot{m}_\mathrm{w} v^2 + k_\mathrm{a0}\frac{1}{2}\dot{M}_\mathrm{a} v^2 \quad (3)$$

The constants k_a0 and k_w0 represent the translation and rotation kinetic energy saved after the AWJ/material interaction, energy of elastic waves in the material, stress energy in the material that can be estimated as proportional to the term $\mu P \pi D_\mathrm{f}^2/4$ in Watt units, heat energy. In the case of k_a0, the energy is used for the abrasive grain fracture. The parameter γ is the surface energy, or better the fracture energy, according to the energetic and tension principles of a new surface area creation defined by Griffith and his subsequent works [12]. Generally, in the literature, the approximation for a perfect crystals is $\gamma \approx r_\mathrm{m}^2 \cdot b/e$. The erosion E is expressed by the surface energy and is obtained as follows

$$E = \frac{\mathrm{d}M_\mathrm{m}}{\mathrm{d}t} = \dot{M}_\mathrm{m} = \frac{\mathrm{d}(Na^3\rho_\mathrm{m})}{\mathrm{d}t} \quad (4)$$

$$\gamma \cdot \Delta\dot{A} = \gamma \dot{N} 6a^2 = \gamma \frac{\dot{M}_\mathrm{m}}{a^3\rho_\mathrm{m}} 6a^2 = \frac{6\gamma \dot{M}_\mathrm{m}}{a\rho_\mathrm{m}} = \frac{6\gamma E}{a\rho_\mathrm{m}} \quad (5)$$

So the Eq. 3 can be re-written to the form

$$(1 - k_\mathrm{w0})\frac{1}{2}\dot{m}_\mathrm{w} v^2 + (1 - k_\mathrm{a0})\frac{1}{2}\dot{M}_\mathrm{a} v^2 = \frac{6\gamma E}{a\rho_\mathrm{m}} \quad (6)$$

The value of the surface energy is typically greater than 1 J/m^2 and less than 10 J/m^2 for brittle materials. However, in case of more plastic materials, the value can be greater than 1000 J/m^2 [1, 9–12]. When expressing v from the Eq. 1, $v = \dot{m}_w \cdot v_w / (\dot{m}_w + \dot{M}_a)$ is obtained and the erosion equation changes to the form

$$E = \left[\left(k_w \frac{1}{2} (\dot{m}_w - \dot{m}_{w0}) + k_a \frac{1}{2} (\dot{M}_a - \dot{M}_{a0}) \right) \cdot \left(\frac{\dot{m}_w v_w}{\dot{m}_w + \dot{M}_a} \right)^2 \right] \cdot \frac{a \rho_m}{6 \gamma} \cdot f(\alpha) \quad (7)$$

where the function $f(\alpha)$ represents the effect of the impact angle. Its approximation here is made according to the general harmonic function experimentally obtained for brittle materials, $f(\alpha) = [1 + \sin(2\alpha + 90°)]/2$. In case of brittle rock materials, the best erosion is achieved at $\alpha = 90°$, i.e. by the perpendicular impact [1, 6]. Since the fraction $a \cdot \rho_m/(6\gamma)$ contains only material constants, it can be replaced by the constant K. The parameters $\dot{m}_{w0}$ and $\dot{M}_{a0}$ are added in order to describe the boundary conditions and the experimental fact that a weak AWJ is not able to cause the erosion. Therefore, a particular flow rate is needed to overcome the strength of the material [1]. The coefficients $k_w = 1 - k_{w0}$ and $k_a = 1 - k_{a0}$ correspond to the dimensionless erosion (impact) in the interval from 0 to 1 for both, the water jet/material and abrasive jet/material interaction. Unfortunately, the coefficients depend on the depth of cut h. If an exponential damping factor is considered, the estimation of the damping coefficients could be $k_w = \exp(-\lambda_1 \cdot h - k_{01})$ and similarly $k_a = \exp(-\lambda_2 \cdot h - k_{02})$, since the effectivity of erosion should decrease with the depth of cut and AWJ stand-off distance. However, solution of the damping phenomena requires a more complex approach exceeding the content of this article.

Based on the technical parameters used by operators during the AWJ machining, the Eq. (7) can be transformed to

$$E = K \cdot f(\alpha) \cdot \left[\left(\frac{k_w}{8} \pi D^2 \sqrt{2\rho_w} \cdot \left(\sqrt{\mu P} - \sqrt{P_0} \right) + \frac{k_a}{2} (\dot{M}_a - \dot{M}_{a0}) \right) \cdot \left(\frac{0.5 \cdot \mu P \pi D^2}{0.25 \cdot \pi D^2 \sqrt{2\rho_w \mu P} + \dot{M}_a} \right)^2 \right] \quad (8)$$

according to the laws of hydrodynamics, where $v_w = (2\mu \cdot P/\rho_w)^{0.5}$ or $\dot{m}_w \approx \pi \cdot D^2 \cdot v_w \cdot \rho_w/4 \approx (2\mu \cdot P/\rho_w)^{0.5} \pi \cdot D^2 \cdot \rho_w/4$, so $\dot{m}_w \cdot v_w = \mu \cdot P \cdot \pi \cdot D^2/2$. The value of P_0 should be the tensile strength of the material r_m (Pa), E is expressed in kg/s units. For the pulsating liquid jet, one can expect a quasi-sinusoidal function of pressure and an effect of the impact pressure [13]. The E function depends on four unknown parameters that should be determined experimentally, i.e. k_w, k_a, $\dot{m}_{w0}$ and $\dot{M}_{a0}$. However, for an intensive AWJ, the $\dot{m}_{w0}$ and $\dot{M}_{a0}$ can be approximated by zero and the Eq. (8) have thus simpler form without the coefficients. The Eq. (8) can be transformed to the relative erosion $E_{rel} = E/\dot{M}_a$ or to the material resistance $R = 1/E_{rel}$. The dependence between the quantities for a model case is shown in Fig. 2. The dependence between the speed of erosion and the pressure of water that is not presented in Fig. 2 shows an increasing trend.

The results show that the abrasive particles enhance the AWJ process. This dependence is increasing and then decreasing to zero (for $\dot{M}_a \to \infty$) with a peak that represents the optimal mass flow rate of the abrasive matter in the mixing chamber. For example, the speed of erosion is the same in $\dot{M}_a \approx 250$ g/min and $\dot{M}_a \approx 3250$ g/min, but the abrasive matter consumption is 5 g/g and 45 g/g. The regime of 250 g/min dosage of the abrasive matter is thus more effective. However, some technical problems occurred during testing, since the mixing chamber could not run at greater amount of the abrasive

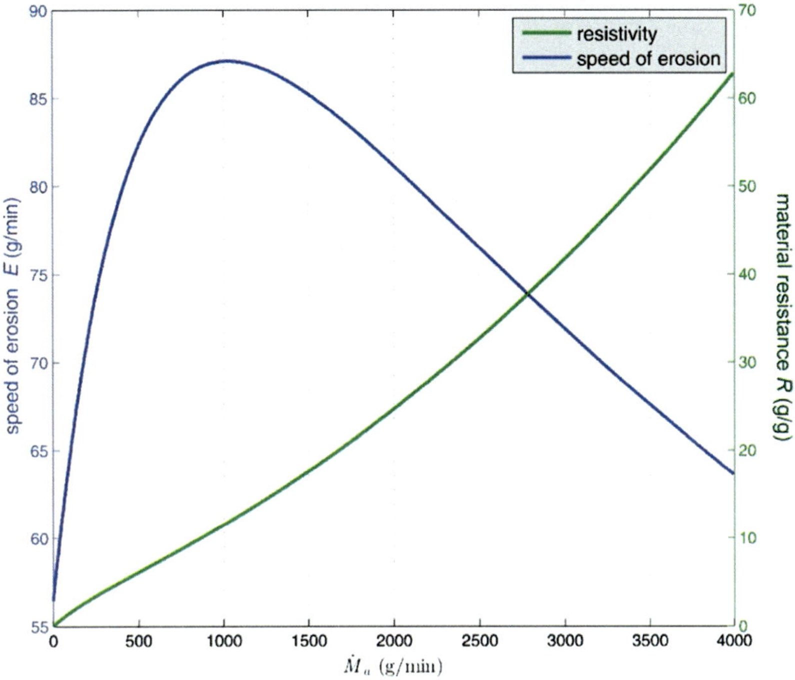

Fig. 2. An example of the theoretical model according to the Eq. 8 for the dependence between the speed of material erosion E and the material resistance $R = 1/E_{\mathrm{rel}}$ at the abrasive mass flow rate $\dot{M}_{\mathrm{a}}$. The model parameters are $k_{\mathrm{w}} = 5{\cdot}10^{-7}$, $k_{\mathrm{a}} = 1.7{\cdot}10^{-6}$, $D = 0.2$ mm, $\rho_{\mathrm{m}} = 2500\ \mathrm{kg/m^3}$, $\rho_{\mathrm{w}} = 1000\ \mathrm{kg/m^3}$, $P = 700$ MPa, $P_0 = r_{\mathrm{m}} = 40$ MPa, $\dot{M}_{\mathrm{a0}} = 100$ g/min, $a = 1$ mm, $\gamma = 2\ \mathrm{J/m^2}$, i.e. $\dot{m}_{\mathrm{w}} = 1.73$ L/min, $v_{\mathrm{w}} = 917$ m/s. The extreme value for E function is in $\dot{M}_{\mathrm{a}}$ point of 1156 g/min. The shape of curves is set by the change of parameter values.

mass flow rate. Very surprising fact is that k_{w} and k_{a} have to be set at about 10^{-6}–10^{-7} to reach the rational value of the E. It means that only a small part of the AWJ kinetic energy is used for the erosion process according to Eq. (8). The erosion with the condition $\dot{M}_{\mathrm{a}} = 0$ g/min is caused by the water jet, the model of which is presented in the ref. [14]. The parameters $\dot{m}_{\mathrm{w0}}$ and $\dot{M}_{\mathrm{a0}}$ were added to the model based on a clear physical derivation in order to describe the experimental erosive effect. If the model shows any difference in comparison with the experiment, it can be solved by using the effect of material tension instead of $\dot{m}_{\mathrm{w0}}$ and $\dot{M}_{\mathrm{a0}}$. For example, the tension energy caused by the water jet can be $\mu P{\cdot}\Delta V$ where ΔV is an elementary volume under stress. It is estimated here as $\Delta V = a\pi D_{\mathrm{f}}^2/4$.

Optimal Abrasive Mass Flow Rate

It can be simply derived from the Eq. 8 (from the condition $\mathrm{d}E/\mathrm{d}\dot{M}_{\mathrm{a}} = 0$) that the optimal mass flow rate of abrasive grains is

$$\text{optimal } \dot{M}_{\mathrm{a}} = \frac{2}{k_{\mathrm{a}}}\left(k_{\mathrm{w}}\dot{m}_{\mathrm{w0}} + k_{\mathrm{a}}\dot{M}_{\mathrm{a0}} + \frac{1}{2}k_{\mathrm{a}}\dot{m}_{\mathrm{w}} - k_{\mathrm{w}}\dot{m}_{\mathrm{w}}\right) \tag{9}$$

and in case of the intensive AWJ machining, the equation $\dot{M}_a \approx \dot{m}_w \cdot (1 - 2k_w/k_a)$ is valid.

It is clear that an ideal setting of an AWJ machine is reached at the maximal E and minimal R. Economically optimal function could be $eo = E/E_{max} - R/R_{max}$, valid in the interval from -1 to 1, where the maximal value of the "eo" function expresses the most optimal machining process with the lowest abrasive matter consumption and the fastest time of machining. In Fig. 2, the economical optimum is at the point of $\dot{M}_a =$ 600 g/min, with $eo_{max} = 0.85$.

Depth of Cut h and Surface Quality *q*

If the main Eqs. (7) and (8) are recalculated to the depth of cut h, the relation is

$$h = \frac{E}{\rho_m u D_f} \tag{10}$$

Now, when expressing the speed of the nozzle movement u from the Eq. (10) as the main AWJ parameter set by operators, in accordance with the paper written by Zeng and Kim [6], the effect of the quality parameter q is

$$u = \frac{E}{q^n \rho_m h D_f} \tag{11}$$

Using the Eq. (11), the AWJ machine operator can determine the quality of material surface by setting a particular speed of the nozzle movement, where n is the exponent approximately equal to 1 (the authors Zeng and Kim presented the relation $u \sim q^{1.15}$; however, in a regression model). As presented in [6]:

$q = 1$ Criteria for separation cuts. Usually, $q > 1.2$ should be used.
$q = 2$ Rough surface finish with striation marks at the lower half surface.
$q = 3$ Smooth/rough transition criteria. Slight striation marks may appear.
$q = 4$ Striation free for most engineering materials.
$q = 5$ Very smooth surface finish.

In the recent time, technical standards for the surface evaluation q are discussed. Table 1 represents the quality assessment according to the Swiss Norm 214001 performed on the aluminium group of materials. It should be noted that individual values can be influenced by various factors. The standard is still in development.

Comparison of Erosion Model with Original Zeng and Kim Model

Finally, the comparison of the Eq. (10) with the regression model of Zeng and Kim shows differences. For the depth of cut in brittle material, Zeng and Kim published a general Eq. $h = n_0 \cdot P^{n1} \cdot \dot{m}_w^{n2} \cdot \dot{M}_a^{n3} / (u^{n4} \cdot D_f^{n5})$, where n_0–n_5 are the regression coefficients. Their equation was calculated on the basis of an original equation they derived – however, the derived equation is of poor validity. Additionally, their equation is applicable and generally used, despite the fact that it does not fulfill the boundary conditions. For example, if $\dot{M}_a = 0$ g/min, then $h = 0$ mm. It cannot be true since the pure water jet is able to erode a material. Linear dependence of the h on $\dot{M}_a$ can be valid only for

Table 1. Quality levels: R_a - average roughness, u - perpendicularity of angularity deviation, j - corner defect, t - range of thickness.

q_5	q_4	q_3	q_2	q_1
	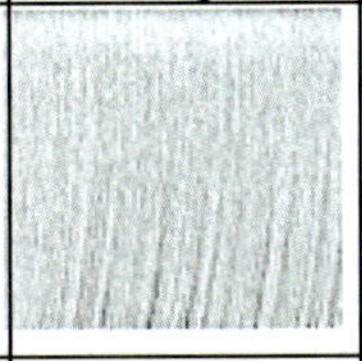	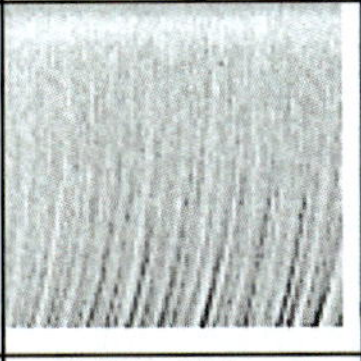	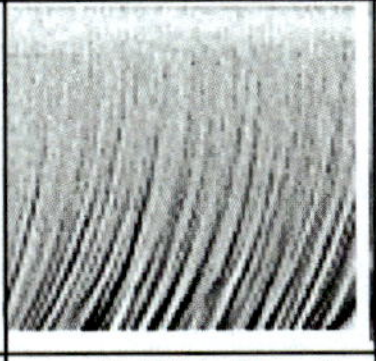	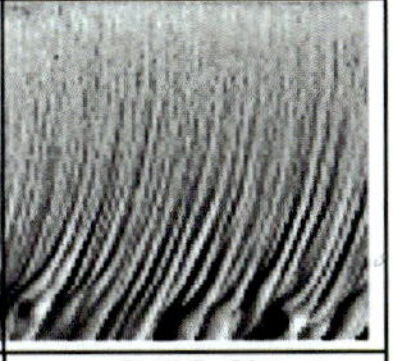
$R_a < 3.2$ μm	R_a (3.2 -6.3 μm)	R_a (6.3 -12.5 μm)	R_a (12.5 -25 μm)	R_a (25 -50 μm)
$u < 0.05$ mm	$u < 0.1$ mm	$u < 0.2$ mm	$u < 0.3$ mm	$u > 0.3$ mm
$j < 0.1$ mm	$j < 0.25$ mm	$j < 0.5$ mm	$j < 1$ mm	$j > 1$ mm
t (0-20 mm)	t (0-30 mm)	t (0-40 mm)	t (0-150 mm)	t (0-300 mm)

small $\dot{M}_a$, since it is well known that greater dosage of abrasives in the mixing chamber diminishes the energy of grains and water jet. It is an experimental fact that very weak AWJ cannot cause erosion, see the term of an incubation time in the work of Sodomka [1]. However, the model of Zeng and Kim includes the erosion. In our model, the boundary conditions are described. The mentioned objections to the Zeng and Kim model cannot reduce its general validity, since it is applicable for an intensive AWJ where the boundary conditions do not play any significant role. Validity of our model, derived in this paper, must be further tested.

3 Conclusion

During the AWJ process in a rock material, the abrasive particles enhance the erosion – this is an experimental fact. On this basis, we made a physical model (Eq. 7 or 8) containing four experimental coefficients. If the mass flow rate of abrasive grains entering in the mixing chamber is increasing, the erosion effect increases accordingly, then reaches its maxima and finally decreases to zero. The point where the abrasive mass flow rate reaches the maxima represents the optimal setting of an AWJ machine from the point of view of the cutting process. However, the economical optimum for AWJ machining is presented by a dimensionless equation, taking into account the consumption of abrasives. The model does not take into account the size of abrasive grains; however the smallest particles do not have enough kinetic energy to overcome the surface energy of a material. From this point of view the model is incomplete. The same also applies for the shape of grains. From the model derived, it is not clear why the dependency of erosion for brittle materials should be $E \sim v^{2.5}$, as presented in previous works, see the introduction section. In the future, the effect of attenuation should be included into the model in better form. The model should be compared with an experiment. The first objective is to test the Eq. (8) for the erosion speed using some standard garnet abrasive matter and changing the abrasive mass flow rate, and to measure the speed of erosion and relative erosion.

Acknowledgements. This work has been supported by:

1. the Project Institute of Clean Technologies for Mining and Utilisation of Raw Materials for Energy Use – Sustainability Program, Reg. No. LO1406 financed by the Ministry of Education, Youth, and Sports of the Czech Republic.

2. the Czech Science Foundation, project Reg. No. 18-250355.

3. the Grant Agency of the Slovak Republic, project Reg. No. 2/0142/19 and No. 1/0599/18.

References

1. Sodomka, L.: Eroze materiálu proudem částic. Jemná mechanika a optika, pp. 77–80 (1976)
2. Martinec, P., Hlaváček, P., Ruppenthalová, L., Sitek, L.: Properties and characteristics of recycled garnet abrasives for AWJ. In: Water Jet 2017, pp. 101–125. Ústav Geoniky AVČR, Ostrava (2017)
3. Summers, David A.: Waterjetting Technology. E & FN Spon, New York (1995)
4. Kovacevic, R., Hashish, M., Mohan, R., Ramulu, M., Kim, T.J., Geskin, E.S.: State of the Art of the Research and Development in Abrasive Waterjet Machining. Trans. ASME, 776–785 (1997)
5. Martinec, P., Foldyna, J., Sitek, L., Ščučka, J., Vašek, J.: Abrasives for AWJ Cutting. Academy of Science of the Czech Republic, Ostrava (2002)
6. Zeng, J., Kim, T.J.: Parameter prediction and cost analysis in abrasive waterjet cutting operations. In: 7th American Water Jet Conference, Seattle (1993)
7. Hlaváč, L.M., et al.: Comminution of material particles by water jets - influence of the inner shape of the mixing chamber. Int. J. Mineral Process. **95**, 25–29 (2010)
8. Henning, A., Miles, P., Schubert, E.: Effect of particle fragmentation on performance of the abrasive waterjet. In: Water Jetting 2018. BHR Group, pp. 89–102 (2018)
9. Jandačka, P., Hlaváč, L.M., Mádr, V., Šancer, J., Staněk, F.: Measurement of specific fracture energy and surface tension of brittle materials in powder form. Int. J. Fract. **1**(1), 103–110 (2009)
10. Jandačka, P., Mádr, V.: Fracture energy of abrasive materials. In: Klichová, D., Sitek, L. (eds.) Water Jet 2015. Institute of Geonics of the CAS, Velké Losiny (2015)
11. Jandačka, P., et al.: Fracture energy of selected brittle silicates. Ceramics - Silikáty **55**, 355–361 (2011)
12. Griffith, A.A.: The phenomena of rupture and flow in solids. Philos. Trans. R. Soc. London Ser. A **221**, 163–198 (1921)
13. Foldyna, J., Sitek, L., Švehla, B., Švehla, Š.: Utilization of ultrasound to enhance high-speed water jet effect. Ultrason. Sonochem. **11**, 131–137 (2004)
14. Jandačka, P., Hlaváč, L., Uhlář, R., Mádr, V., Burkovič, J.: Regression model for depth of cut of water jet into rock materials, pp. 117–124. Transaction of the VŠB - Technical University of Ostrava, Mechanical Series (2009)

Submerged Abrasive Water Jet Piercing/Drilling: Preliminary Tests

Jiri Klich[1,2(✉)], Dagmar Klichova[1,2], and Guoyi Peng[2]

[1] Institute of Geonics of the Czech Academy of Sciences, Ostrava, Czechia
jiri.klich@ugn.cas.cz

[2] College of Engineering, Nihon University, Koriyama, Fukushima, Japan

Abstract. In the technological process of cutting by means of an abrasive water jet, the so-called piercing is often used when starting the operation of cutting. As the piercing is a time-important part of the process, it is advantageous to reduce the time by using different piercing strategies. This article discusses a nozzle circling method that reduces the time required for piercing. The tested circling diameters varied from 0 to 1.55 mm. It was found that the time required for piercing decreased to one sixth with the biggest circling diameter. Another issue discussed in this article is the piercing in submerged water. This method is mainly used to ensure comfort and safety of operators using the water jet. The advantage is the reduction of dust, noise and jet reflection. However, the jet efficiency decreases. An experiment was carried out at five standoff distance settings. The experiment has shown that the resulting times differed only minimally from the piercing in air. Using a digital microscope and subsequent image analysis, the jet behavior as it passed through the material was examined and recommendations for using the nozzle circling method were designed.

Keywords: Abrasive water jet · Submerged · Piercing · Drilling

1 Introduction

In the abrasive water jet cutting technology, at first, the so-called material piercing is largely used. This is an initial process in which the water jet acts as a drill and gradually removes material until the jet reaches the opposite side of a workpiece and a through hole is created. This process is followed by material cutting [1, 2]. There are productions where only this part of the process, i.e. the material piercing, is used. These are, for example, productions of various dies, sieves, vents, etc. In industrial production, technologists try to optimize this process in order to reduce the time required to create a hole, reduce technological demands on the equipment and achieve the required size and quality of a hole. In some cases, the piercing time represents a significant part of the component manufacturing time. Thus, it should be minimized as much as possible [3].

D. Klichová et al. (Eds.): Water Jet 2019, LNME, pp. 91–98, 2021.
https://doi.org/10.1007/978-3-030-53491-2_10

The thicker the material, the more problematic the piercing is for the abrasive water jet. A backflow of the liquid with abrasives occurs damaging the nozzle head material, which decreases the efficiency of water jet cutting. Therefore, other possibilities how to deflect the backflow from the jet space are considered. Various methods of moving the water nozzle relative to the machined material have been used, for example, the straight line motion, circular motion, spiral motion, swinging motion, their combination, etc. [4].

One of the easiest methods is to move the nozzle in a straight line parallel to a material surface. The rebound jet has space to escape into the created kerf behind the jet. This is a very effective way how to create a hole in the shortest possible time. Unfortunately, it requires a sufficiently large space on the workpiece. Another and frequently used motion is the circular motion. In this case, a hole with a diameter larger than a direct piercing is created. The reflected jet always has an escape space on the opposite side of the hole diameter than the direct jet [5].

Another problems of the piercing process, and WJ cutting in general, are the high level of acoustic noise, dustiness and jet reflection. All these phenomena negatively affect/complicate the operation of the equipment. The reflected jet is reduced by various damping covers on the water nozzle, or the entire machine is completely covered. The operator is also required to wear hearing protection (earmuffs) against the noise. One way how to limit the negative processes is the commonly used method of submerged operation.

According to this method, the workpiece and the nozzle are submerged in the catch tank water. This is called the underwater submerged cutting. The noise of the jet is thus significantly reduced as the water jet does not pass through the air before it hits the workpiece. In addition, it dampens the remaining energy of the reflected jet. When operating under water, the water jet is affected by passing through an environment with properties which vary from those of the air. The water jet efficiency thus also decreases [6, 7].

The experiment aimed at obtaining general information on the water jet behavior during the submerged piercing compared to the piercing in the air at different standoff distances. Furthermore, it was observed how the piercing time varies in relation to the diameter d of the circular motion of the nozzle.

2 Experimental Setup

20 mm thick aluminum alloy A5052 was chosen as a test material. The parameters of the abrasive water jet are as follows: water pressure - 300 MPa, abrasive material - Australian garnet 80 MESH, abrasive mass flow rate - 250 g/min, water nozzle diameter - 0.33 mm, focusing tube diameter - 1.0 mm.

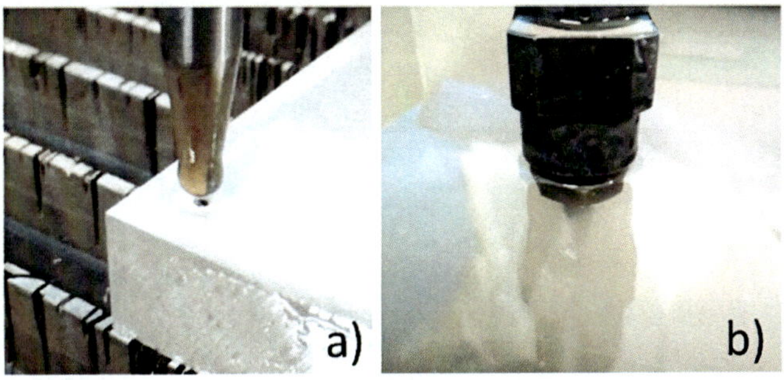

Fig. 1. Position of the nozzle in relation to the sample a) in the air b) submerged in water.

The variable parameters were the standoff distance of 2–10 mm, jet position in air or submerged (Fig. 1), circling nozzle speed of $v_t = 4$ mm/s with a diameter of 0, 0.4, 0.8, 1.2 and 1.55 mm (Fig. 2a, b). The time duration needed for creating a hole was measured using two stopwatches from the moment of jet activation to an audible material breakage. In the case of submerged conditions, the water level was 55 mm above the material.

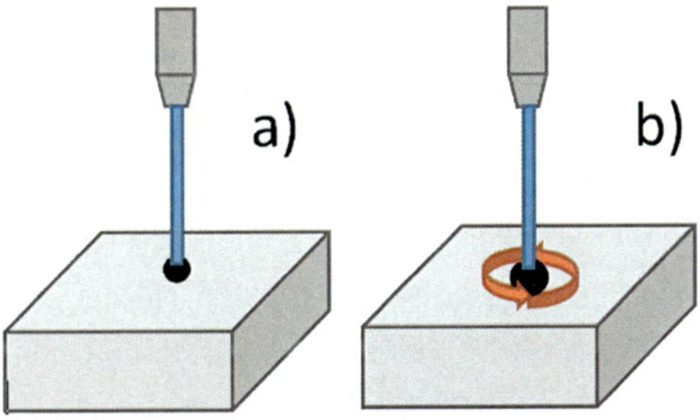

Fig. 2. Jet position a) without circulation $d = 0$ mm b) circulation $d = 0.4, 0.8, 1.2$ and 1.55 mm.

3 Results

The experimental results are shown in Fig. 3 and Fig. 4. In case of the piercing in the air, the effect of standoff distance on the duration time t does not significantly change. In case of the submerged piercing, the duration time t increases slightly with greater standoff distances.

The diameter of the nozzle circulation greatly influences the piercing time. In the air, the highest value of time $t = 31.6$ s is reached by the piercing at a diameter $d = 0$, which represents the static penetration. The piercing time decreases with an increase in the circulation diameter. The shortest piercing time $t = 5.3$ s is achieved at the diameter $d = 1.55$ mm. In the submerged condition, the highest time $t = 40.5$ s is achieved when piercing without circulation at $d = 0$ mm. As the circulation diameter increases, the piercing time decreases accordingly. The shortest piercing time $t = 5.55$ s is achieved at the diameter $d = 1.55$ mm.

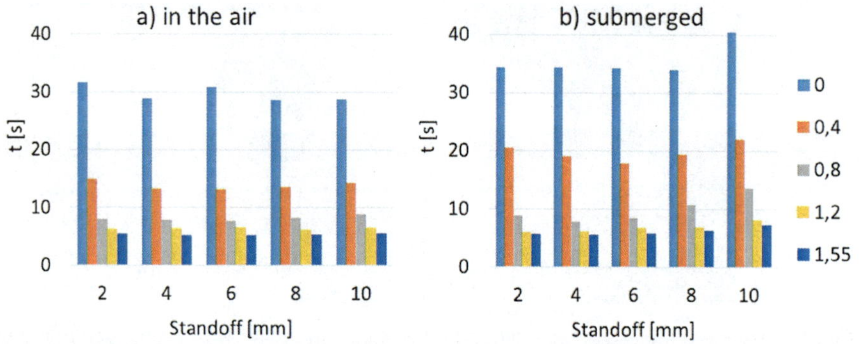

Fig. 3. Time duration of the piercing process a) in the air b) submerged at various stand-off distances and with different circulation diameters.

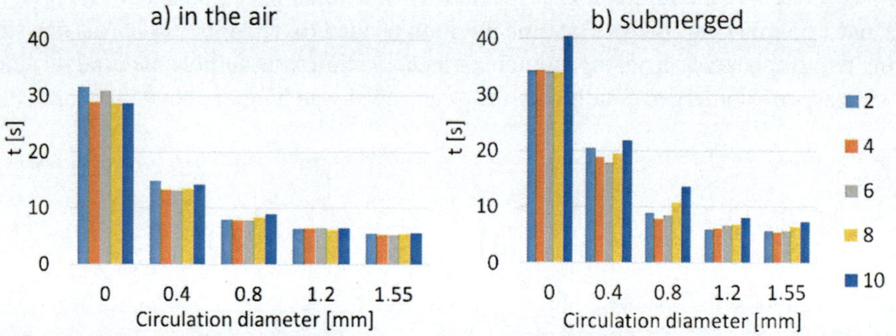

Fig. 4. Time duration of the piercing process a) in the air b) submerged with different circulation diameters and at various standoff distances.

4 Discussion

The obtained results show the influence of individual parameters on the time needed for a hole creation. One of the parameters studied in this article is the effect of gaseous (air) and liquid (water) environments. The results confirmed the assumption that when the jet passes through a liquid medium, the energy of the jet is more attenuated and piercing times are thus higher. In case of a standoff distance of 2 mm, the difference was about 10%. This difference further increased with an increasing standoff distance. At a 10 mm standoff distance, the difference was about 25 to 35%. It was caused by the increasing resistance of the surrounding liquid.

In addition, the effect of the surrounding environment on the water jet can be monitored at the hole input (see Fig. 5a, b). When the jet is used in the air environment, it can expand into the space, thus influencing the surroundings of the hole input. Abrasive particles deflected from the mainstream as they pass through the focusing tube are not sufficiently slowed down by the air and impinge the material surface around the hole input being created. The surface is thus eroded, which is, in some cases, an undesirable process. On the other hand, when the water jet is immersed in a liquid medium, the deflected abrasive particles are more attenuated. Therefore, the diameter of an affected area is smaller.

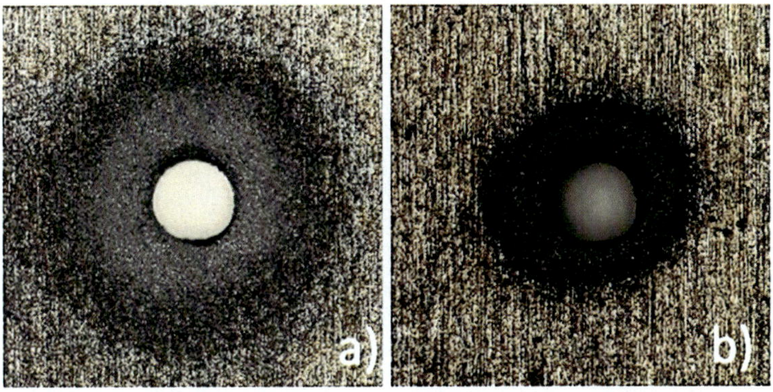

Fig. 5. Areas of the surface erosion by a water jet at the same operating parameters: circular motion diameter $d = 0$ mm, standoff distance of 10 mm. a) Air, b) submerged.

Another studied parameter that affects the time required to create a hole is the diameter of nozzle circulation. The results clearly show that the piercing time is shorter as the circulation diameter increases. The biggest difference is observed between 0 mm and 0.4 mm; the time used is shortened by almost half. When comparing 0 mm and 1.55 mm, it is one-sixth of the time. This is due to the way how the reflected beam exits from the hole. The larger space for the jet escape, the less the impinging jet is attenuated by the reflected jet. Accordingly, the piercing efficiency of the water jet is increased and the piecing times are reduced.

However, it cannot be easily applied in all cases. There are cases where it is necessary to create a hole with the smallest possible diameter or to affect the area around the hole as little as possible. The nozzle circulation causes a side effect, i.e. erosion of the inner hole, see Fig. 7. This occurs when the jet is reflected in a direction tangent to the diameter of circulation. Formations of irregular shape are created on a downward spiral from a particular point of entry to the water jet exit. This phenomenon does not occur by the static piercing (Fig. 6a). With a circulation diameter of 1.55 mm, these formations reach the depth of about 1 mm (Fig. 6b). Therefore, when using this piercing strategy, it is necessary to provide more space around the hole, otherwise the work piece may be degraded.

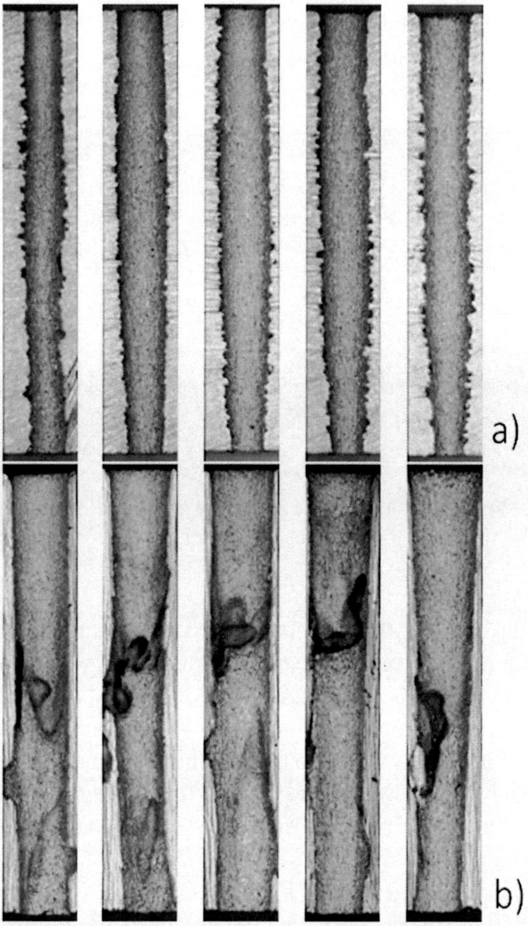

Fig. 6. Visual comparison of holes from the side view. From left to right – increasing standoff distance from 0 to 1.55 mm. a) Circulation diameter of 0 mm (static). b) Circulation diameter of 1.55 mm.

Another thing needs to be considered. That is the shape of the inside of a hole. Looking at Fig. 6a, it is apparent that the middle part of the hole depth has a larger diameter than the inlet and outlet of the hole. This phenomenon is affected, in particular, by the type of material to be pierced.

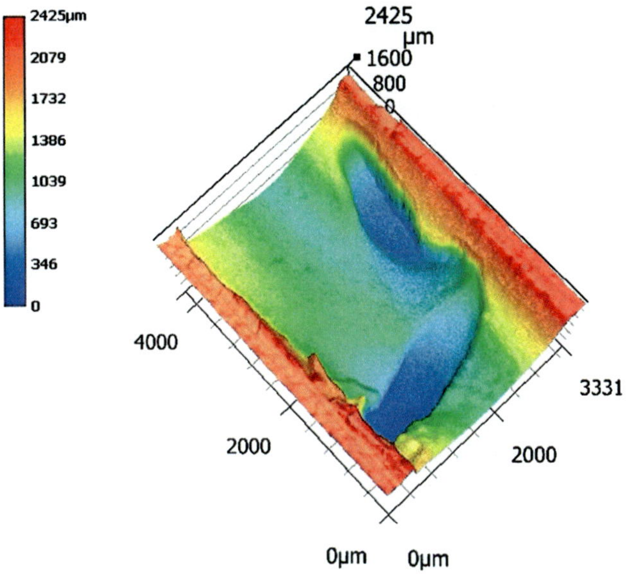

Fig. 7. Detail of an eroded formation inside a hole.

5 Conclusion

It has been found that when using the method of a circulating nozzle, the time required for material piercing is several times shorter than the time of static piercing. As the circulation diameter increases, the piercing time decreases accordingly.

The effect of submerged conditions on the water jet efficiency based on the time of material piercing was verified.

The time loss is minimal and its advantages can be exploited, i.e. noise and dust reduction, reflected jet attenuation and affected zone minimization. The measurement performed shows that if the situation permits and piercing at the diameter of 1.55 mm is selected, the submerged option can be used with great advantages.

The preliminary tests give us an idea of the abrasive water jet behavior when piercing a submerged material. This issue needs to be further examined. Another objective is, therefore, to deepen the knowledge in this area by studying the ongoing processes. In addition, a greater number of parameters should be investigated and the significance of their effects on the piercing process should be determined.

Acknowledgements. This work was supported by the Japan Society for the Promotion of Science, ID No. P18778. The authors also wish to acknowledge the support of Nihon University.

References

1. Hashish, M.: A model for abrasive-waterjet (AWJ) machining. J. Eng. Mater. Technol. **111**, 154–162 (1989)

2. Paul, S., Hoogstrate, A.M., van Lutterwelt, C.A., Kals, H.J.J.: Analytical and experimental modelling of the abrasive water jet cutting of ductile materials. J. Mater. Process. Technol. **73**, 189–199 (1998)
3. Zeng, J., Kim, T.J.: Parameter prediction and cost analysis in abrasive waterjet cutting operations. In: Proceedings of 7th American Water jet Conference, Seattle, pp. 175–189 (1993)
4. Ohlsson, L., Powell, J., Ivarson, A., Magnusson, C.: Optimisatoin of the piercing or drilling mechanism of abrasive water jets. In: Proceedings of 11th International Conference on Jet Cutting Technology, pp. 359–370, St Andrews (1992)
5. Fredin, J., Jonsson, A.: Experimentation on piercing with abrasive waterjet. Int. J. Mech. Aerosp. Ind. Mech. Manuf. Eng. **5**, 2400–2406 (2011)
6. Haghbin, N., Spelt, J.K., Papini, M.: Abrasive waterjet micro-machining of channels in metals: comparison between machining in air and submerged in water. Int. J. Mach. Tools Manuf **88**, 108–117 (2014)
7. Shimizu, S., Peng, G., Oguma, Y.: Air coated abrasive suspension jets under submerged condition. Mod. Mach. Sci. J. **3**, 2214–2217 (2018)

Evaluation of Surface Topography Created by Abrasive Suspension Jet Under Submerged Condition

Dagmar Klichová[1(✉)], Jiří Klich[1], and Guoyi Peng[2]

[1] Institute of Geonics of the Czech Academy of Sciences, Ostrava, Czechia
dagmar.klichova@ugn.cas.cz

[2] College of Engineering, Nihon University, Koriyama, Fukushima, Japan

Abstract. The paper deals with the technology of abrasive suspension jet (ASJ) with a ventilation nozzle system for cutting a metal material (aluminum alloy A5052) submerged under water. SN 214001 standard is used for the quality evaluation of a cut. The standard describes several shape parameters according to which cuts can be compared under different technological conditions. One of the monitored parameters is the air ventilation, i.e. the jet is coated in an air coating. This changes the efficiency and shape of the water jet, and thereafter the shape of the resulting cut. The shapes were scanned using a digital microscope and further analyzed. Furthermore, the article deals with the measurement of the surface roughness according to the ISO 4287 standard in relation to the amount of ventilation air. Roughness values were obtained using an optical profilometer.

Keywords: Surface quality · Standard parameters · Abrasive suspension jet · Submerged jet

1 Introduction

If it is necessary to cut a device, structure or something submerged under water, it is a great advantage to use a high-speed abrasive suspension jet (ASJ). Unlike an abrasive injection jet, ASJ is easier to be applied and more effective at the same time. [1, 2] To ensure an easy handling and control during the cutting operation, it is necessary to determine an appropriate working distance of the nozzle from the surface of the material to be cut. However, as the distance increases, jet damping occurs due to the resistance of the surrounding environment, i.e. water [2, 3]. To eliminate this negative phenomenon, a method of air ventilation forming a kind of sheath around the jet was developed [4].

The article focuses on the quality evaluation of the of cut surfaces created by a submerged abrasive suspension jet (ASJ) according to the Swiss SN 214001 standard [5] Contact-free cutting – Water jet cutting – Geometrical product specification and quality. In 2010, a working draft of ISO/TC 44 N 1770 [6] was formulated. This standard determines new parameters and methods of parameter measurements used in order to categorize a cut surface into a particular quality level Q [7].

D. Klichová et al. (Eds.): Water Jet 2019, LNME, pp. 99–104, 2021.
https://doi.org/10.1007/978-3-030-53491-2_11

Cutting capability of submerged abrasive suspension jets decreases with the increase in the standoff distance [8]. The effect of ventilation was investigated. The results showed that the cutting ability of the ventilated ASJ in the submerged condition was enhanced [9, 10].

The main objective of the presented research is to describe the effect of the submerged condition of the ventilated abrasive suspension jet on the surface topography.

2 Experimental Part

Experiments were realized by means of an ASJ system. The system consisted of a high-pressure pump, an abrasive storage tank of approximately 10 l and an abrasive mixture unit. Abrasive concentration in the suspension jet was measured to be 23 w%. Commonly available Indian garnet with MESH 120 was used as the abrasive material. The maximum working pressure of the system was 35 MPa. The cutting speed was set at 3 mm·min^{-1}. The depth of a nozzle head *H* was 100 mm. Air was sucked to the sheath nozzle by a plastic tube. The airflow rate was regulated by a valve.

The samples were evaluated by a digital microscope. Figure 1 shows parameters of the cut determined in the SN 214001 standard [5] which were measured on samples. The following terms are related to the cut: *g* – burr, *hf* - fine cut, *hr* - remaining surface, *rk* - edge radius, *sb* - jet affected zone, *t* - workpiece thickness, *u* - perpendicularity or angularity tolerance.

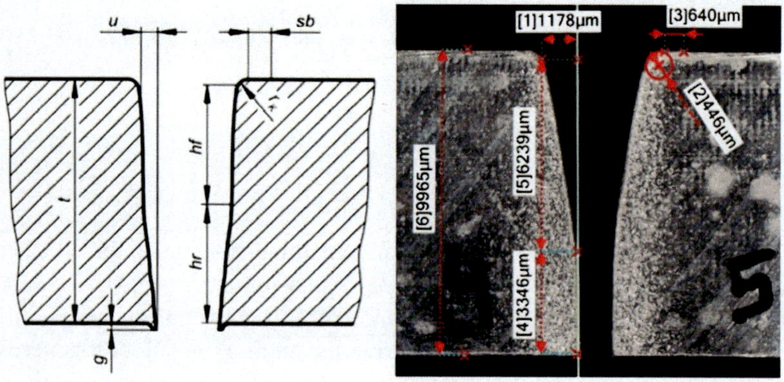

Fig. 1. Location of measured parameters on the sample [5].

The quality of samples was then assessed by an optical MicroProf FRT profilometer. The evaluated surface profile was measured in the middle of the cut surface for thicknesses of less than 2 mm. For 2 mm and more, the measured area was located at the bottom of the cut (within 10% of workpiece thickness), more precisely at least 1 mm above the bottom surface (see Fig. 2).

Roughness is one of the most important parameters which enables to define characteristics of a cut surface. It is influenced mainly by the cutting speed, type and thickness of material and, to a lesser extent, by the cutting medium and process control. In the SN

214001 standard [5], the surface roughness is determined by the *Ra* parameter, which is defined in the ISO 4287 standard [11].

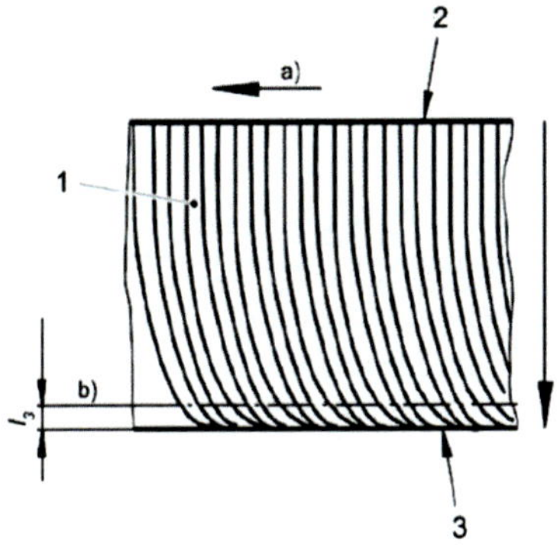

Key:
1 Workpiece
2 Upper surface
3 Bottom surface
l_3 Zone with greatest roughness
a) Advance direction
b) Line of R_a measurements

Fig. 2. Location of the measured line on cut surface of a sample [5].

3 Results

All standardized parameters were measured using a digital Keyence VHX-6000 microscope. The parameter *u* enabled the analysis determining the quality level *Q*. In addition, the width of the formed groove at the jet input w_1, the narrowest point of the groove w_2 and the groove width at the jet outlet w_3 were monitored.

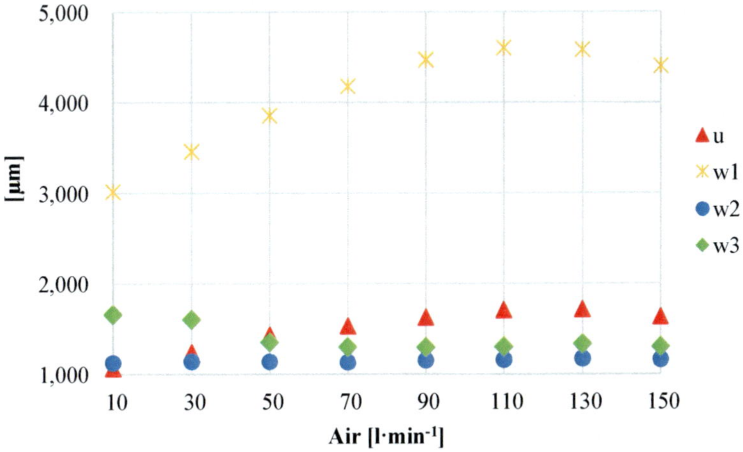

Fig. 3. Effect of ventilation on the monitored parameters.

Figure 3 graphically shows the effect of ventilation on the monitored parameters *u*, w_1, w_2 and w_3. As the ventilation increased, the efficiency of the jet improved accordingly, which was reflected in the upper part of the groove where u and w_1 parameters were monitored. The values of parameter w_2 were constant.

Roughness profile parameters provide a quantified form of surface topography description. The surface structure is assessed based on a surface profile which is created as the intersection of the actual surface and a suitably selected plane. In practice, a plane perpendicular to the plane parallel to the actual surface in a suitable direction is chosen.

The surface profile is then analyzed by separation of individual components by filtering. A profile filter divides profiles into long-wave and short-wave components [12]. Rules and procedures for assessing the surface structure are specified in more detail in the ISO 4288 standard [13]. This standard determines the basic roughness length *lr* and the evaluated roughness length *ln* necessary for measurement of the *R*-parameters of both periodic and non-periodic surfaces. *Ra*, i.e. the mean arithmetic deviation of the profile, is an arithmetic mean of absolute values of the $Z(x)$ ordinates in the range of the basic length *lr*. It is one of the most commonly used surface roughness characteristics in the engineering practice; although the predicative ability of the *Ra* parameter is rather low as it does not respond sensitively to extreme tip heights and groove depths of the profile. [11].

The cut surface was measured at a distance of 1 mm from the edge of the jet exited from the material using a MicroProf FRT optical profilometer. The measured data were analyzed using the SPIP software. The measurement and subsequent evaluation of cut surfaces was performed according to the standards [11–13].

Figure 4 shows the effect of an increasing volume of air-bubble layer on the quality of a cut surface. For this purpose, surface photographs were taken using a digital microscope. It enabled to observe in detail changes in the surface topography which varied depending on the air pressure.

The changes are highly visible on the surface profile measured at the bottom of the sample and they determine the *Ra* roughness profile parameter. As the volume of air increased, the efficiency of the suspension jet improved accordingly and the value of the *Ra* parameter decreased. A cut surface with better surface quality was thus obtained. The change of the surface profile proved that greater volume of air caused creation of a more stable coating. The jet efficiency increased and the *Ra* roughness parameter was reduced. On the other hand, the values of w_1 and *u* parameters showed that at volume of 110, 130, 150 l·min^{-}1, the air capsule is not stable enough and the performance of the submerged jets is getting worse, as confirmed by Peng's studies.

However, if the cut surface is only evaluated according to the *Ra* parameter, it can be erroneously concluded that the jet is the most effective when using the air volume of 150 l·min^{-1}. Lower values of the *Ra* parameter may be affected by a change in the jet flow during cutting that may be caused by an unstable air coating.

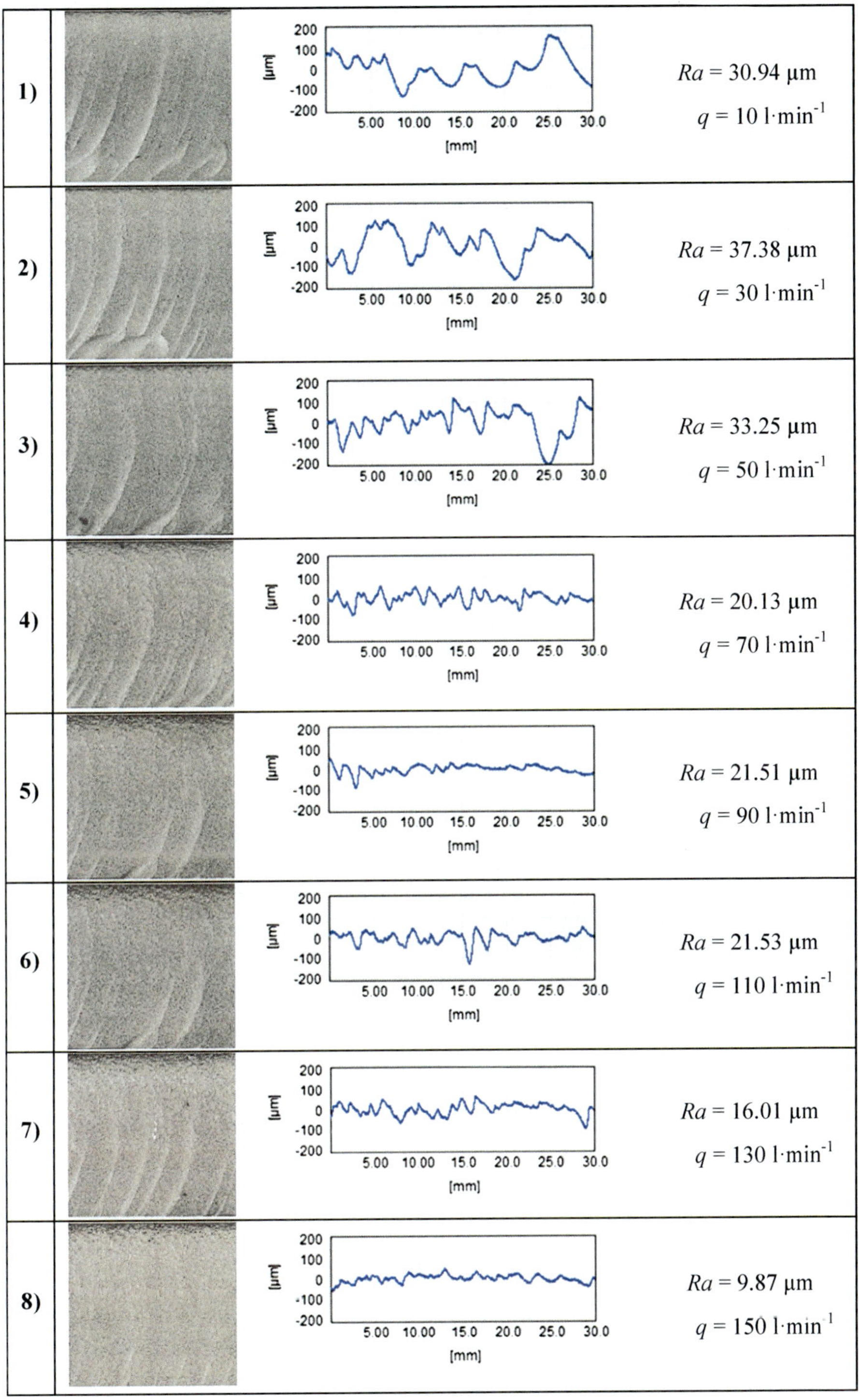

Fig. 4. Effect of an increasing volume of air-bubble layer on the quality of a cut surface.

4 Conclusion

Various parameters can be monitored in order to evaluate the effectiveness of a submerged ventilated ASJ. It has been proven [5, 6] that an air coating increases the jet efficiency.

The study of a groove shape by means of standardized parameters allows investigating the behavior of the jet passing through a material (input, output). The previous research has mainly focused on the jet efficiency. The surface topography analysis and groove shape provide with valuable information about the jet behavior when passing through a material.

Acknowledgments. This work was supported by the Japan Society for the Promotion of Science, ID No. P18778. The authors also appreciate the support of the Nihon University.

References

1. Summers, D.-A.: Water Jetting Technology, pp. 1–25. CRC Press, Boca Raton (2003)
2. Hung, T.N., et al.: Submerged application of abrasive water suspension jets (AWSJs). In: Proceedings of the International Conference on Water Jetting, BHR Group, pp. 205–216 (2004)
3. Shimizu, S., et al.: Submerged cutting by abrasive suspension jet issuing from sheathed nozzle with ventilation. In: Proceedings of the International Conference on Water Jetting, BHR Group, pp. 435–441 (2010)
4. Peng, G., et al.: Periodic behavior of cavitation cloud shedding in submerged water jets issuing from a sheathed pipe nozzle. J. Flow Control Measur. Vis. **6**(1), 15–26 (2018)
5. SN 214001. (Contact-free cutting – Water jet cutting – Geometrical product specification and quality). Schweizerische Normen-Vereinigung. 16 s (2010)
6. ISO/TC 44 N 1770. Contact-free cutting – Water jet cutting – Geometrical product specification and quality. (2010)
7. Klichová, D., Gurková, L.: Evaluation of quality of cut surfaces created by abrasive water jet according to swiss standard SN 214001. In: Proceedings of the Conference on Water Jetting Technology. Water Jet 2017, Lednice, pp. 81–89 (2017)
8. Shimizu, S., Nishiyama, T., Shimura, T., Omote, T.: Drilling capability of submerged abrasive water suspension jet. In: Water Jetting 16, BHR Group, pp. 509–520 (2002)
9. Shimizu, S., Peng, G., Oguma, Y.: Air coated abrasive suspension jets under submerged condition. M.M. Sci. J. **2018**, 2214–2217 (2018)
10. Peng, G., Tamurab, Y,. Ogumac, Y., Klich, J., Quan, H.: Effect of ventilation on the performance of abrasive suspension jet for submerged cutting. In: The 12th Pacific Rim International Conference on Water Jet Technology, 2019, Chiba, Japan, pp. 1–7 (2019)
11. ISO 4287. 1999. Geometrical product specifications (GPS) – Surface texture: Profile method – Terms, definitions and surface texture parameters, 24 p. (1999)
12. ISO 16610-21. 2012. Geometrical product specifications (GPS) – Filtration – part 21: Linear profile filters: Gaussian filters, 28 p. (2012)
13. ISO 4288. 1999. Geometrical product specifications (GPS) – Surface texture: Profile method – Rules and procedures for the assessment of surface texture, 16 p. (1999)

Creating a Database for Turned Surfaces

Dagmar Klichová[1](✉), Jiří Klich[1], Dominika Lehocká[2], Petr Hlaváček[1], Libor Sitek[1], and Vladimír Foldyna[1]

[1] Institute of Geonics of the Czech Academy of Sciences, Ostrava, Czechia
dagmar.klichova@ugn.cas.cz

[2] Faculty of Manufacturing Technologies with a seat in Presov, The Technical University of Kosice, Presov, Slovakia

Abstract. This article deals with the evaluation of surface topography created by Abrasive Water Jet (AWJ) technology. The effects of machining parameters on the turned surface was analyzed by roughness parameters. Further attention is paid to the influence of technological factors on the surface quality created by abrasive water jet turning processes. Based on an input and output system, a database design was created providing the evaluation of the surface quality. Various materials were used for the database which should serve to optimize the abrasive water jet turning processes in order to increase the technological efficiency and contribute to its automation.

Keywords: Abrasive water jet · Turning · Roughness · Surface quality

1 Introduction

High-speed abrasive water jet (AWJ) has a wide range of applications in the area of material machining [1–5]. In particular, the AWJ turning technology [6, 7] is suitable to cut materials that are difficult to process by conventional methods [1, 8–11]. Conditions of material removal significantly vary from conditions of the conventional machining. When a material is processed using the AWJ technology, a workpiece is not machined by a tool with a defined cutting edge geometry [12] but by abrasive particles without a defined cutting edge geometry which are unevenly distributed in the high-speed water jet [13]. Materials which perform a defined rotary motion are machined using the AWJ [14]. The workpiece clamped in a chuck rotates at a constant rotation speed n [min^{-1}]. The main dimension of the workpiece is characterized by a diameter d [mm].

In recent years, we have noticed a shift from increasing technical parameters of cutting machines (pump performance, applied pressure values) to the introduction of new machines and high-pressure water jet cutting technologies [15] to optimize and standardize all cutting operation activities according to the latest knowledge acquired from the basic and applied research and general economic requirements [16]. In addition to the high standard of machines, equipment, tools, management and control systems, the requirements include the optimization and standardization of cutting conditions for a wide range of materials to be cut [17–20]. The current production of turned parts

D. Klichová et al. (Eds.): Water Jet 2019, LNME, pp. 105–114, 2021.
https://doi.org/10.1007/978-3-030-53491-2_12

depends primarily on the expertise of operators. Solution of this particular problem is vital for increasing the cutting performance and achieving operational savings. Research into the conditions and dependencies of the high-pressure water jet turning technology has enabled to create a material database in order to determine the impact of AWJ cutting parameters on the surface quality [21–23]. The conceptual structure of the database is illustrated in Fig. 1. For equipment suppliers, the database represents a significant comparative advantage. For customers, it enables an easy orientation when determining the cutting conditions of a particular material.

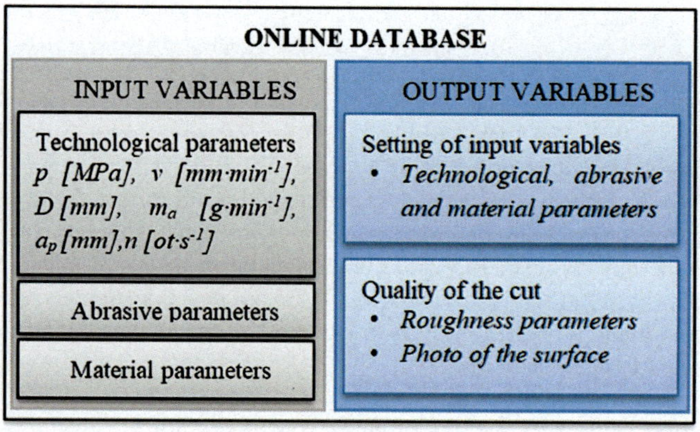

Fig. 1. Scheme of database.

The main aim of presented article is describing to conceptual structure of the database of various materials surfaces machined by AWJ tugning. Database contains overview of AWJ turned materials (disintegration of rotating parts) according to workpiece material, abrasives and technological factors of AWJ turning. Using presented database is possible to predict required surface quality based on surface topography visualization and numerical values of surface roughness parameters (defined in the standard ISO 4287 [24]) of turned (machined) surface.

2 Experimental Setting

Extensive laboratory experiments were conducted for a wide range of materials. The setting of experimental input parameters was determined in connection with the choice of experimental materials. The factors employed in the experiment, i.e. traverse speed v [mm·min^{-1}], revolutions per minute n, material diameter D [mm], depth of cut a_p [mm], were selected according to experimental materials. A standard cutting head for the abrasive water jet generation (PTV 301022-X) with a water nozzle diameter of 0.33 mm, focusing tube diameter of 1.02 mm and focusing tube length of 76.2 mm was used. Experimental equipment consisted of the PTV 75-60 high pressure pump and the X-Y cutting table PTV WJ2020-1Z-D. The studied surfaces were created at a pressure of 400 MPa. The commercially available Australian garnet and Olivine garnet with MESH 80 were used as abrasive materials. The sample was fixed in the clamping jaws mounted in the chuck device, which enabled the rotation of the workpiece in variable direction of rotation, see Fig. 2a) the workpiece rotating in the direction of the abrasive water jet, Fig. 2b) the workpiece rotating against the AWJ.

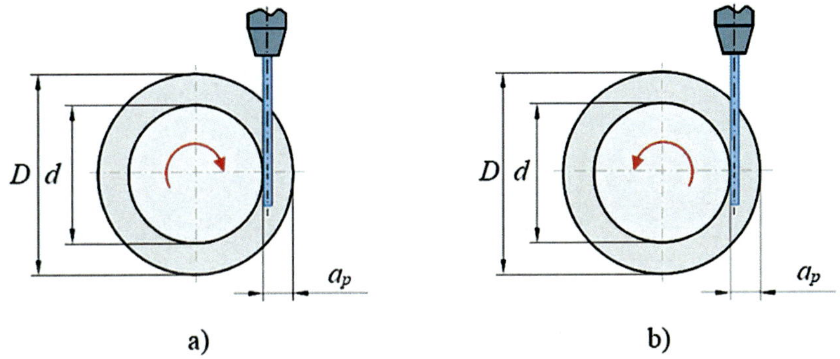

Fig. 2. Abrasive water jet (AWJ) turning methods used during tests: a) turning away from the jet (TAJ), b) turning towards the jet (TTJ), where *D* – diameter of material before machining; *d* – machined diameter, *ap* – depth of cut.

Topography of the surface [25–27] created by the abrasive water jet technology was precisely studied by means of an optical profilometer and a digital microscope. The obtained data were further analyzed with the SPIP software according to the standard ISO 4287 [24].

3 Methodology of Measurement

Test samples were measured using an optical profilometer in accordance with the European surface quality measurement standards. A photo documentation of turned surfaces using a digital microscope was made. The following section focuses on the profile standardization. The basic source for assessing the surface structure is a surface profile. The surface texture is assessed based on the surface profile which results from the intersection of the real surface by an appropriately selected plane. Usually, the plane perpendicular to the plane parallel to the real surface in an appropriate direction is selected. In the engineering practice, parameters of the profile roughness (so-called *R*-parameters) are used for the evaluation of the surface quality. *R*-parameters are defined in the standard ISO 4287 [24]. The evaluation of the turned surfaces was performed using the following profile roughness parameters:

- *Ra* - arithmetical mean deviation of the assessed profile,
- *Rq* - root mean square deviation - *RMS*,
- *Rz* - maximum height of profile,
- *Rt* - total height of profile,
- *Rp* - maximum profile peak height,
- *Rv* - maximum profile valley depth,

- *Rsk* - skewness,
- *Rku* - kurtosis,
- *Rc* - mean height of profile elements,
- *RSm* - mean width of profile elements,
- *RPc* - peak count number.

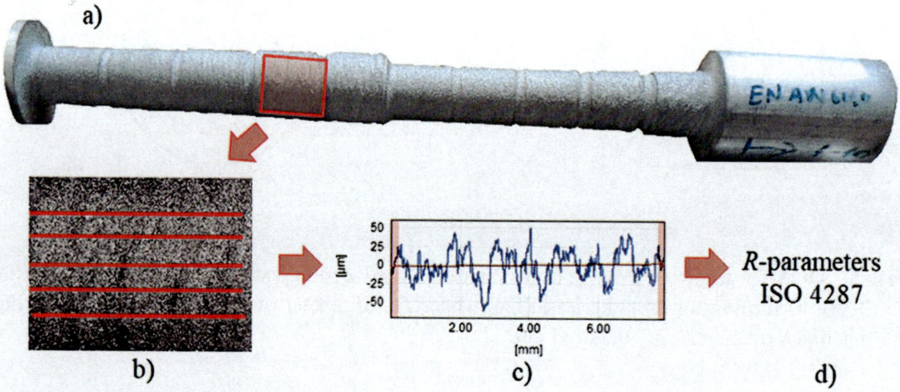

Fig. 3. Schematic representation of a sample measurement procedure: a) picture of the turned sample, b) arrangement of lines measured by the optical profilometer, c) obtained roughness profile, d) analysis of the profile *R*-parameters according to the standard ISO 4287.

Measurement of the turned surface topography of the experimentally prepared sample was realized by an optical profilometer in 15 lines (3 measurements with approx. 60° rotation and 5 lines in each measurement). From the measured lines, roughness profiles were obtained after filtering irregularities and waviness of the surface. Subsequently, roughness parameters were calculated according to the standard ISO 4287 [24]. The measurement procedure is shown in Fig. 3. The acquired profile roughness parameters then quantify the surface texture of the turned sample.

4 Results and Discussion

Conceptual structure of the database is shown in Fig. 1. The input parameters include the technological parameters (pressure p, cutting head traverse speed v, diameter of turned material D, material removal a_p, abrasive mass flow rate m_a, rotational speed n, direction of rotation), abrasive parameters (type and grain size of abrasives) and material parameters (physical and chemical properties of the material used). Setting of the input variables affects the values of the output variables. The surface quality thus informs about the topographic state, as well as about the input variables of the technological process used for surface creation and the physical and mechanical character of the turned material.

Companies using AWJ have already developed their own databases of testing materials. However, the surface quality assessment is qualitative and description of individual quality levels is therefore only indicative. The designed database should help to increase the AWJ efficiency by quantifying the resulting quality of the cut.

Setting of input parameters include the selection of material to be cut and determination of process parameters used for a cutting edge creation by the abrasive water jet technology. An exemplary record of the input parameters for consequent turning by the AWJ technology is presented in Table 1.

Table 1. Input technological parameters.

Abrasive material	GMA 80 MESH	
Turned material	Titan (Ti6Al4V)	
Diameter of turned material	D [mm]	20
Traverse speed	v [mm·min^{-1}]	5; 10; 15; 20; 25
Pressure	p [MPa]	400
Material removal	a_p [mm]	5
Mass flow rate	m_a [g·min^{-1}]	400
Rotational speed	n [r·min^{-1}]	200
Direction of rotation	s [-]	Consecutive

Table 2. Output parameters at changing traverse speed v [mm·min^{-1}].

Output parameters		Traverse speed v [mm·min^{-1}]				
		5	10	15	20	25
d	[mm]	10.49	11.48	12.71	13.78	14.80
Ra	[μm]	6.13	13.49	22.64	21.59	26.79
Rq	[μm]	7.76	16.65	28.21	26.49	33.42
Rz	[μm]	43.34	76.04	117.07	104.38	136.26
Rt	[μm]	49.81	90.29	145.92	127.17	166.16
Rp	[μm]	19.32	37.66	57.75	52.49	73.70
Rv	[μm]	24.02	38.38	59.32	51.89	62.55
$R\Delta q$	[-]	0.60	0.63	0.59	0.54	0.57
Rsk	[-]	−0.49	−0.17	−0.18	−0.09	0.18
Rku	[-]	3.20	2.75	2.84	2.64	2.74
Rc	[μm]	25.81	54.18	91.09	84.42	108.05
RSm	[μm]	386.46	869.43	1324.82	1285.61	1411.60
RPc	[-]	26.01	11.83	7.81	8.02	7.31

Design of tables for an exact description of the cutting edge is based on parameters quantifying the sample surface. The created database of tested materials contains an overall surface roughness analysis in the form of an evaluation of all *R*-parameters defined by ISO 4287 [24]. Table 2 shows the processing of evaluated *R*-parameters for a Titanium material (Ti6Al4V).

Another output element is a database of photographs of studied materials. Every created sample was photographed. The quality of produced surfaces can be thus compared even visually, as shown in Fig. 4. The surface of a material machined by the AWJ demonstrates a specific structure.

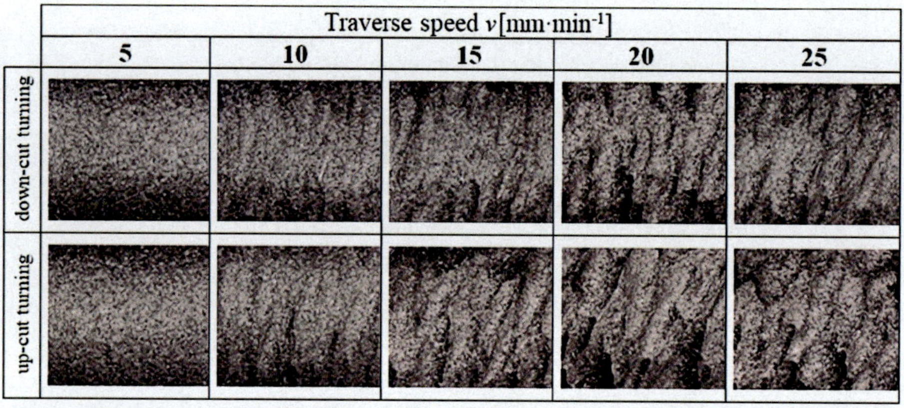

Fig. 4. An example from the database of photographs of turned surfaces, material AISI 304. Down-cut and up-cut surface turning.

Surface quality determination requires analysis of multiple standardized R-parameters. Use of a single parameter only provides with a partial analysis of the surface quality, which can lead to wrong conclusions about the overall quality of the workpiece. Multi-parameter assessment of the surface quality gives a more comprehensive picture of the surface behavior (state). The Fig. 5 graphically represents the dependence of selected roughness parameters on the traverse speed for the down-cut and up-cut surface turning of aluminum alloy EN AW 6060 using the AWJ.

In case of both types of chuck rotation, material removal (Fig. 5a) as well as the quality of the turned surface decreases (Fig. 4) with an increasing traverse speed. The up-cut turning is more efficient, however the surface quality is worse. This shows the roughness parameter *Ra* achieving higher values during the up-cut turning (Fig. 5b).

Ra is a very stable and repeatable parameter. However, it does not react sufficiently to local height differences of the studied surface profile. For these reasons, another height parameters of the profile *Rz* is graphically represented, serving as a control parameter. Figure 5c shows that effects of the traverse speed cause the same character of changes of the *Ra* and *Rz* parameters. Therefore, it can be stated that the measured samples do not demonstrate any local height unevenness.

The skewness of the assessed profile provides with a specific measure comparing the degree of concentration of height values with the degree of density of other values.

For all traverse speeds, *Rsk* values are <0 (Fig. 5d). This proves occurrence of greater and deeper depressions on the surface, which indicates good load bearing properties of the surface.

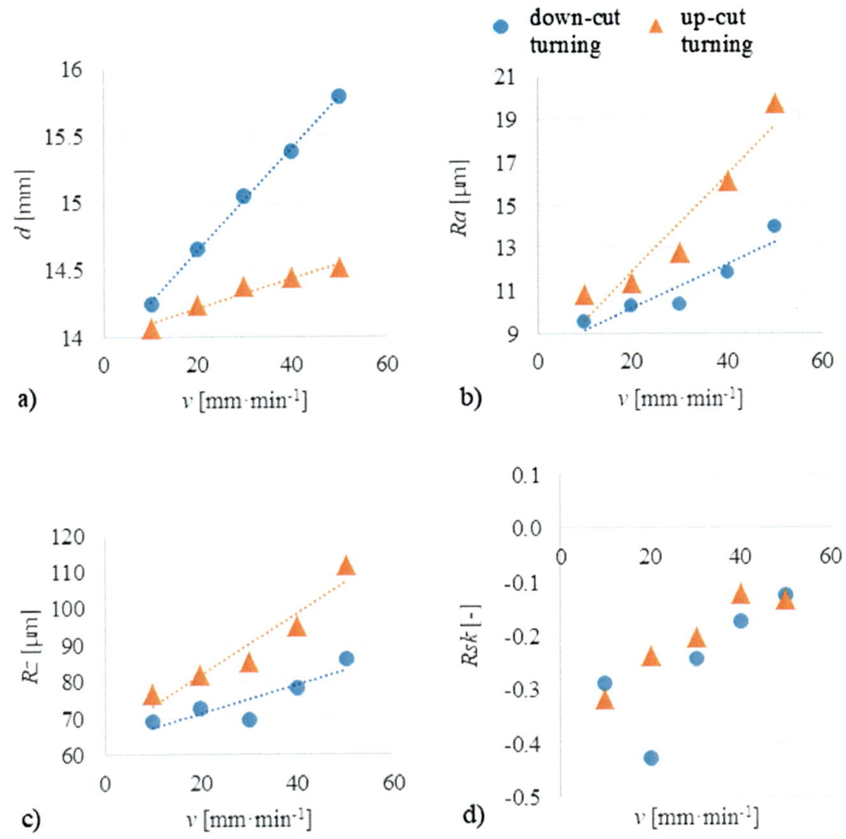

Fig. 5. Graphical representation of turning results of aluminum alloy EN AW 6060.

The created database quantifies the surface quality of studied materials turned by the abrasive water jet based on tables used for an exact determination of the cutting edge quality. The elaborated database of materials provides information enabling a productivity improvement and achievement of the desired topographic properties of cutting surfaces.

5 Conclusion

A database of turned materials has been created and is being intensively developed based on the research into applications of the advanced high-speed water jet machining technology carried out at the Institute of Geonics of the Czech Academy of Sciences. The created Internet Knowledge Portal (IKP) supports significantly the innovation activity by gathering necessary knowledge for research teams, creating an electronic library where the state of the art of the high-pressure water jet machining technology is regularly updated. The developed IKP will be applied as a project output for customer/user activities. The online dialogue of the user (customer) with IKP will enable to determine appropriate cutting conditions for specific technological needs. The knowledge-based part of the IKP system will guide the customer (user) through the structure of possible solutions based on the user's technological requirements (requirements for cutting surface topography and cutting edge quality). The IKP system offers the customer (user) the optimal water jet cutting technology, cutting parameters, selection of suitable abrasives and other production options.

The results can be used to optimize the abrasive water jet technology in order to produce the required surface topography of various materials. In particular, a correlation between the selected AWJ technological parameters and the resulting surface topography is studied. A database was designed to obtain analytical information about the geometry of the surface topography. This knowledge provides with important information to increase the productivity of production and to ensure the required topographic properties of the cutting surface. The results contribute to enhance the knowledge of possible applications of the AWJ technology. In addition, the database can be used in educational processes. In general, the main objective is to significantly extend the area of industrial applications of the advanced high-speed water jet machining technology in industry.

Acknowledgments. This article was written within the scope of a project of the Czech Ministry of Industry and Trade No. FV30233. The authors are very thankful for the support.

References

1. Srivastava, A.K., Nag, A., Dixit, A.R., Tiwari, S., Srivastava, V.S.: Parametric study during abrasive water jet turning of hybrid metal matrix composite. In: Advances in Manufacturing Engineering and Materials, ICMEM 2018, pp. 72–84 (2019). https://doi.org/10.1007/978-3-319-99353-9_9
2. Ali, Y.M., Wang, J.: Impact abrasive machining. In: Machining with Abrasives, pp. 385–419 (2011). https://doi.org/10.1007/978-1-4419-7302-3_9
3. Sitek, L., Foldyna, J., Soucek, K.: Shaping of rock specimens for testing of uniaxial tensile strength by high speed abrasive water jet: first experience. In: EUROCK 2005: Impact of Human Activity on the Geological Environment, pp. 545–549 (2005)
4. Mlynarczuk, M., Skiba, M., Sitek, L., Hlavacek, P., Kozusnikova, A.: The research into the quality of rock surfaces obtained by abrasive water jet cutting. Arch. Min. Sci. **59**(4), 925–940 (2014). https://doi.org/10.2478/amsc-2014-0064
5. Mardi, K.B., Dixit, A.R., Mallick, A., Pramanik, A., Ballokova, B., Hvizdos, P., et al.: Surface integrity of Mg-based nanocomposite produced by Abrasive Water Jet Machining (AWJM). Mater. Manuf. Process. **32**(15), 1707–1714 (2017). https://doi.org/10.1080/10426914.2017.1279306
6. Hutyrova, Z., Scucka, J., Hloch, S., Hlavacek, P., Zelenak, M.: Turning of wood plastic composites by water jet and abrasive water jet. Int. J. Adv. Manuf. Technol. **84**(5–8), 1615–1623 (2016). https://doi.org/10.1007/s00170-015-7831-6

7. Kartal, F., Gokkaya, H.: Effect of abrasive water jet turning process parameters on surface roughness and material removal rate of AISI 1050 steel. Mater. Test. **57**(9), 773–782 (2015). https://doi.org/10.3139/120.110777
8. Carach, J., Hloch, S., Hlavacek, P., Scucka, J., et al.: Tangential turning of incoloy alloy 925 using abrasive water jet technology. Int. J. Adv. Manuf. Technol. **82**(9–12), 1747–1752 (2016). https://doi.org/10.1007/s00170-015-7489-0
9. Yue, Z.B., Huang, C.Z., Zhu, H.T., Wang, J., Yao, P., Liu, Z.W.: Optimization of machining parameters in the abrasive waterjet turning of alumina ceramic based on the response surface methodology. Int. J. Adv. Manuf. Technol. **71**(9–12), 2107–2114 (2014). https://doi.org/10.1007/s00170-014-5624-y
10. Li, W.Y., Zhu, H.T., Wang, J., Ali, Y.M., Huang, C.Z.: An investigation into the radial-mode abrasive waterjet turning process on high tensile steels. Int. J. Mech. Sci. **77**, 365–375 (2013). https://doi.org/10.1016/j.ijmecsci.2013.05.005
11. Axinte, D.A., Stepanian, J.P., Kong, M.C., McGourlay, J.: Abrasive waterjet turning-An efficient method to profile and dress grinding wheels. Int. J. Mach. Tools Manuf **49**(3–4), 351–356 (2009). https://doi.org/10.1016/j.ijmachtools.2008.11.006
12. Hreha, P., Radvanska, A., Hloch, S., Perzel, V., Krolczyk, G., Monkova, K.: Determination of vibration frequency depending on abrasive mass flow rate during abrasive water jet cutting. Int. J. Adv. Manuf. Technol. **77**(1–4), 763–774 (2015). https://doi.org/10.1007/s00170-014-6497-9
13. Hlaváček, P., Cárach, J., Hloch, S., Vasilko, K., Klichová, D., Klich, J., Lehocká, D.: Sandstone turning by abrasive waterjet. Rock Mech. Rock Eng. **48**(6), 2489–2493 (2015). https://doi.org/10.1007/s00603-015-0719-9
14. Ganovska, B., Molitoris, M., Hosovsky, A., Pitel, J., Krolczyk, J.B., Ruggierio, A., Krolczyk, G.M., Hloch, S.: Design of the model for the on-line control of the AWJ technology based on neural networks. Indian J. Eng. Mater. Sci. **23**(4), 279–287 (2016)
15. Molitoris, M., Pitel, J., Hosovsky, A., Tothova, M., Zidek, K.: A review of research on water jet with slurry injection. In: International Conference on Manufacturing Engineering and Materials, ICMEM 2016, vol. 149, pp. 333–339 (2016)
16. Tekaut, I.: Theoretical evaluation of the effect of surfaces processed with abrasive water jet on fatigue life. Trans. Famena **43**(2), 85–98 (2019). https://doi.org/10.21278/TOF.43207
17. Hou, R.G., Huang, C.Z., Zhu, H.T., Wang, J.: Study on experiment device of abrasive water jet micro-turning. In: Advances in Materials Processing X, vol. 500, pp. 339–344 (2012). https://doi.org/10.4028/www.scientific.net/AMR.500.339
18. Hatala, M., Duplak, J., Duplakova, D., Botko, F.: Effect of traverse speed on surface roughness parameters after laser cutting of non-alloy structural steel. TEM J.-Technol. Educ. Manag. Inform. **8**(2), 402–408 (2019). https://doi.org/10.18421/TEM82-12
19. Sitek, L.: High-pressure abrasive jet cutting using linear stepping movement. In: Geomechanics 93: Strata Mechanics/Numerical Methods/Water Jet Cutting/Mechanical Rock Disintegration, pp. 311–315 (1994)
20. Foldyna, J., Zelenak, M., Klich, J., Hlavacek, P., Sitek, L., Riha, Z.: The measurement of the velocity of abrasive particles at the suction part of the cutting head. Tehnicki Vjesnik-Technical Gazette **22**(6), 1441–1446 (2015)
21. Perec, A., Pude, F., Kaufeld, M., Wegener, K.: Obtaining the selected surface roughness by means of mathematical model based parameter optimization in abrasive waterjet cutting. Strojniski Vestnik-J. Mech. Eng. **63**(10), 606–613 (2017). https://doi.org/10.5545/sv-jme.2017.4463
22. Duplak, J., Hatala, M., Duplakova, D., Botko, F.: Prediction model of surface roughness parameters of structural steel created by plasma arc cutting via full factor experiment. Materialwiss. Werkstofftech. **50**(10), 1207–1220 (2019). https://doi.org/10.1002/mawe.201900032

23. Cep, R., Malotova, S., Lichovnik, J., Hatala, M., Legutko, S.: The influence of cutting conditions on the selected parameters of the surface integrity. Acta Polytech. **58**(6), 334–338 (2018). https://doi.org/10.14311/AP.2018.58.0334
24. ISO 4287. Geometrical Product Specifications (GPS) - surface texture: profile method - terms. Definitions and surface texture parameters (1997)
25. Hutyrova, Z., Zajac, J., Michalik, P., Mital, D., Duplak, J., Gajdos, S.: Study of surface roughness of machined polymer composite material. Int. J. Polymer Sci. (2015). https://doi.org/10.1155/2015/303517. Article no. 303517
26. Hreha, P., Radvanska, A., Knapcikova, L., Krolczyk, G.M., Legutko, S., Krolczyk, J.B., Hloch, S., Monka, P.: Roughness parameters calculation by means of on-line vibration monitoring emerging from AWJ interaction with material. Metrol. Measur. Syst. **22**(2), 315–326 (2015). https://doi.org/10.1515/mms-2015-0024
27. Vikram, G., Babu, N.R.: Modelling and analysis of abrasive water jet cut surface topography. Int. J. Mach. Tools Manuf **42**(12), 1345–1354 (2002). https://doi.org/10.1016/S0890-6955(02)00064-0

The Use of High-Pressure Water Assistance in the Rock Mining Process Using Cutting Tools

Krzysztof Kotwica(✉)

AGH University of Science and Technology, Cracow, Poland
kotwica@agh.edu.pl

Abstract. The article presents the possibilities of using high-pressure water assistance in the rock mining process using cutting tools. Currently applied methods are based on the external spraying, in the form of water or air-water curtains, sprinkling whole mining head or internal, the so-called sectoral spraying, sprinkling of individual cutting tools, being in contact with the rock. The advantages and disadvantages of the described methods were presented. An innovative solution for the holder of tangential-rotary picks was proposed, in which high-pressure lubrication with water or emulsion was used. The results obtained during laboratory, field and industrial tests were described.

Keywords: Rock mining · Cutting tool · High-pressure water assistance lubrication

1 Introduction

Currently in driving process of the tunnels and roadways in underground mines most often are used mechanical methods applying roadheaders with mining heads equipped with cutting tools, mainly rotary-tangential picks. The gallery excavation is carried out very often in difficult mining and geological conditions. It relates to the big hardness, compactness and homogeneous of mined rock, which uniaxial compressive strength exceeds even 120 MPa. Rock contents also a lot of minerals and inclusions which results abrasion and tool wear and in the case of sphaerosiderites generates sparking [8–10].

The drilling speed of excavations strongly depends on the condition of the tools, mainly cone picks, used for mining. If the mining pick is very wear, it causes bigger dustiness and sparking but also limits mining advance and increases energy consumption. In consequence gallery drilling costs are also increasing.

When the cone tool wedge enters into the contact with the rock it starts developing the triaxial hydrostatic stress, which plastifies the rock [1]. The constant factors of a proper operation of tangential-rotary picks are as follows: their right setting (selection of the optimum cutting angles), the way they are mounted and the kind of bearing in the tool holders. The tools are set up in the holders at the required angles in order to maintain their proper operation [5]. These angles are shown on the scheme in Fig. 1.

D. Klichová et al. (Eds.): Water Jet 2019, LNME, pp. 115–125, 2021.
https://doi.org/10.1007/978-3-030-53491-2_13

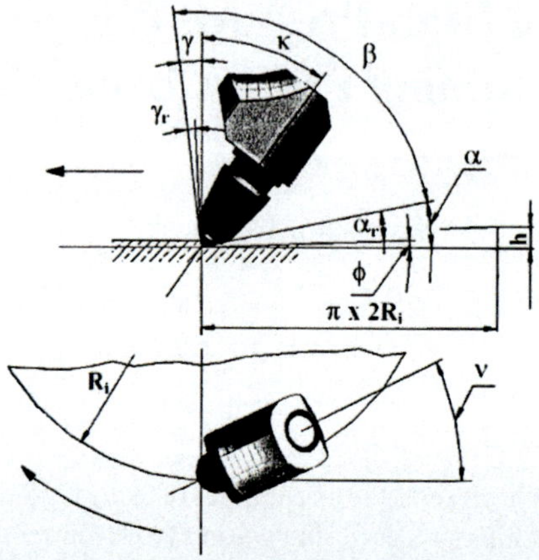

Fig. 1. Parameters of cutting and tool setting angle of the rotary tangential tool [5]: α – tool clearance, α_r – tool auxiliary clearance, β – blade angle, γ – tool rake angle, γ_r – auxiliary tool rake angle, ν – tool side setting angle, κ – tool setting angle, h – depth of cut, R_i – radius of i – tool.

When the conditions for mounting the rotary tangential pick in the holder are correct, it allows free and regular turnover of the pick in the holder. It causes symmetrical and very even wear of the pick and increases tool life. Such ideal wear of the cutting tool - rotary tangential pick, is shown in Fig. 2.

Fig. 2. A view of regular and very even wear of rotary tangential pick obtained in laboratory conditions by the Krupp-Widia company [8, 10].

Dry friction usually occurs in the pick holders' constructions mentioned above and it is characterized by a relative displacement of collaborating elements with no participation of a lubricating medium and by cutting down the tops of uneven surface. Tangential-rotary picks tend to operate under quite difficult conditions, especially when during hard rock mining. Additionally, some substances (fine fraction of the rock or polluted water used for spraying) reach and contaminate the zone of "bearing" of tool holders and make it difficult for cone picks to rotate in the holders. These working conditions of cone picks lead in consequence to decay their rotations in the holders and cause their premature asymmetric wearing. This situation applies for ca. 60% of tools used in the mining industry. On the Fig. 3 a very unregular wear of cone pick edge is shown. It was obtained on a laboratory stand after 2 min of cutting process in incorrectly conditions, in the absence of rotation of the tool in the holder. [8, 10].

Fig. 3. An influence of a tangential rotary pick rotation decay on the edge wear, a laboratory tests result [10].

For assurance the required working conditions in holders for tangential-rotary picks, few ways of high-pressure water assistance have been developed and tested. They will be presented in next chapters.

2 Application of High-Pressure Water Jets

In this chapter the possibilities of application of high-pressure water jets in the rock mining process using cutting tools for edge wear decreasing were described. Present applied methods are based on the external spraying, in the form of water or air-water curtains, sprinkling whole mining head or internal, the so-called sectoral spraying, sprinkling of individual cutting tools, being in contact with the rock. The benefits and faults of these methods were shortly presented.

An example of external spraying are the so-called water curtains and their improved version aimed at reducing water consumption in the face, air-water curtains. The water-air curtain of the "Jet - Rohr" system used in roadheaders from Sandvik is shown in Fig. 4 [11]. It blows the water-air mixture above the mining head into the mining area. A sprayed water-air mixture surrounds the organ and prevents ignition of methane in two ways: as a result of continuous spraying of the mining head and as a result of creating a non-ignition atmosphere. This external water-fog spraying system consists of 19 pairs of nozzles: 19 air nozzles (nozzle diameter 5.0 mm) and 19 water nozzles (diameter 1.0 mm). The use of such a system can almost eliminate the danger of ignition of methane and reduce dust production by over 90%.

However, it works only on their effect of hazards, not directly on the cause. In this case it is also not possible to occur significantly reduce the wear of cutting picks.

During rock mining using cone picks, there are two ways of high-pressure water jet positioning in relation to the cone pick. On the Fig. 5 high-pressure nozzle may be used in the front or behind the cutting tool. In both cases it is mounted in the tool axis [8, 10]. This solution of water jet operation in the area of contact of the pick with mined rock allows to obtain significant benefits.

The high-pressure water jet from a nozzle located in front of the pick is supplied into the area of cracked rock and moistens it which makes the mining process easier by decreasing the rock strength. Other benefit is easier removal of mined rock outside the cone pick using hydro transport. But it is possible with more water flow.

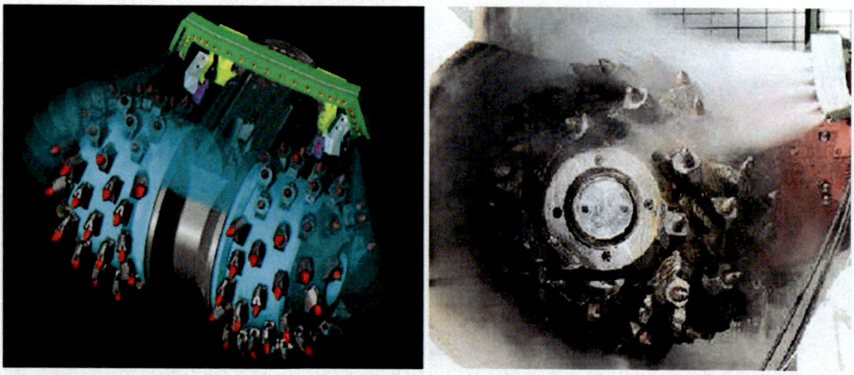

Fig. 4. Constructional solution and working method of the external water-air spraying system "Jet-Rohr" [11].

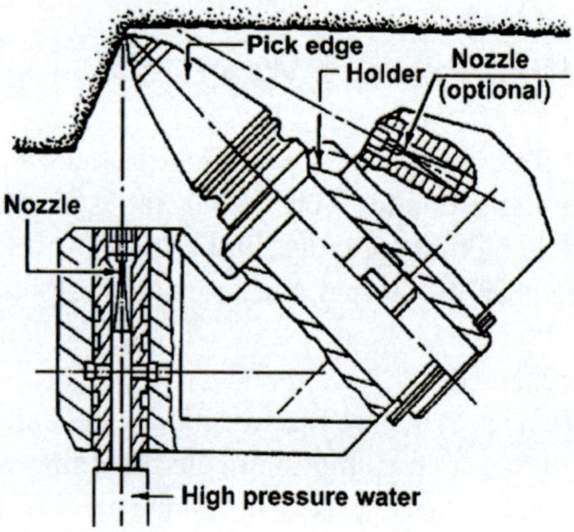

Fig. 5. A diagram of nozzles distribution at high-pressure water jet assistance of mining process with a tangential rotary pick [8, 10].

The slot, which has been mined it in the tool axis by high-pressure water jet, is capable to decrease the rock mining resistance. In addition, it causes the lowering of energy needed for rock mining. The pick edge does not penetrate compact rock, but it is pressed into the slot and breaks the rock massive.

The high-pressure water jet located in the front of the pick is not able for limiting the friction between tool and mined rock. It is also lower possibility for spark eliminating.

The situation is better when high-pressure water nozzle is located behind the pick. Water jet is directed directly into cutting zone and behind them (Fig. 6). Water stream moistens the bottom and sides of the cut, preventing a spark. The watered rock allows also friction coefficient value lowering, what has big influence into cutting tool wear. Beyond the tool wear limiting, this way of water assistance decreases also dust production.

The results of performed tests confirmed the influence of high-pressure water jet introduction into cutting process, especially into tool wear limiting. But test results shown mainly quantitative rather than qualitative type of tool edge wear. It could not be determined whether the nature of the wear was even and uniform or random.

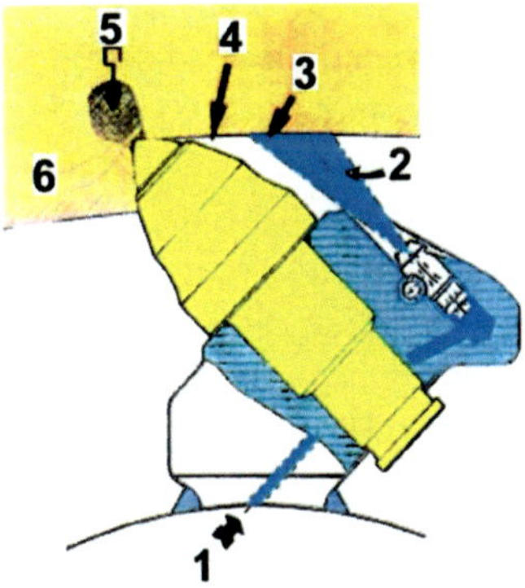

Fig. 6. Scheme spraying of the cutting pick spraying behind the pick [11]: 1 - water supply, 2 - high pressure-water jet, 3 - the highest temperature zone, 4 - the zone with the largest dust formation, 5 - crumpling zone, 6 - crushing zone.

During several performed tests it was also found that rotary-tangential picks have been blocked in the holders. It was caused by wet dust and small particles of mined rock which got between pick shank and inner surface of the holder. It was the reason for start in the AGH UST Cracow with the researches for limitation or prevention this phenomenon. The results of performed tests are described in next chapter.

3 New Solutions of Tangential-Rotary Pick's Holders

There were few attempts for replacement of dry or semi-dry friction between pick shank and inner surface of the holder. Mixed or semi-fluid friction should allow for pick's rotations in the holder and its even and uniform wear.

Due to the limited constructional space, which is determined by the size of tools, their holders and bodies of the units, as well as low rotational speed, fluid friction cannot be acquired in any case. Such a situation is also related to the fact that the requirements for Sommerfeld number *So* are not likely to be met, even when very viscous lubricating media are applied [4].

$$So = \frac{\eta \cdot \omega}{p \cdot \Psi^2} \tag{1}$$

where: η – viscosity of a lubricating medium, [Pa·s]

ω – tool angular velocity, [1/s]

p – surface pressure in the bearing sleeve, [MPa]

Ψ – relative clearance for slide bearings on fluid friction should range from $\Psi =$ 0.0006 to 0.015 as depends on the bearing load and pin rotational speed.

Thus, there have been made some efforts to apply exchangeable slide bearing sleeves. Due to the difficulty in sealing of bearing joint operating in very tough conditions, standard abrasive wear index was taken into account. It is defined as a relation of the volume of ground material to the friction road [6].

$$W = K_A \cdot S \cdot \rho \left[\mathrm{m}^2 \right] \tag{2}$$

where: K_A – standard abrasive wear index, so-called Archard constant for a given material, which is a function of pressure (thrust) according to the Fig. 7, [MPa^{-1}]

S – friction face (contact surface) [m^2],

ρ – pressure (thrust) [MPa].

The bearings of rotary-tangential picks tend to operate under unfavourable conditions, where the sliding speed is very low and thrust – high. It is then difficult to provide proper lubrication, and additionally the output and abrasion products press themselves between frictional surfaces [2, 3]. Therefore, the alternative solution for bearing of rotarytangential picks was prepared in the AGH UST Cracow. Dry friction was replaced by semi-dry or mixed friction. The figure of the modernized solution of the RM8 system holder with forced lubrication of the sleeve is shown in Fig. 7 [6, 10]. In such a solution a lubricating agent (low-percentage oil-water emulsion) is distributed under the pressure through the hole of ferrule 2 to four grooves made in the sleeve 3, mounted in the tool holder 1.

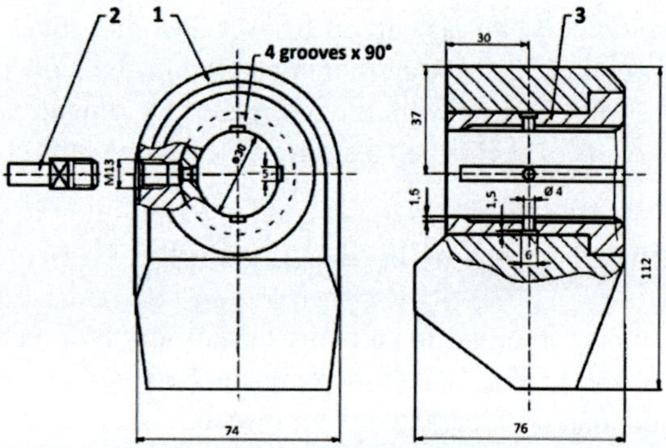

Fig. 7. Modernized, lubricated holder for the tangential rotary picks [6, 10]: 1 – tool holder, 2 – delivery ferrule, 3 – sleeve.

In order to assess the effectiveness of new constructional solutions for the holders of rotary-tangential picks, there have been made copies of both holders with forced lubrication and with rolling bearing. Such solutions were examined on a special laboratory stand designed for testing individual mining tools. For comparison, the first tests on this laboratory stand were performed during the mining of artificial samples of rock with selected types of tangentialrotary picks with and without high-pressure water jest assistance.

4 Results of Laboratory and Industrial Tests Performed in the Department of Machinery Engineering and Transport

Using special laboratory stand for single mining tools testing, the comparative tests were carried out. On the test stand it was possible to carry out mining tests of the ring concrete sample on the side surface at defined mining conditions as cutting depth and spacing, cutting speed, pick setting in the holder. The same tests were performed for standard holder and new solution of lubricated holder.

For all tests tangential rotary, the same cone picks with the diameter of carbide insert of 20 mm were used. The number of pick rotations was measured using markers on the tool's shanks or heads. It allowed counting the number of rotations per 1 min. The tests were performed for determined conditions of the tool operation. During the tests the depth of cutting $g = 9$ and 12 mm was changed as well as side deflection angle ν within the range from 0 to 45°. Constant were the rock compressive strength $Rc = 65$ MPa, cutting speed v at about 1.25 m/s and the cutting spacing $t = 12$ mm. The concrete sample was mined at its whole height. The lubricated holder was supply in one part of the tests with 1.5% oil water emulsion in one of the cases and with clean water in the second one. In both cases the value of feed pressure $p = 1.5$ MPa [8, 10].

In last part of the tests, for selected mining parameters - cutting depth $g = 9$ mm, cutting speed of 3 m/s, cutting spacing $t = 10$ mm, pick side deflection angle at about 9° there were tests for tool edge wear measurement carried out. Rotary-tangential pick was cut the concrete sample without and with high-pressure water jet assistance in front of and in the rear of the tool and using pick mounted in low grade emulsion lubricated holder. High-pressure water jet was generated using nozzle diameter of 0.8 mm and at water pressure of 45 MPa. A view of cone picks during the tests, assisted with high-pressure water jet is shown in Fig. 8. Figure 9 presented a view of pick mounted in lubricated holder at intervals of every 0.5 s from the assistance inclusion.

Fig. 8. A view of mining a concrete sample with a rotary tangential pick: a) without high-pressure water jet assistance, b) with front assistance, c) with rear assistance [8, 10].

The influence of pick assembly method in the holder and the parameters of cutting process on the number of picks revolutions is shown in the diagram in the Fig. 10. The results showed that in the case of lubricated holders, even at small value of side deflection angle of the pick, it rotates in its holder. For the standard holder even when the tool side deflection angle is as much as 45%, it did not cause its lower and irregular rotations [10].

The pick and holder surfaces after the tests have been clean and free from impurities, both in the case when the holder was fed by emulsion or clean water. The view of tool edge and holder surfaces after tests of rock sample cutting at the cutting length of app. 1000 m is shown in Fig. 11. The holder was powered up with clean water. The pick edges' wear after the tests was also compared. Using lubricated holder, more or lower regular rotation of the picks was observed. It caused desirable, regular and very even, small wear of the tool's edges [10].

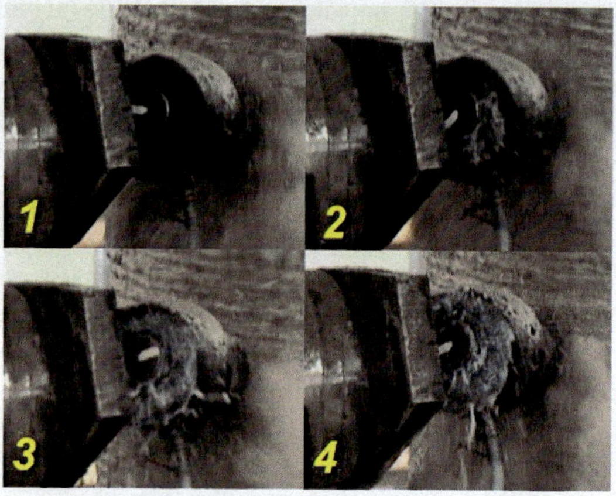

Fig. 9. A view of a holder with open sleeve with straight grooves assisted with low-grade emulsion at intervals of every 0.5 s from the moment of assistance addition [10].

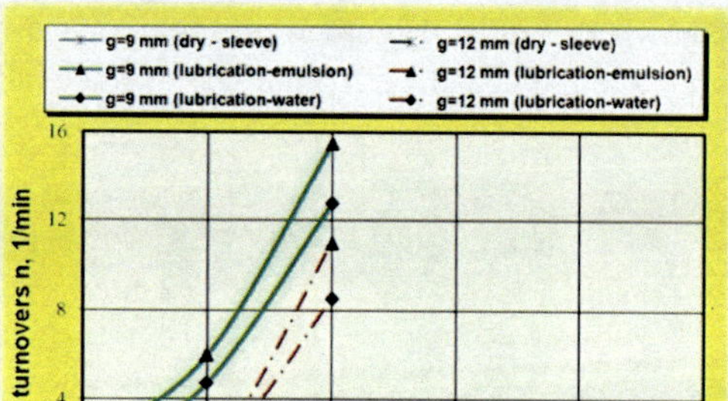

Fig. 10. The number of pick rotations in dependence of the tool side deflection angle υ and assembly method in the holder [6, 7].

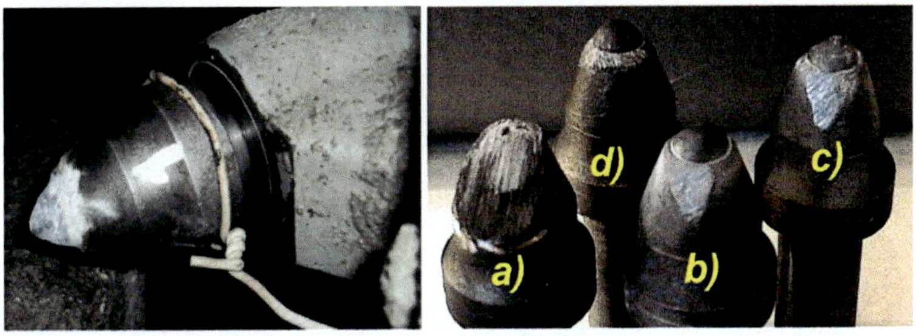

Fig. 11. At the left - a view of a pick edge mounted in holder powered up with clean water, At the right - pick edges wear after mining tests at the cutting length of 1000 m: a) without water assistance, b) with high-pressure water jet assistance from the front, c) with high-pressure water jet assistance behind the tool, d) mounted in clean water lubricated holder [8, 10].

The volumetric wear results of rotary-tangential picks, which were shown in Fig. 11, are presented in Fig. 12. There were also wear measurement conducted using optical microscope. It allows to compare the shape of pick edge and to calculate not only of volumetric tool edge wear, but also to determine the character of wear (regular and even or asymmetrical and large). For rotary-tangential pick which was mounted in standard holder catastrophic wear of not only edge but also a rest of working part of the pick was observed (Fig. 13). With application of high-pressure water jet assistance, the degree of the wear was much lower and regardless the water jet location was comparable. For pick mounted in lubricated low-grade emulsion holder, the degree of the wear was much lower - more than 25–30%, reduction of wear. Using water jet assistance, on the pick edge signs of uneven and irregular wear were observed. It may cause fast tool destruction. For pick which was mounted in lubricated holder the wear was very evenly and small.

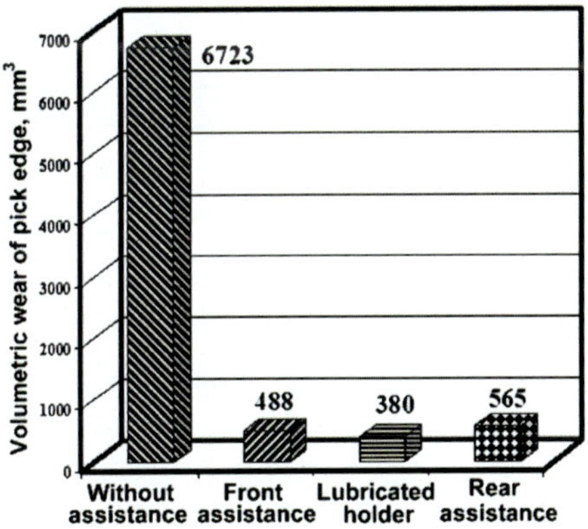

Fig. 12. Volumetric wear of tools edges after mining tests without and with high-pressure water jet assistance and in assisted holder [8, 10].

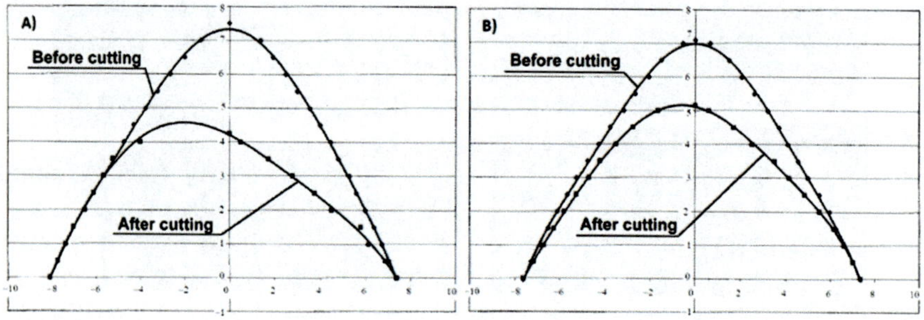

Fig. 13. Exemplary profiles of tangential rotary picks edges wear at mining with high-pressure water jet assistance from rear (on left) and with lubrication (on right) [8, 10].

5 Summary

The results of laboratory, field and industrial tests allow to state that new solution of lubricated holder can be able to increase durability of the cutting tools, mainly rotary-tangential picks, in comparison to standard holders. Based on above described laboratory and field test, in AGH UST Cracow, in collaboration with REMAG Factory, new design solution of mining head for roadheaders with new lubricated holders was developed [8, 10]. The underground test performed in hard coal mine, using the AM-50 roadheader, confirmed the effectiveness of the proposed solution. New holder for rotary-tangential picks, lubricated with only pure water, was fully useful for increasing of tool's rotation. The lifetime of picks mounted on the mining head was over twice increased, with a significant reduction in dust. The mining head equipped with lubricated holders after the whole working shift was additionally clean, without dust and impurities. Mining head with standard holder, after the same working time was dirty and polluted. Both mining head are compared in Fig. 14.

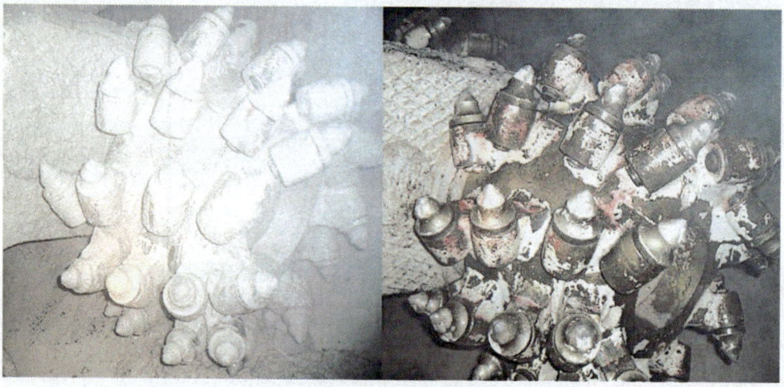

Fig. 14. View of AM-50 roadheader mining head after working shift: in the left- equipped with standard holders and polluted; in the right - equipped with lubricated holders and clean [8, 10].

References

1. Evans, T.: Basic Mechanics of the Point-Attack Pick. Colliery Guardian, London (1984)
2. Fuller, D.D.: Theory and practice of lubrication. PWT Warszawa (1960)
3. Hirst, W.: Wear of unlubricated metals. In: Proceeding of Conference on Lubrication and Wear. Institution of Mechanical Engineers, London (1955)
4. Janecki, J., Hebda, M.: Friction, lubrication and wear of machine parts. WNT Warszawa (1969)
5. Jonak, J.: Theoretical basics of cutting rocks with conical and rotary picks. Wydawnictwa Uczelniane Politechniki Lubelskiej, Lublin (1998)
6. Kotwica, K., Maziarz, M.: Impact of the mounting of tangential rotary tools on their proper operation. Arch. Min. Sci. **49**(1), 85–100 (2004)
7. Kotwica, K.: Atypical and innovative tool, holder and mining head designer for roadheaders used to tunnel and gallery drilling in hard rock. Tunnel. Underground Space Technol. **82**, 493–503 (2018)
8. Kotwica, K.: Effect of selected working conditions of cutting picks on their wear during the mining of hard rocks. Quarterly Mechanics and Control, vol. 29/3, AGH Kraków, pp. 110–119 (2010)
9. Kotwica, K.: Results of laboratory investigations into operating conditions of cutting tool. J. Min. Sci. **39**(2), 168–173 (2003)
10. Kotwica, K.: The influence of water assistance on the character and degree of wear of cutting tools applied in roadheaders. Arch. Min. Sci. **56**(3), 353–374 (2011)
11. Sandvik Company - Brochures and information materials

Effect of the Ultrasonically Enhanced Water Jet on Copper Surface Topography at a Low Traverse Speed

Dominika Lehocká[1,2(✉)], Jiří Klich[2], Vladimír Simkulet[1], František Botko[1], Karol Kovaľ[3], Ján Kepič[3], Zuzana Mitaľová[1], and Michal Hatala[1]

[1] Faculty of Manufacturing Technologies with a seat in Presov, Technical University of Kosice, Presov, Slovakia
dominika.lehocka@tuke.sk
[2] Institute of Geonics of the Czech Academy of Sciences, Ostrava, Czechia
[3] Institute of Materials Research, Slovak Academy of Sciences, Kosice, Slovakia

Abstract. The article describes the maximum erosion efficiency of an acoustically enhanced pulsating water jet on CW004A copper surface. The copper surface was disintegrated at a traverse speed $v = 0.1$ mm.s^{-1}, water pressure of 39 MPa and circular nozzle diameter of 1.321 mm. The effect of the PWJ on the copper surface was evaluated based on a surface topography evaluation using an optical profilometry and a scanning electron microscopy analysis. In the experimental investigation, it was found that a low traverse speed rate has destructive effects on the material with significant weight and volume loss of the material. Surface topography was characterized by a formation of protrusions, depressions and craters. Subsurface material failure of up to 200 μm was detected.

Keywords: Copper · Surface topography · Disintegration · Pulsating water jet erosion

1 Introduction

Water jet technology is widely used in various industrial areas for cutting [1], disintegration and peening of materials [2]. Various effects on the material surface can be achieved by modifying the water flow. The continuous water jet can be used for hydrodemolition [3]. The abrasive water jet [4–7] can be applied for machining of metals such as aluminium alloys, titanium [8]. In addition, the water jet can be used for machining of rocks such as stone [9], limestone [10], graphite [11] and machining of composites [12–15] and ceramics [16]. Micromachining of the Quartz glass using the ultrasonic vibration-assisted abrasive water jet is also being tested [17].

Main advantages of the water jet application are dust-free cut and small cutting gap. Water jet cutting is also not limited by material conductivity. The disadvantage is the limited material thickness, which depends on the material type and properties.

D. Klichová et al. (Eds.): Water Jet 2019, LNME, pp. 126–134, 2021.
https://doi.org/10.1007/978-3-030-53491-2_14

The pulsating water is another modification of the water flow, which finds application in practice. This type of the water jet technology uses "a water hammer effect" for material disintegration and properties influencing. "The water hammer effect" is characterized by a cyclical impact [18] of clusters of liquid on the surface [19, 20]. Clusters of fluid can be generated in a variety of ways, i.e. by an ultrasound or using specially designed self-resonating nozzles. For the experiment described in this article, the water flow was enhanced using an acoustic generator.

The principle of an ultrasonically modulated pulsating water jet and its effects (Fig. 1) are described in more detail in works [21–33, 41–43].

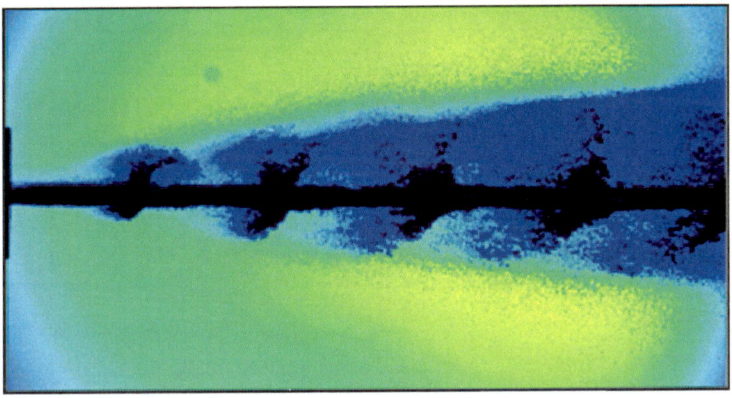

Fig. 1. Visualization of the pulsating water jet using a circular nozzle.

Various types of the PWJ can be used to influence and disintegrate the surface layers of different materials in the engineering, mining and construction industries. In the construction and mining industry, PWJs are applied for concrete structures remediation [34], cutting of concrete composites [35], rocks [36–39] and treatment of ornamental stones. In the engineering industry, the PWJ is used for descaling [40], material removal [41] and surface roughening [42]. Surface peening [30, 43] is also being tested. PWJ is also used for cleaning in the food industry [44, 45].

The main objective of the presented research is.to describe the maximum erosion efficiency of the acoustically enhanced water jet on the surface of CW004A copper based on a surface topography evaluation and a scanning electron microscopy (SEM) analysis. The described experimental research is a part of a larger research study [25–30, 33, 41, 43] aimed at describing the erosive effects of the acoustically enhanced PWJ on various types of metal surfaces and encouraging potential of its further application in practice.

2 Experimental Procedure

A sample of CW004A copper was selected as an experimental material (Table 1). This copper is electroconductive and can be hot and cold worked. It is produced by electrolysis of metallurgical copper. The material shows high electroconductivity, relatively high erosion resistance and a high-temperature coefficient. CW004A is commonly used for electronic and electrical applications [46] and for production of thermocouples, bimetallic heating elements and compensating cables.

In the experimental investigation, the following technology set-up was used: a cutting head with a StoneAge nozzle with a circular equivalent diameter of 1.321 mm, a hydraulic high-pressure Hammelmann.HDP 253 pump, an ultrasound Ecoson WJ-UG_630-40 device and an industrial ABB IRB 6640-180 robot.

Table 1. Material properties and material composition of CW004A copper.

Chemical composition [%]		Mechanical properties			
Cu + Ag	O	Tensile strength R_m [MPa]	Yield strength $R_{p0,2}$ [MPa]	Min. elongation L_0 = 100 mm A [%]	Hardness HB
>99,9	<0,04	395	365	4	114

The sample was disintegrated under the technological conditions shown in Table 2. The low traverse speed was characterized by high mass removal of the material and high disintegration effect.

Table 2. Experimental technological conditions.

Technological conditions of the experiment					
Nozzle diameter d [mm]	Pressure of pump p [MPa]	Traverse speed v [mm. s^{-1}]	Stand-off distance z [mm]	Frequency f [kHz]	Power of ultrasound P[W]
1.321/circular	39	0.10	35	20.31	340

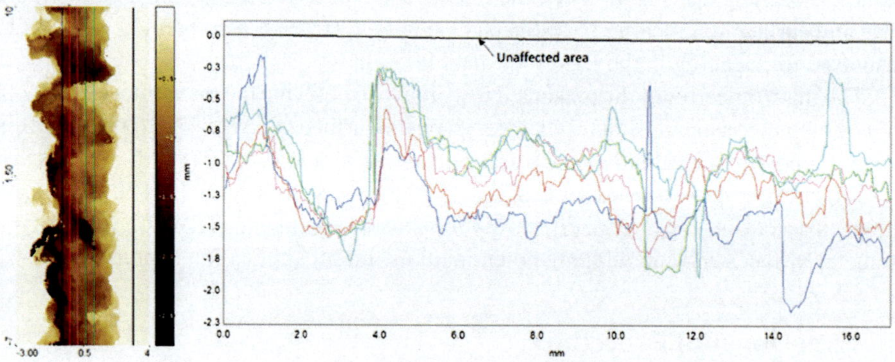

Fig. 2. Surface profile evaluation of the disintegrated CW004A copper.

The topography of the disintegrated groove surface was measured with a MicroProf FRT.optical profilometer. Values of the profile roughness parameters [47] R_z [μm].

(average maximum height of the profile), R_p [μm] (maximum profile peak height) and R_v [μm] (maximum profile valley depth) were evaluated in the SPIP software. The measured values of the profile roughness parameters represent an average of values from five measurements at different locations of the disintegrated groove, see Fig. 2. The scanning electron microscopy (SEM) analysis was used for an observation of the eroded groove surface (Figs. 4 and 5) in crosscut and viewed from the top. The grooved surface was investigated using an electron JEOL JSM 7000F microscope.

3 Results

Surface topography of the CW004A copper was evaluated based on the profile roughness parameters R_z [μm], R_p [μm] and R_v [μm]. These parameters properly characterized the surface disintegrated by means of the PWJ. Figure 3 shows a 3D view and a real view of a groove with values of the roughness parameters measured in the unaffected area and average values measured in the disintegrated area. The disintegration by a circular nozzle shows a significant volume (61.873 mm^3) and mass material (810 mg) removal, representing a loss of 4.03 $mg.s^{-1}$ and 40.30 $mg.mm^{-1}$. The maximum measured groove depth was 2.7 mm. This shows a significant depth efficiency of the acoustically enhanced PWJ. The average value of the measured profile parameters is presented in Fig. 3. The created profile differs significantly from the unaffected area. The groove surface is irregularly eroded at the maximum PWJ erosion and is characterized by important height differences. The results show a small difference between the formation of peaks (protrusions) (R_p = 222, 73 μm) and valleys (depressions) (R_v = 211.09 μm). The overall average maximum height of the profile is significantly higher in the disintegrated area (R_z = 433, 82 μm) as in the unaffected area (R_z = 2.17 μm). The groove edges could not be clearly determined and thus the shape of the groove by a repeated disintegration could not be predicted.

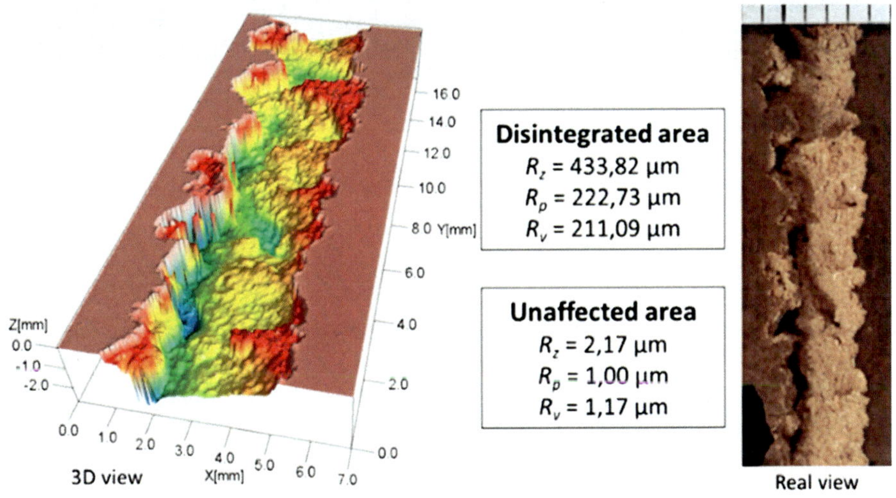

Fig. 3. The topography of the groove surface with values measured in the affected and unaffected area.

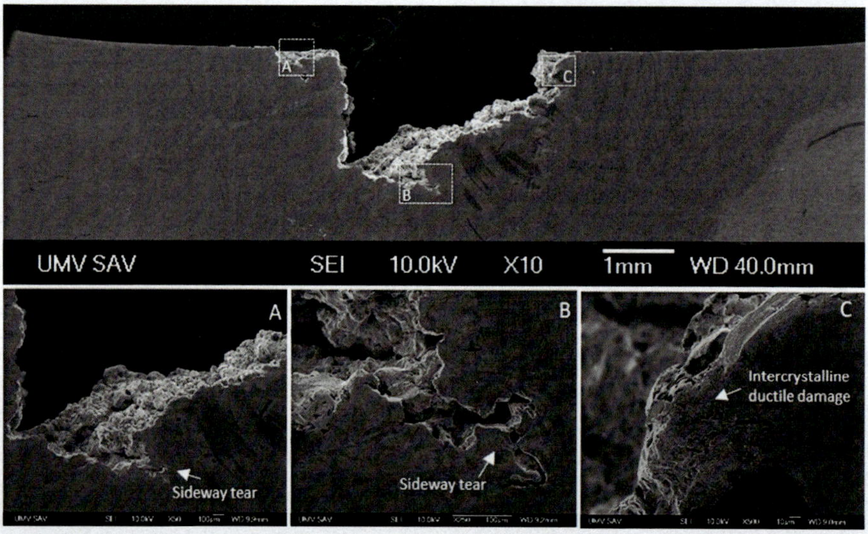

Fig. 4. SEM analysis of the cross-cut of the sample (magnification: 10x, 1–250x, 2–250x and 3–500x).

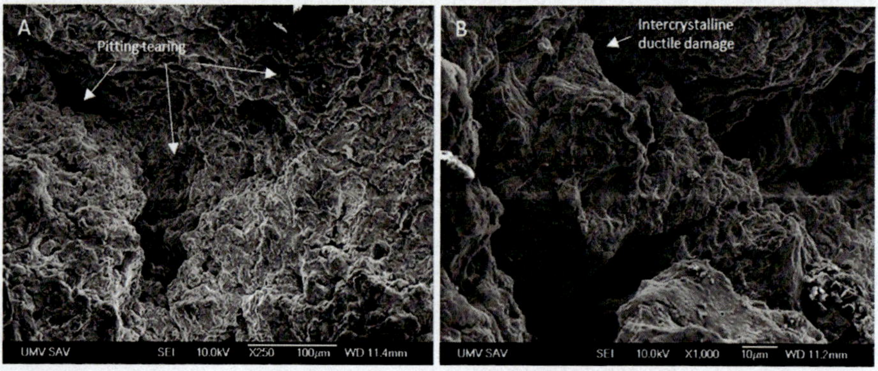

Fig. 5. SEM analysis of the top of the sample (magnification: A – 250x, B – 1000x).

Figures 4 and 5 show the surface of CW004A copper after the disintegration by the PWJ scanned using an electron microscopy at various magnifications. The material was eroded at the water pressure $p = 39$ MPa with high efficiency. The water jet impact on the material caused the material to tear sideways from the disintegration area (Fig. 4), which is visible at a larger magnification in Fig. 4, approximately to 200 μm depth. Side tearing can also be observed from the top view, see Fig. 5, where the pitting mechanism of material tearing took place. In the studied materials, the manifestation of intracrystalline ductile damage at grain boundaries can be observed.

4 Conclusions

Effects of the acoustically enhanced pulsating water jet during disintegration of CW004A copper at the traverse speed $v = 0.1$ mm.s^{-1} using a circular nozzle were investigated. The PWJ disintegration resulted in a significant volume (61.873 mm^3) and mass material (210 mg) removal. The maximum groove depth was 2.7 mm. Values of the profile roughness parameters showed similar results when evaluating peaks (protrusions) and valleys (depressions). Similar values were measured for both parameters ($R_p = 222.73$ μm and $R_v = 211.09$ μm). Significant erosion efficiency of the PWJ on CW004A copper was proved. However, the edge of the groove formed by the circular nozzle could not be clearly determined and thus the shape of the groove by a repeated disintegration could not be predicted.

The SEM analysis showed the formation of subsurface cracks in materials up to about 200 μm. In the studied material, pitting material tearing and intercrystalline ductile fracture across grain boundaries was observed.

Based on results from this part of a larger experimental study, a high erosion efficiency of the acoustically enhanced water jet in materials at low traverse speed rate was proved. The studied material was significantly eroded on the surface and disrupted in its subsurface layers. The disintegration process with the maximum efficiency is thus not suitable for technological operations requiring high accuracy.

The results can be used in areas where a massive material erosion is desirable. The acoustically enhanced PWJs should not be applied in the areas where material structure disruption is not required. The PWJ application potential is mainly in industrial areas focusing on the removal of bulk material layers, concrete structures remediation and treatment of ornamental stone.

Acknowledgements. This work was supported by the Slovak Research and Development Agency under the contract no. APVV-17-0490. This work was further supported by projects: KEGA č. 030TUKE-4/2018, VEGA č. 1/0492/16 - Research of the Possibilities of Elimination Deformations of the Thin Components with the Use of High Speed Machining and by the Ministry of Industry and Trade of the Czech Republic, project No. FV 30233. This publication is the result of the Project implementation: University Science Park TECHNICOM for innovative applications with the support of knowledge technologies - Phase II, ITMS2014+: 313011D232, supported by the European Regional Development Fund.

References

1. Zajac, J., Mital', D., Radchenko, S., Kokuľa, P., Olexa, L., Cep, R.: Short-term testing of cutting materials using the method of interrupted cut. Oper. Diagn. Mach. Prod. Syst. Oper. States II **616**, 236–243 (2014). https://doi.org/10.4028/www.scientific.net/AMM.616.236
2. Liu, X.C., Liang, Z.W., Wen, G.L., Yuan, X.F.: Waterjet machining and research developments: a review. Int. J. Adv. Manuf. Technol. **102**(5–8), 1257–1335 (2019). https://doi.org/10.1007/s00170-018-3094-3
3. Hyland, C.W.L., Ouwejan, A.: Fatigue of reinforcing bars during hydro-demolition. In: 6th International Conference on fracture fatigue and wear (FFW), Journal of Physics Conference Series, IOP Publishing, vol. 843, article number 0102033 (2017). https://doi.org/10.1088/1742-6596/843/1/012033

4. Perec, A., Pude, F., Grigoryev, A., Kaufeld, M., Wegener, K.: A study of wear on focusing tubes exposed to corundum-based abrasives in the waterjet cutting process. Int. J. Adv. Manuf. Technol. **104**(5–8), 2415–2427 (2019). https://doi.org/10.1007/s00170-019-03971-0
5. Molitoris, M., Pitel, J., Hosovsky, A., Tothova, M., Zidek, K.: A review of research on water jet with slurry injection. Procedia Engineering **2016**(149), 333–339 (2016). https://doi.org/10.1016/j.proeng.2016.06.675. International Conference on Manufacturing Engineering and Materials, ICMEM
6. Hreha, P., Radvanska, A., Hloch, S., Perzel, V., Krolczyk, G., Monkova, K.: Determination of vibration frequency depending on abrasive mass flow rate during abrasive water jet cutting. Int. J. Adv. Manuf. Technol. **77**(1–4), 763–774 (2015). https://doi.org/10.1007/s00170-014-6497-9
7. Pitel, J., Matiskova, D., Marasova, D.: A new approach to evaluation of the material cutting using the artificial neural networks. TEM J. Technol. Educ. Manag. Inform. **8**(2), 325–332 (2019). https://doi.org/10.18421/TEM82-02
8. Hloch, S., Hlavacek, J., Vasilko, K., et al.: Abrasive waterjet (AWJ) titanium tangential turning evaluation. Metalurgija **53**(4), 537–540 (2017)
9. Sitek, L., Foldyna, J., Soucek, K.: Shaping of rock specimens for testing of uniaxial tensile strength by high speed abrasive water jet: First experience. In: Eurock 2005: Impact of Human Activity on the Geological Environment, 545–549. A A Balkema Publisher (2005)
10. Perec, A.: Investigation of limestone cutting efficiency by the abrasive water suspension jet. In: Lecture Notes in Mechanical Engineering, pp. 124–134. Springer (2019). https://doi.org/10.1007/978-3-319-99353-9_14
11. Carach, J., Lehocka, D., Legutko, S., et al.: Surface roughness of graphite and aluminium alloy after hydro-abrasive machining. In: Advances in Manufacturing, Lecture Notes in Mechanical Engineering, pp. 805–813. Springer. (2018). https://doi.org/10.1007/978-3-319-68619-6_78
12. Bodnarova, L., Valek, J., Sitek, L., Foldyna, J.: Effect of high temperatures on cement composite materials in concrete structures. Acta Geodynamica et Geomaterialia **10**(2), 173–180 (2013)
13. Hutyrova, Z., Scucka, J., Hloch, S., Hlavacek, P., Zelenak, M.: Turning of wood plastic composites by water jet and abrasive water jet. Int. J. Adv. Manuf. Technol. **84**(5–8), 1615–1623 (2016). https://doi.org/10.1007/s00170-015-7831-6
14. Yue, Z., Huang, C., Zhu, H., et al.: Optimization of machining parameters in the abrasive waterjet turning of alumina ceramic based on the response surface methodology. Int. J. Adv. Manuf. Technol. **71**(9–12), 2107–2114 (2017). https://doi.org/10.1007/s00170-014-5624-y
15. Hutyrova, Z., Zajac, J., Michalik, P., Mital, D., Duplak, J., Gajdos, S.: Study of surface roughness of machined polymer composite material. Int. J. Pol. Sci. (2015). article number 303517. https://doi.org/10.1155/2015/303517
16. Popan, I.A., Contiu, G., Campbell, I.: Investigation on standoff distance influence on kerf characteristics in abrasive water jet cutting of composite materials. In: MATEC Web of Conference, vol. 137 (2017). article number 01009. https://doi.org/10.1051/matecconf/201713701009
17. Hou, R., Wang, T., Lv, Z., Liu, Y.: Experimental study of the ultrasonic vibration-assisted abrasive waterjet micromachining the quartz glass. Adv. Mat. Sci. Eng. (2018). article number 8904234. https://doi.org/10.1155/2018/8904234
18. Luiset, B., Sanchette, F., Billard, A., Schuster, D.: Mechanisms of stainless steels erosion by water droplets. Wear **303**(1–2), 459–464 (2013). https://doi.org/10.1016/j.wear.2013.03.045
19. Foldyna, J., Klich, J., Hlavacek, P., et al.: Erosion of metals by pulsating water jet. Tech. Gazette **19**(2), 381–386 (2012)
20. Thomas, G.P., Brunton, J. H.: Drop impingement erosion of metals. In: Proceedings of the Royal Society of London, Series A, Mathematical and Physical Sciences, pp. 549–565. Royal Soc London (1970). https://doi.org/10.1098/rspa.1970.0022

21. Zelenak, M., Foldyna, J., Scucka, J., Hloch, S., Riha, Z.: Visualisation and measurement of high-speed pulsating and continuous water jets. Measurement **72**(1–8) (2015). https://doi.org/10.1016/j.measurement.2015.04.022
22. Říha, Z., Foldyna, J.: Ultrasonic pulsations of pressure in a water jet cutting tool. Tech. Gazette **19**(3), 487–491 (2012)
23. Riha, Z., Foldyna, J.: Ultrasonic pulsations of pressure in a tool for water jet cutting. In: Vodni Paprsek 2011: Vyzkum, Vyvoj, Aplikace, pp. 245–254. Inst Geonics Cas, Ostrava-Poruba (2011)
24. Kušnerová, M., Foldyna, J., Sitek, L., et al.: Innovative approach to advanced modulated waterjet technology. Tech. Gazette **19**(3), 475–480 (2012)
25. Lehocka, D., Klich, J., Foldyna, J., et al.: Copper alloys disintegration using pulsating water jet. Measurement **82**, 375–383 (2016). https://doi.org/10.1016/j.measurement.2016.01.014
26. Lehocká, D., Klich, J., Foldyna, J., et al.: Surface integrity evaluation of brass CW614N after impact of acoustically excited pulsating water jet. In: ICMEM 2016, International Conference on Manufacturing Engineering and materials, Book Series: Procedia Engineering, vol. 149, pp. 236–244. Elsevier Science BV (2016). https://doi.org/10.1016/j.proeng.2016.06.662
27. Lehocka, D., Klichova, D., Foldyna, J., et al.: Comparison of the influence of acoustically enhanced pulsating water jet on selected surface integrity characteristics of CW004A copper and CW614N brass. Measurement **110**, 230–238 (2017). https://doi.org/10.1016/j.measurement.2017.07.005
28. Lehocka, D., Simkulet, V., Legutko, S.: Assessment of deformation characteristics on CW004A copper influenced by acoustically enhanced water jet. In: MANUFACTURING 2017, 5th International Scientific-Technical Conference on Advances in Manufacturing, pp. 717–724. Springer (2018). https://doi.org/10.1007/978-3-319-68619-6_69
29. Lehocká, D., Botko, F., Simkulet, V., et al.: Study of surface topography of CW004A copper after PWJ disintegration. In: MMS Conference, 2nd EAI International Conference on Management of Manufacturing Systems. ACM (2018). https://doi.org/10.4108/eai.22-11-2017.2274347
30. Srivastava, M., Hloch, S., Tripathi, R., et al.: Ultrasonically generated pulsed water jet peening of austenitic stainless-steel surfaces. J. Manuf. Process. **32**, 455–468 (2018). https://doi.org/10.1016/j.jmapro.2018.03.016
31. Li, D., Kang, Y., Ding, X.L., Liu, W.C.: Effects of the geometry of impinging surface on the pressure oscillations of self-resonating pulsed waterjet. Adv. Mech. Eng. **9**(8) (2017). article number 1687814017720081. https://doi.org/10.1177/1687814017720081
32. Klich, J., Klichova, D., Foldyna, V., Hlavacek, P., Foldyna, J.: Influence of variously modified surface of aluminium alloy on the effect of pulsating water jet. Strojniski Vestnik J. Mech. Eng. **63**(10), 577–582 (2017). https://doi.org/10.5545/sv-jme.2017.4356
33. Lehocka, D., Klich, J., Botko, F., Simkulet, V., Foldyna, J., Krejci, L., Storkan, Z., Kepic, J., Hatala, M.: Comparison of ultrasonically enhanced pulsating water jet erosion efficiency on mechanical surface treatment on the surface of aluminum alloy and stainless steel. Int. J. Adv. Manuf. Technol. **103**(5–8), 1647–1656 (2019). https://doi.org/10.1007/s00170-019-03680-8
34. Sitek, L., Foldyna, J., Martinec, P., et al.: Use of pulsating water jet technology for removal of concrete in repair of concrete structures. Baltic J. Road Bridge Eng. **6**(4), 235–242 (2011). https://doi.org/10.3846/bjrbe.2011.30
35. Foldyna, V., Foldyna, J., Klichova, D., Klich, J., Hlavacek, P., et al.: Effects of continuous and pulsating water jet on CNT/concrete composite. Strojniski Vjesnik J. Mech. Eng. **63**(10), 583–589 (2017). https://doi.org/10.5545/sv-jme.2017.4357
36. Foldyna, J., Sitek, L., Martinec, P., et al.: Rock cutting by pulsing water jets. In: Eurock 2005: Impact of Human Activity on the Geological Environment, pp. 129–134. A A Balkema Publisher (2005)

37. Dehkhoda, S., Hood, M.: An experimental study of surface and sub-surface damage in pulsed water-jet breakage of rocks. Int. J. Rock Mech. Min. Sci. **63**, 138–147 (2013). https://doi.org/10.1016/j.ijrmms.2013.08.013
38. Liu, Y., Wei, J.P., Ren, T., Lu, Z.H.: Experimental study of flow field structure of interrupted pulsed water jet and breakage of hard rock. Int. J. Rocks Min. Sci. **78**, 253–261 (2015). https://doi.org/10.1016/j.ijrmms.2015.06.005
39. Lu, Y.Y., Xiao, S.Q., Ge, Z.L., Zhou, Z., Ling, Y.F., Wang, L.: Experimental study on rock-breaking performance of water jets generated by self-rotatory bit and rock failure mechanism. Powder Technol. **346**, 203–216 (2019). https://doi.org/10.1016/j.powtec.2019.01.078
40. Hnizdil, M., Raudensky, M.: Descaling by pulsating water jet. In: METAL 2010 - 19th International Conference on Metallurgy and Materials, pp. 209–213. Tanger LTD (2010)
41. Lehocka, D., Klich, J., Pitel, J., Krejci, L., Storkan, Z., Duplakova, D., Schindlerova, V., Sajdlerova, I.: Analysis of the pulsating water jet maximum erosive effect on stainless steel. In: Lecture Notes in Mechanical Engineering. Advances in Manufacturing II, Mechanical Engineering, vol 4, pp. 233–241 (2019). https://doi.org/10.1007/978-3-030-16943-5_21
42. Klich, J., Klichova, D., Hlavacek, P.: Effects of pulsating water jet on aluminium alloy with variously modified surface. Tech. Gazette **24**(2), 341–345 (2017). https://doi.org/10.17559/TV-20140219100749
43. Hloch, S., Srivastava, M., Krolczyk, J.B., et al.: Strengthening effect after disintegration of stainless steel using pulsating water jet. Tech. Gazette **25**(4), 1075–1079 (2018). https://doi.org/10.17559/TV-20170327134630
44. Stoye, H., Koehler, H., Mauermann, M., Majschak, J.P.: Investigations to increase cleaning efficiency with pulsed liquid jet. Chem. Ing. Tech. **86**(5), 707–713 (2014). https://doi.org/10.1002/cite.201300047
45. Augustin, W., Fuchs, T., Foste, H., Scholer, M., Majschak, J.P., Scholl, S.: Pulsed flow for enhanced cleaning in food processing. Food Bioprod. Process. **88**(C4), 384–391 (2010). https://doi.org/10.1016/j.fbp.2010.08.007
46. Teliskova, M., Torok, J., Duplakova, D., Kascak, J., Mezencevova, V., Bircak, J.: Non-destructive diagnostics of hard-to-reach places by spatial digitization. TEM J. Technol. Educ. Manag. Inform. **7**(3), 612–616 (2018). https://doi.org/10.18421/TEM73-18
47. Hreha, P., Radvanska, A., Knapcikova, L., Krolczyk, G.M., Legutko, S., Krolczyk, J.B., Hloch, S., Monka, P.: Roughness parameters calculation by means of on-line vibration monitoring emerging from AWJ interaction with material. Metrol. Measur. Syst. **22**(2), 315–326 (2015). https://doi.org/10.1515/mms-2015-0024

Particle Velocity in Abrasive Waterjets

Michael Lo(✉), Axel Henning, Kevin Hay, and Peter Miles

OMAX Corporation, 21409 72nd Ave S, Kent, WA 98032, USA
Michael.Lo@omax.com

Abstract. Abrasive particle velocity plays a key role in material removal in abrasive waterjet (AWJ) machining. Understanding how certain operating conditions effect the speed of abrasive particles is important for the improvement of the AWJ process. Without a method of measuring abrasive speed, it is difficult to evaluate the cutting performance of an AWJ using theoretical particle velocity modeling. To measure the velocity of single abrasive particles at a variety of different conditions, a magnetic particle sensor system was designed. In order to increase the accuracy and versatility of the sensors, it was constructed with the ability to capture a large amount of velocity data which allows for outlying data to be filtered and the velocity distribution to be analyzed. The results were used to examine how the velocity data can be used to improve the theoretical particle speed models, the influences of varied operating conditions on velocity spread, and the correlation between kinetic cutting power and material erosion.

Keywords: Abrasive waterjet cutting · Particle velocity · Magnetic sensor

Nomenclature

A = *Particle area*
a = *Accelaration*
C_d = *Coefficient of drag*
M_a = *Mass flow abrasive*
ρ_{fluid} = *Density of fluid*
m = *Particle mass*
P = *Water pressure at nozzle*
P_h = *Hydraulic power*
P_p = *Particle cutting power*
t = *Time*
V_p = *Particle velocity*
V_w = *Water velocity*
Y_p = *Particle position*

D. Klichová et al. (Eds.): Water Jet 2019, LNME, pp. 135–144, 2021.
https://doi.org/10.1007/978-3-030-53491-2_15

1 Introduction

An AWJ uses the energy of high-pressure water to accelerate abrasive particles and remove material through a process of enhanced erosion. Using a small orifice, the high-pressure water is directed into a section of the AWJ cutting head called the mixing chamber. Depending on the pressure, the water velocity can range between 500 m/s and 1300 m/s. Since the water is traveling at high speeds, a Venturi vacuum is created that draws air into mixing chamber. The air pulls abrasive into the cutting head where it is mixed with the high velocity jet. The mixing tube both focusses the stream while allowing the abrasives to be accelerated by the air and water mixture. This high velocity jet can then be used to cut virtually any material at a range of different thicknesses. The velocity of the particle is an important factor in the jet's ability to remove material effectively and understanding the particle behavior occurring in the AWJ nozzle is extremely important. Precise measurements of particle velocity allow for the development of analytical modelling approaches of the particle's acceleration and material erosion. In this paper, the examination of particle velocity measurements will be discussed.

2 Speed Measurement

The measurement of the velocity of abrasive particles presents many inherent difficulties. The abrasive particles are small (diameter 63–300 μm), traveling at high speeds (500–1300 m/s), and mixed with water droplets that are travelling at similar speed. Therefore, a specialized approach is needed to distinguish the particles from the mixture of air and water to reliably measure the abrasive velocity. To detect the speed of the abrasive, a variety of methods have been described in prior paper such as the dual disk anemometer (DDA) [4, 5], florescent particle tracking [7], laser Doppler [1], and magnetic particle sensor [2, 3]. In this study, the magnetic particle sensor was chosen because it allows for precise measurements of single particles speeds.

The magnetic sensor measures the amount of time required for a particle to travel a known distance which allows for a single particle's velocity to be calculated. Instead of capturing only an average speed value, multiple individual velocity readings are taken at a given operating condition. This allows for the analysis of the velocity distribution and velocity variation at different operating parameters.

Due to limitations in previous magnetic measurement devices [2, 3], a maximum of three data points were collected at a specific operating condition (i.e. pressure, feed rate, nozzle geometry, etc.). The new sensor design is capable of collecting a large number of velocity data points at a given condition. Gathering more data points provides more robust results because outlying data points can be filtered based on the variance from the median or average velocity values. Increasing the number of data points also increases the resolution of the data distribution.

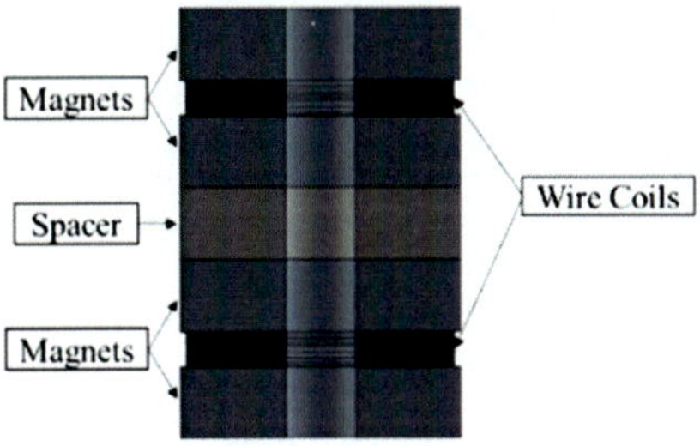

Fig. 1. Cross sectional view of the magnetic particle velocity sensor.

The basic principle for the magnetic velocity device is to detect when a particle passes through two known points in the sensor. The device senses particles using a layered magnet and wire coil design illustrated in Fig. 1. When a particle passes through the magnets, it creates a disturbance in the magnetic field that generates an electrical pulse in the wire coils. The pulses are used to measure the time needed for the particle to pass from top to bottom coil. The velocity can be calculated using the length between the two coils divided by the time required to travel the distance.

During the process of AWJ cutting, a large number of high-speed particles are used to cause sufficient erosion at the material. As it is desired to identify the speed of single particles, ferromagnetic particles were mixed into the garnet supply at 0.5% of the total mass contained in the hopper. This allows typical conditions of varying abrasive feed rates while still measuring the velocity of single particles. With the small number of stainless-steel particles in the abrasive flow, the characteristics of the garnet feed were not significantly affected by the mass difference.

A test was performed to verify that the difference in particle mass between garnet and the tracer particles (density of 4.3 g/cm^3 and 7.7 g/cm^3 respectively) had no effect on the velocity measurement. Stainless steel tracer particles were classified into different weights and subjected to the AWJ process. As can be seen in Fig. 2, particle mass had

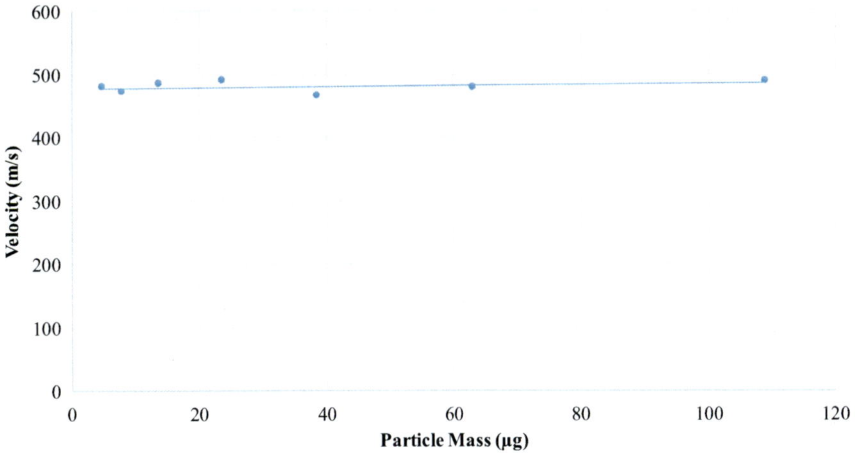

Fig. 2. Experimentally determined velocities for different size particles (P = 279 MPa, M_a = 340 g/min)

no effect on final particle velocity. With little velocity difference measured between the sizes, the point at which the particles reach a terminal velocity occurs before exiting the mixing tube. This means that the garnet particles are traveling roughly the same velocity as the stainless steel at the location of the sensor.

3 Particle Speed Distribution

With the ability to collect a large amount of velocity data, a distribution of 1500 points was collected at a single operating condition. The histogram in Fig. 5 shows the velocity distribution that follows rough Gaussian distribution trends. The speed values were normalized around the median to show deviation percentages.

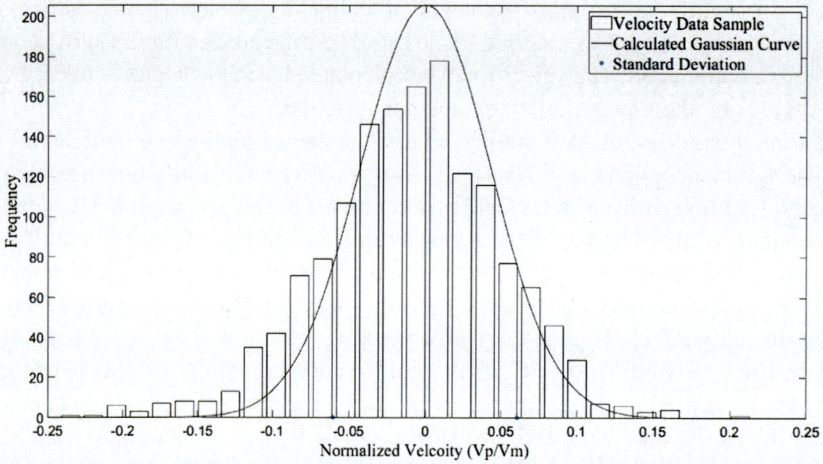

Fig. 3. Experimentally determined velocity distribution for stainless steel abrasive mixed with garnet abrasive ($P = 379$ MPa, $M_a = 340$ g/min).

To examine the effect of different operating conditions on velocity distribution, speed data was collected at a variety of pressures and abrasive flow rates. The amount of deviation in a sample demonstrates the consistency of the acceleration process. If a distribution has more variations, the particles are not accelerating uniformly in the mixing tube. A non-uniform acceleration causes the particles to reach a range of different final velocities at the exit of the mixing tube. Samples that achieve less variation have more consistent acceleration and the particles reach similar final velocities.

Figure 4 shows there is little effect of abrasive feed rate on the relative spread of particle velocity. The spread remains around ± 16% of the median velocity. However, with a smaller ratio of abrasive to water, the acceleration process of the particles becomes more consistent and the final velocity spread decreases. A large abrasive feed rates causes the particles to achieve a wider range of velocities at the exit of the mixing tube. As the mean velocity increases at higher pressure, the absolute spread increases as well.

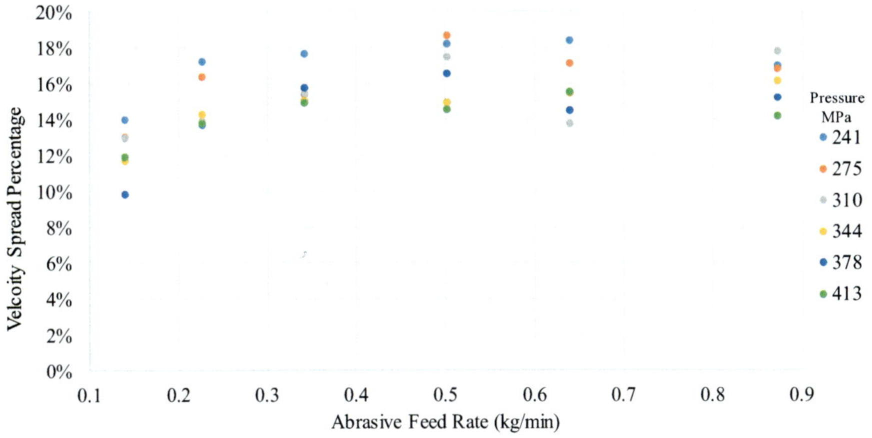

Fig. 4. Experimentally determined velocity variation at different abrasive feed rates and pressures.

4 Theoretical Particle Velocity Derivation

To compare the measured data from the velocity sensor to theoretical values, a model was derived to describe the velocity of an abrasive particle. This model utilizes the drag force equation and Newton second law of motion to describe velocity behavior of a particle accelerating through the nozzle. To calculate the velocity at the exit of the orifice, the Bernoulli equation was used.

$$V_w = \sqrt{\frac{2P}{\rho_{fluid}}} \tag{1}$$

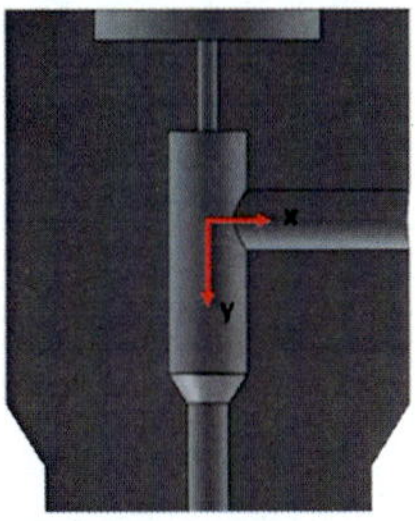

Fig. 5. Cross sectional view of the waterjet nozzle.

Using the drag equation and Newton's second law, the velocity and position functions can be found. Assuming the frictional losses are small in comparison to drag, only drag is the substituted into the force equation.

$$\frac{C_d A \rho_{fluid} V^2}{2} = ma \tag{2}$$

Solve the equation for the acceleration.

$$a = \frac{C_d A \rho_{fluid} V^2}{2m} \tag{3}$$

As the particle accelerates, the amount of drag decreases because the relative velocity between the fluid and particle decreases. Substitute this relationship into the Eq. 3.

$$a = \frac{C_d A \rho_{fluid} \left(V_w - V_p\right)^2}{2m} \tag{4}$$

Rewrite in ordinary differential equation form.

$$\ddot{y} = \frac{C_d A \rho_{fluid} (V_w - \dot{y})^2}{2m} \tag{5}$$

Solving Eq. 5 provides the following position and velocity equations.

$$Y_p(t) = \frac{V_w\left(\frac{C_d A \rho_{fluid}}{2m} t + \frac{1}{v_w}\right) - \ln\left(\frac{C_d A \rho_{fluid}}{2m} t + \frac{1}{v_w}\right) + \ln\left(\frac{1}{v_w}\right) - 1}{\frac{C_d A \rho_{fluid}}{2m}} \tag{6}$$

$$V_p(t) = V_w - \frac{1}{\frac{C_d A \rho_{fluid}}{2m} t + \frac{1}{v_w}} \tag{7}$$

5 Experimental Data and Theoretical Results Comparison

Particle speed data was collected at a variety of different pressures and abrasive flow rates to provide a wide range of data points. Each point on Fig. 6 represents the median

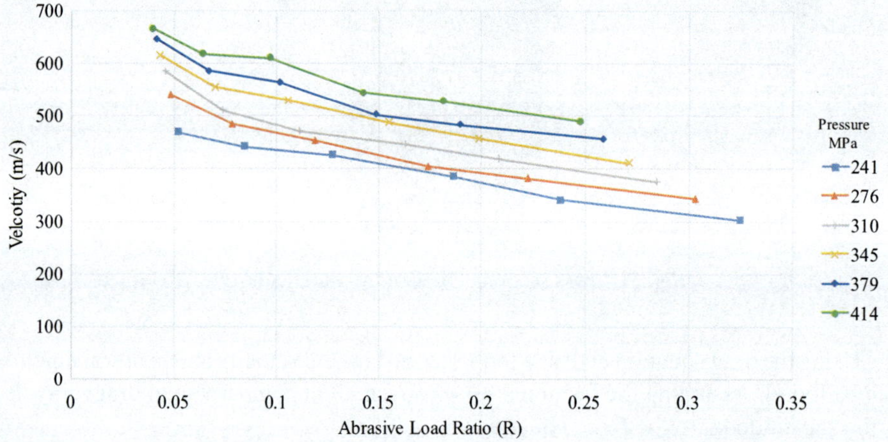

Fig. 6. Experimentally determined median particle velocity at different pressures and abrasive load ratios.

value of a 200 sample velocity distribution at the given operating condition. The abrasive feed rate was converted to abrasive loading using Eq. 8.

$$R = \frac{\dot{m}_p}{\dot{m}_w} \tag{8}$$

Figure 6 shows that when the abrasive loading is increased, the velocity of the particles decreases at a constant pressure. Using the theoretically derived particle velocity model (Eq. 7), a comparison was made between the predicted and measured particle velocity at the location of the sensor. The following assumptions were made to solve Eq. 7; the density of the fluid is the ratio of the total fluid mass (air and water) divided by total volume in the mixing tube, the coefficient of drag was $C_d = 1.05$, and the particle was assumed to be traveling only in the y-direction (see Fig. 3) with no initial velocity.

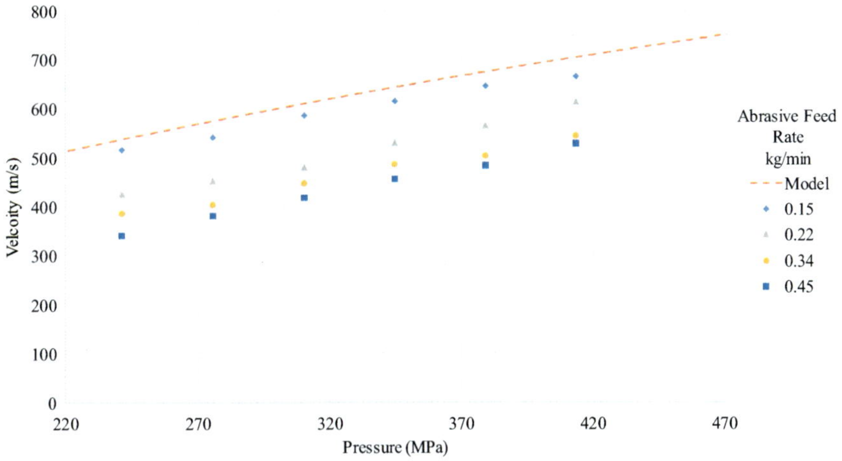

Fig. 7. Experimentally determined abrasive velocity at different pressures compared to theoretical particle velocity model.

The modeled particle speed follows the trends presented by the measured velocity data (see Fig. 7). The error in the model is due to the effect of feed rate on particle velocity. When the abrasive feed rate is increased, the velocity of the particles decreases. This parameter is not accounted for in the derivation of the particle speed model which causes an over prediction (higher value) in the calculated velocity.

6 Kinetic Cutting Power and Material Separation Speed

Measuring particle velocity at the impact location between the abrasives and material allows for the comparison between kinetic cutting power and material separation. Figure 8 is generated by translating the abrasive velocity (Fig. 6) into abrasive speed ratio (see Eq. 9).

$$\Psi_p = \frac{V_p}{V_w} \tag{9}$$

A regression analysis was performed and the velocity ratio was written as a function of abrasive loading using Eq. 10 [4, 6], where a = .343 and b = .435.

$$\Psi_p = \frac{a}{b + R} \tag{10}$$

The kinetic cutting power of the particles is given by Eq. 11.

$$P_p = \frac{1}{2}\dot{m}_p V_p^2 \tag{11}$$

Use the velocity ratio and abrasive loading correlations (Eqs. 8 and 9 respectively).

$$P_p = \frac{1}{2}R\dot{m}_w \Psi_p^2 V_w^2 \tag{12}$$

Simplify Eq. 13 using the hydraulic power.

$$P_p = R\Psi_p^2 P_h \tag{13}$$

Substitute Eq. 10 into abrasive kinetic cutting power (Eq. 13). The kinetic cutting power is written in terms of abrasive loading and hydraulic power.

$$P_p = R\left(\frac{a}{b + R}\right)^2 P_h \tag{14}$$

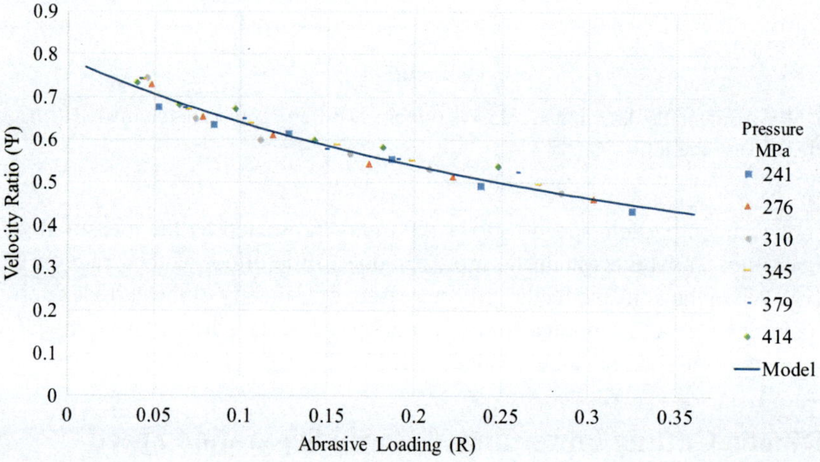

Fig. 8. Experimentally determined particle velocity ratio at different pressures and abrasive load ratios (model with a = .343 and b = .435).

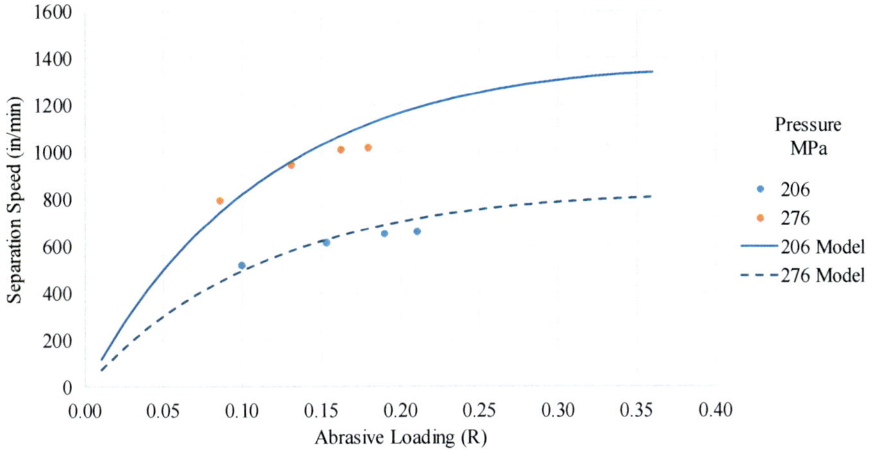

Fig. 9. Experimentally determined separation speed (6.25 mm A36 steel) at different abrasive load ratios compared to theoretical abrasive kinetic cutting power model (model with a = .343 and b = .435).

The kinetic cutting power model is scaled to fit the separation data to show the correlation between separation speed and kinetic cutting power. As can be seen in Fig. 9, the model follows the trends of the separation speed data. When the kinetic power of the cutting stream increases, the material separation speed increases as well.

7 Conclusions

Measuring the velocity of the abrasive particles accelerated by the AWJ nozzle is an extremely challenging process. The particles are mixed with a high velocity mixture of water and air which makes it difficult to detect the abrasives. The magnetic sensor was able to measure the individual velocity of single particle and sense the particle speed through the mixing to the AWJ cutting head. With the goal of increasing the reliability and resolution of the velocity distribution, an improved data acquisition system was used to collect a large number of data points.

Stainless steel tracer particles were mixed with the garnet because a magnetic abrasive had to be used. A test was conducted to determine if the density difference between stainless steel and garnet would affect particle velocity. It was found that mass and size did not influence the particle's velocity after exiting the mixing tube.

The measured particle velocity was then compared to a theoretically derived particle speed model. The predicted speed values were higher than the measured because the model did not take into consideration the effect of abrasive feed rate on velocity.

Particle velocity was tested at a variety of different pressures and garnet flow rates. This data was translated into terms of velocity ratio and abrasive loading. Using a regression analysis, the velocity ratio can be written as a function of abrasive loading. Substituting this relationship into the kinetic cutting power provides an equation that describes the particle's kinetic power as a function of abrasive loading. Comparing these results to material separation data found that the kinetic power results followed the trends presented by the separation values. When the kinetic cutting power was increased, the material separation increased as well.

The magnetic sensor measured the velocity of individual particles accelerated by an AWJ very well. The sensor provided data that was examined in a variety ways to improve the understanding of the acceleration process and material erosion behavior. To improve upon the collected results, more data is needed to build a stronger model of the AWJ cutting process. With a larger variety of tested operating conditions, the predictive quality of the theoretical calculations would improve. In conclusion, this paper provided an analysis of measured particle velocity data obtained using a magnetic particle speed sensor.

References

1. Annoni M.; Monno M.: A repeatable set-up for water jet laser Doppler velocity measurements. In: Proceedings of the 2008 19th International Conference on Water Jetting, Nottingham, UK, 15–17 October (2008)
2. Dong, Y., Tyler, L., Summers, A., Johnson, M.: Experimental study of particle velocity measurement in abrasive water jet cutting in the nozzle and stream using multiple sensing elements. In: Proceedings of the 2004 17th International Conference on Water Jetting, Mainz, Germany, September 7–9 (2004)
3. Dorle, A., Tyler, L., Summers, D.: Measurement of particle velocities in high speed Waterjet technology. In: Proceedings of the 2003 WJTA American Waterjet Conference, Houston, Texas, 17–19 August (2003)
4. Henning, A., Miles, P.: Effect of pressure on abrasive Waterjet cutting performances. In: Proceedings of 2014 22nd International Conference of Water Jetting, Haarlem, Netherlands, 3–5 September (2014)
5. Liu, P., Miles, P., Cooksey, N., Hibbard, C.: Measurement of water-droplet and abrasive speeds in ultrahigh-pressure abrasive waterjets. In: Proceedings of 1999 Abrasive Waterjet Conference, Houston, Texas, August 14–17 (1999)
6. Miles, P., Henning, A.: Pump efficiency's effect on cutting performance. In: Proceedings of the 2013 WJTA-IMCA Conference and Expo, Houston, Texas, 9–11 September (2013)
7. Roth, P., Looser, H., Heiniger, K.C., Bühler, S.: Determination of abrasive particle velocity using laser induced fluorescence and particle tracking methods in abrasive water jets. In: Proceedings of the 2005 WJTA American Waterjet Conference, Houston, Texas, 21–23 August (2005)

Multi-criteria Optimization of the Abrasive Waterjet Cutting Process for the High-Strength and Wear-Resistant Steel Hardox®500

Andrzej Perec[1](✉), Wojciech Musial[2,3], Jaroslaw Prazmo[4,5], Ryszard Sobczak[4,6], Aleksandra Radomska-Zalas[1], Anna Fajdek-Bieda[1], Slawomir Nagnajewicz[2], and Frank Pude[7]

[1] Faculty of Technology, Jacob of Paradies University, Gorzów Wielkopolski, Poland
aperec@ajp.edu.pl
[2] Faculty of Mechanical Engineering, Koszalin University of Technology, Koszalin, Poland
[3] Regional Research and Development Center in Walcz, Walcz, Poland
[4] Institute of Engineering and Management, University of Applied Sciences in Walcz, Walcz, Poland
[5] ProWaterjet, Koszalin, Poland
[6] RS WaterJet, Koszalin, Poland
[7] Steinbeis Consulting Center High-Pressure Waterjet Technology, Horgau, Germany

Abstract. This paper presents the results of the Abrasive Water Jet (AWJ) cutting process optimization for Hardox® steel. This type of steel makes it possible to reduce the weight of a work piece and increases the service life of steel elements in comparison with common steel applications. The special combination of features, such as hardness, strength, and toughness, allows using elements with smaller cross-sections. The effect of different pressure levels, traverse speeds and abrasive flow rates, on cutting depth, kerf width and roughness on the cut surface of the Hardox® 500 steel was experimentally tested. In order to determine the most beneficial control parameters of the cutting process, multi-criteria optimization by means of the Gray Relation Analysis (GRA) combined with the Taguchi approach was carried out.

Keywords: High-pressure abrasive water jet · Cutting · Hardox® steel · Gray relational analysis

1 Introduction

Common methods of treatment for hard to machining materials do not give sufficient results in terms of its efficiency and quality [3]. AWJ [16, 19] and pulsating water jet (PWJ) [9, 10] machining are not limited by such disadvantages. Even AWJ machining of very hard materials such as polycrystalline diamond (PCD) is possible, but corundum-based abrasives [1, 14] are recommended.

In order to examine the multiple attributes related to properties of Hardox® - wear resistance steel, cut by AWJ, one decision-making technique, GRA has been chosen.

D. Klichová et al. (Eds.): Water Jet 2019, LNME, pp. 145–154, 2021.
https://doi.org/10.1007/978-3-030-53491-2_16

It's one of the main techniques, together with TOPSIS method [20], used when the information is uncertain and incomplete.

Wojciechowski et al. [23] published research of the minimization of cutting forces and vibrations during precise ball end milling of hardened 55NiCrMoV6 steel. The optimal selection of process parameters which enables the minimization of cutting forces and vibrations in order to improve the machined surface quality was carried out by the minimization of process responses with the application of signal to noise S/N ratio and GRA. Optimal values of process control parameters were achieved and was validated during the tests of the measurements of treated surface quality. Tavodova [4] shows the use of the statistical process control (SPC) method, specifically a Design of Experiment (DoE) with a Taguchi approach, during evaluation of a finished area of aluminum alloy by AWJ. Based on the results from the experiment optimal levels of control parameters for achieving the best surface possible after cutting by the AWJ were established. Another DoE method – response surface methodology (RSM) in modeling of AWJ machining was published by Perec [17] for modelling and optimization of limestone cutting efficiency by the Abrasive Water Suspension Jet. Nair et al. [13] presented research on multi-objective GRA for characteristics of holes in Inconel 617 alloy, drilled by AWJ. The properties of the hole was determined using among others entry and exit hole geometrical dimensions, taper angle of hole, drill rate and surface roughness. The control parameters were pressure of water, stand-off distance, and abrasive flow rate. Analysis of variance (ANOVA) of the individual responses was utilized to identify of each parameter influence on the process efficiency. The effects of the (ANOVA) confirmed the strong correlation of results was obtained. The investigation of the cutting performance of AWJ machining on Aluminum Hybrid Metal Matrix Composites (HMMCs) was published by Senthilkumar et al. [21]. The influence of cutting control parameters such as pressure, stand-off distance and traverse speed was analyzed on the basis of the orthogonal array on the machining effects such as taper angle, surface roughness, and material removal rate. The optimal condition for the better cutting performance was defined by GRA. The stand-off distance was the most contributing control parameter with almost 82% as resolved by ANOVA.

The literature search shows that GRA is an interesting optimization method and can be used to make a wide range of analyzes. Especially for optimizing mutually opposite parameters. That is why GRA was used for multi-criteria optimization of cutting depth and roughness of the cut surface by AWJ.

2 Materials and Experimental Setup

2.1 Cutting Materials and Abrasives

In research, the Hardox® 500 steel was used. It is a bendable and weldable abrasion-resistant steel, with a nominal Brinell hardness of 500 HBW. Hardox® steel has been produced since 1970 in six grades. It is distinguished by high wear resistance, ability to being machined by special tools, good weldability, high mechanical properties, and impact resistance. Steel from the Hardox® group has a tensile strength from 1250 MPa (Hardox 400) to 2000 MPa (Hardox 600). The heat treatment is carried out from the normalized states through quenching in water and tempering in temperature 200–700 °C. It is suitable for applications that demand higher wear resistance [8, 26].

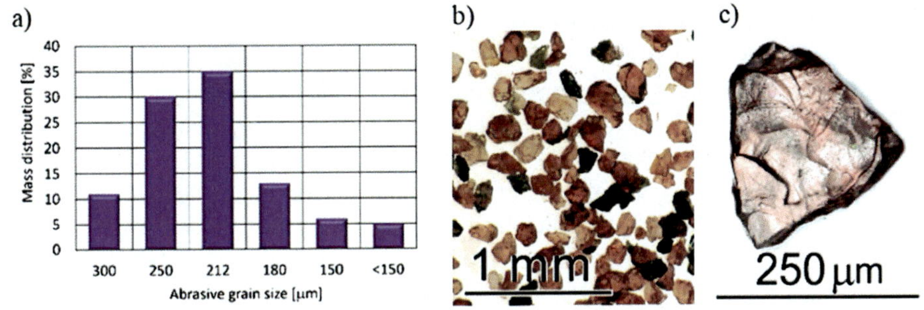

Fig. 1. Garnet #80: a) grains distribution, b) grains view c) detail of sample grain.

As abrasive the crushed rock garnet type J80A from Jinhong Mining located in Jiangsu, China was used. Abrasive grain size distribution and view of grains of this abrasive present Fig. 1. The shape of grains is nearly to isometric. Grains are characterized by sharp edges and corners. Grain's color is from light to deep violet with smaller number dark brown to black. Mineral content, chemical composition and physical characteristics are presented by Table 1.

Table 1. Properties of almandine garnet J80A abrasive [25].

Mineral content [%]							
Amandine	Ilmenite	Omphacite	Rutile	Quartz	Hornblende	Free Silica	
90–96	1.0	1.5	0.6	<0.1	<0.5	<0.5%	
Chemical composition [%]							
Fe_2O_3	SiO_2	TiO_2	Al_2O_3	FeO	CaO	MgO	MnO
17	39	0.05	21	8	9.5	5	0.4
Physical characteristics							
Density [kg/dm^3]	Bulk Gravity [kg/dm^3]	Mohs Hardness	Color	Grain shape	Conductivity	Acid solubility (HCL)	
3.8–4.1	2.3–2.4	7.5–8.0	Dark red	Sub angular	<25 S/m	<1.0%	

2.2 Multi-response Optimization Using GRA

GRA is a section of Grey Theory published by Deng Julong in 1989 [7]. GRA allows an effective and significant inference to a test or model which has incomplete information by creating a relationship among two discrete sequences. The grey relational rank shows the degree of influence that the comparison sequence could exert over the reference sequence. In the question of two or more outputs, with different features, multi-response optimization using GRA is one of the best methods. GRA can also be used to establish the similarity between apparently uneven output data [12, 24].

The objective of this paper is to utilize GRA to gain an optimal combination of control parameters for maximum cutting depth and minimum surface roughness and to uncover the individual result of each control parameter on cutting depth and surface roughness. In this research, multi-response optimization of cutting parameters on the GRA basis was carried out according to the succeeding steps.

Grey Relational Generation

In GRA, data preprocessing is carried out to normalize the primary sequence values to the sequence within the scope of zero to one [11]. The preprocessing of data by normalization, into a group of sequences, is named grey relational generation.

At the stage of preprocessing data using GRA, the response of the converted sequences can be attributed to two quality features, namely, larger-is-better or smaller-is-better. For the smaller-is-better quality feature, as roughness S_a, the sequence can be normalized using the equation:

$$x_i^*(k) = \frac{\max y_i(k) - y_i(k)}{\max y_i(k) - \min y_i(k)} \tag{1}$$

Where:

$x_i^*(k)$ indicates the reference sequence after preprocessing for the *i*-th test

$y_i(k)$ shows the start sequence of the mean of the responses.

For the larger-is-better quality feature, as the cutting depth, the sequence can be normalized using the equation:

$$x_i^*(k) = \frac{y_i(k) - \min y_i(k)}{\max y_i(k) - \min y_i(k)} \tag{2}$$

Computation of GRC and GRG

After normalization of the sequence, the next stage is to compute the deviation sequence of the reference sequence using the equation:

$$\Delta_{0i}(k) = \left|x_0^*(k) - x_i^*(k)\right| \tag{3}$$

Where:

$\Delta_{0i}(k)$, $x_0^*(k)$, and $x_i^*(k)$ are related to the deviation, reference, and comparison sequences. The Grey Relational Coefficient (GRC) is calculated on the basis of the equation:

$$\xi_i(k) = \frac{\Delta_{min} - \zeta\Delta_{max}}{\Delta_{0i}(k) - \zeta\Delta_{max}} \tag{4}$$

where $\xi_i\,(k)$ means the GRC of each response variable calculated as a function of the minimal and maximal deviations of all response variables. The characterization or recognition coefficient exemplified by ζ, designated in the range $\zeta \in [0,1]$, is usually set at 0.5 to appropriate equal weights to every parameter. As indicated in Eq. (4), the composite GRG, is then calculated by averaging the GRC of all response variables:

$$\gamma_i(k) = \frac{1}{n}\sum\nolimits_{k=1}^{n} \xi_i(k) \quad (5)$$

Where:
$\gamma_i\,(k)$ presents the value of GRG established for the i-th test,
n is the aggregate count of efficiency characteristics.

After determining the optimal level of the factors using GRG, the last stage is prediction and checking the quality properties based on the equation:

$$\gamma_{pred} = \gamma_m + \sum\nolimits_{i=1}^{q} \gamma_0 - \gamma_m \quad (6)$$

Where:
γ_0 indicates the maximal of average GRG at the optimal factor level,
γ_m describes the mean GRG,
q denotes the number of factors impacting response values.

Commonly, a big value of the GRG suits to a powerful relation between the reference data sequence and the relative sequence. Then a higher value of the GRG indicates that the corresponding control parameters of the process are nearer to the optimal values.

2.3 Test Rig

The cutting tests were carried out on a WaterJet CNC OMAX 60120 machining center. The experiments were conducted using a full factorial design model in view of three of the control parameters: pressure, traverse speed, and abrasive flow rate (Table 2). Parameters were chosen on the basis of earlier obtained research results [15, 18] and publications of other researchers [2, 5, 6, 22].

Table 2. Control parameters of the cutting process.

Control parameter		L1	L2	L3
Abrasive flow rate	[g/min]	250	350	450
Pressure	[MPa]	350	375	400
Traverse speed	[mm/min]	100	200	300

Surface roughness measurements were carried out in the middle of the cutting zone in a 1.5 mm × 1.5 mm elemental area by using a Sensofar - Model S neox - Five Axis 3D Optical Microscope. The surface structure was observed on a FEI Quanta 200 microscope operating at 25 keV accelerating voltage in the low-vacuum mode.

3 Results of Tests and Discussion

3.1 Grey Relational Analysis

Results of the tests are shown by Table 3, while Table 4 presents the calculation effects of the normalizing, the deviation sequence (Delta), the GRC, the GRG, and their ranks. The overall rating of two responses is evaluated on the basis of the GRG.

Table 3. Tests results of the cutting process.

No.	Control parameters			Responses	
	Abrasive flow rate	Pressure	Traverse speed	Roughness S_a	Cutting depth D_c
	[g/min]	[MPa]	[mm/min]	[μm]	[mm]
1	450	350	100	4.23	20.81
2	450	375	200	5.78	12.07
3	450	400	300	10.64	11.05
4	350	350	200	6.00	12.87
5	350	375	300	7.32	7.96
6	350	400	100	4.21	21.06
7	250	350	300	8.18	7.86
8	250	375	100	5.15	19.79
9	250	400	200	7.49	12.89

Table 4. Grey relational analysis coefficients and grades.

No.	Normalizing		Delta		GRC		GRG	Rank
	S_a	D_c	S_a	D_c	S_a	D_c		
1	1.000	0.981	0.000	0.019	1.000	0.964	0.982	**2**
2	0.758	0.319	0.242	0.681	0.674	0.423	0.549	**4**
3	0.000	0.242	1.000	0.758	0.333	0.397	0.365	**9**
4	0.724	0.380	0.276	0.620	0.644	0.446	0.545	**5**
5	0.518	0.008	0.482	0.992	0.509	0.335	0.422	**7**
6	0.994	1.000	0.006	0.000	0.988	1.000	0.994	**1**
7	0.384	0.000	0.616	1.000	0.448	0.333	0.391	**8**
8	0.856	0.904	0.144	0.096	0.777	0.839	0.808	**3**
9	0.491	0.381	0.509	0.619	0.496	0.447	0.471	**6**

The GRC thus allows optimization of multiple responses by conversion into a single GRG. The calculated GRG has a value between 0 and 1 and values represent a better status the closer they are to the ideal value of 1. Out of all the combinations of selected process variables, one that has the largest GRG value in the frame of the reference sequence is considered to be the best combination of parameters and is thereby recommended. The optimum level for the control factors (process parameters) is chosen respectively to the maximum level of each factor (Fig. 2). The best combination is AFR2-P3-TS1. The total mean value of the GRG (dashed line) is 0.6229.

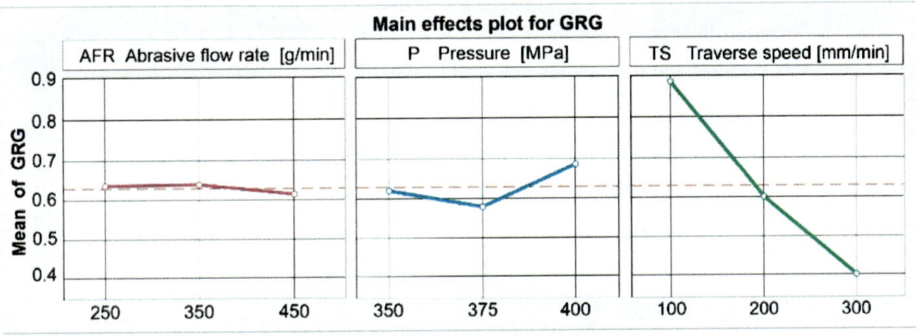

Fig. 2. Dependency of GRG values from control parameters of the cutting process.

The values of GRG corresponding to each control factor at their levels are calculated and presented in Table 5.

Table 5. Optimal level of control parameters.

Parameters	1	2	3	Rank
Abrasive flow rate	0.6298	0.6309	0.6080	3
Pressure	0.6146	0.5746	0.6795	2
Traverse speed	0.8862	0.5938	0.3887	1

3.2 Confirmation Test

The control parameters and effects of the confirmation test are shown by Table 6. For this condition the GRG is calculated on basis of Eq. (5) and equals 0.9938 and the predicted GRG which is calculated on the basis Eq. (6) equals 0.9926. The difference does not exceed 0.0012.

Table 6. Tests results of the cutting process.

Control parameters			Responses	
Abrasive flow rate	Pressure	Traverse speed	Roughness S_a	Cutting depth D_c
[g/s]	[MPa]	[mm/min]	[μm]	[mm]
350	400	100	4.21	21.06

The sample views of the surfaces which were generated by a cutting process at optimal conditions are presented by Fig. 3 and Fig. 4.

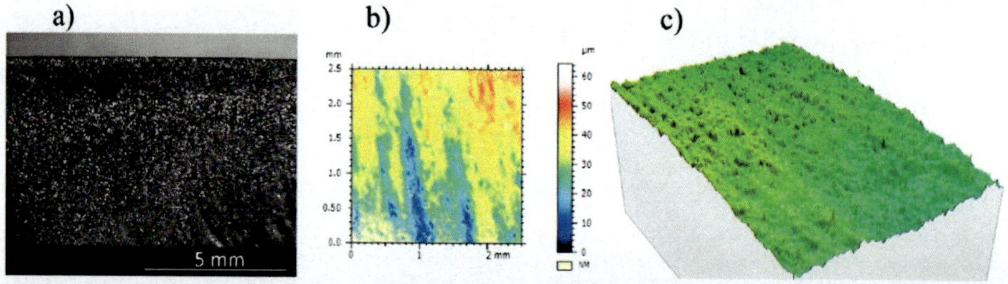

Fig. 3. Sample of cut surface: a) cutting surface view, b) 2D roughness map, c) 3D roughness map.

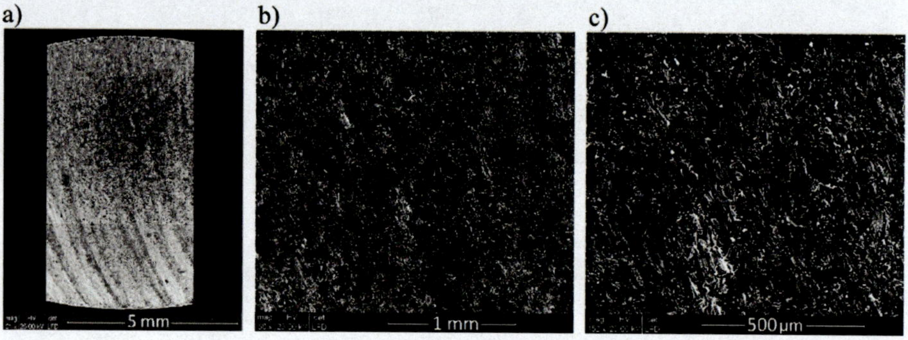

Fig. 4. Sample of EDX microscopy view of cut surface: a) 20x, b) 60x, c) 150x.

4 Conclusions

The primary aim of this research was to obtain the optimal set of control factors that affect the cutting depth and the surface roughness in the presence of multiple responses. The conducted research confirmed the equity of applying the method in multi-criteria optimization of the Hardox® steel cutting process by AWJ. The GRA transforms the optimization of the multiple characteristics into the optimization of an individual function called GRG, which significantly simplifies the computation. The GRA determines the ranks of evidence from computational results by optimal machining parameter combinations.

Based on the GRA graph it can be concluded that the best possible combination of the machining parameters to optimize the output parameters is:

abrasive flow rate = 350 g/min, pressure = 400 MPa, traverse speed = 100 mm/min.

The change in traverse speed causes the biggest changes in effects, followed by the pressure output and the smallest impact was noted for the abrasive flow rate.

References

1. Axinte, D.A., et al.: Abrasive waterjet cutting of polycrystalline diamond: a preliminary investigation. Int. J. Mach. Tools Manuf **49**(10), 797–803 (2009). https://doi.org/10.1016/j.ijmachtools.2009.04.003
2. Bicejova, L.: Abrasive kind and granularity changes affects to water jet technology head vibration during cutting HARDOX material thickness alternation process. Appl. Mech. Mater. **308**, 75–79 (2013). https://doi.org/10.4028/www.scientific.net/AMM.308.75
3. Bohdal, Ł., et al.: Three dimensional finite element simulation of sheet metal blanking process. Appl. Mech. Mater. **474**, 430–435 (2014). https://doi.org/10.4028/www.scientific.net/AMM.474.430
4. Cierna, H., Tavodova, M.: Using the design of experiment method to evaluate quality of cuts after cutting aluminum alloy by AWJ. Manuf. Technol. **13**(3), 303–307 (2013)
5. Filip, A.C., et al.: Research on surface roughness of hardox steels parts machined by Abrasive Waterjet. Stroj. Vestn. – J. Mech. Eng. (2019). https://doi.org/10.5545/sv-jme.2019.5998
6. Hlavac, L., et al.: Influence traverse speed on surface quality after water-jet cutting for hardox steel. In: Proceedings of 24th International Conference on Metallurgy and Materials METAL 2015, pp. 723–728 TANGER, Ostrava, Czech Republic (2015)
7. Julong, D.: Introduction to grey system theory. J. Grey Syst. **1**, 1–24 (1989)
8. Krolczyk, G.M., et al.: Metrological changes in surface morphology of high-strength steels in manufacturing processes. Measurement **88**, 176–185 (2016). https://doi.org/10.1016/j.measurement.2016.03.055
9. Lehocka, D., et al.: Analysis of the pulsating water jet maximum erosive effect on stainless steel. In: Gapiński, B., et al. (eds.) Advances in Manufacturing II. pp. 233–241 Springer, Cham (2019). https://doi.org/10.1007/978-3-030-16943-5_21
10. Lehocka, D., et al.: Comparison of ultrasonically enhanced pulsating water jet erosion efficiency on mechanical surface treatment on the surface of aluminum alloy and stainless steel. Int. J. Adv. Manuf. Technol. **103**, 1647–1656 (2019). https://doi.org/10.1007/s00170-019-03680-8
11. Manoharan, S., et al.: Development and characterization of novel fiber reinforced hybrid friction composites. In: Davim, J.P. (ed.) Green Composites. De Gruyter, Boston (2017). https://doi.org/10.1515/9783110435788-004
12. Mia, M., et al.: Multi-objective optimization and life cycle assessment of eco-friendly cryogenic N2 assisted turning of Ti-6Al-4V. J. Clean. Prod. **210**, 121–133 (2019). https://doi.org/10.1016/j.jclepro.2018.10.334
13. Nair, A., Kumanan, S.: Optimization of size and form characteristics using multi-objective grey analysis in abrasive water jet drilling of Inconel 617. J. Braz. Soc. Mech. Sci. Eng. **40**(3), 1–15 (2018). https://doi.org/10.1007/s40430-018-1042-7
14. Perec, A., et al.: A study of wear on focusing tubes exposed to corundum-based abrasives in the waterjet cutting process. Int. J. Adv. Manuf. Technol. **104**(5–8), 2415–2427 (2019). https://doi.org/10.1007/s00170-019-03971-0
15. Perec, A.: Environmental aspects of abrasive water jet cutting. Ann. Set Environ. Prot. Rocz. Ochr. Srodowiska. **20**, 258–274 (2018)
16. Perec, A.: Experimental research into alternative abrasive material for the abrasive water-jet cutting of titanium. Int. J. Adv. Manuf. Technol. **97**(1–4), 1529–1540 (2018). https://doi.org/10.1007/s00170-018-1957-2
17. Perec, A.: Investigation of limestone cutting efficiency by the abrasive water suspension jet. In: Hloch, S., et al. (eds.) Advances in Manufacturing Engineering and Materials, pp. 124–134. Springer, Cham (2019). https://doi.org/10.1007/978-3-319-99353-9_14

18. Perec, A., et al.: Obtaining the selected surface roughness by means of mathematical model based parameter optimization in abrasive waterjet cutting. SV-J. Mech. Eng. **63**(10), 606–613 (2017). https://doi.org/10.5545/sv-jme.2017.4463
19. Perec, A., Radomska-Zalas, A.: Modeling of abrasive water suspension jet cutting process using response surface method. In: AIP Conference Proceedings 2078, 020051 (2019); pp. 0200511–0200518 (2019). https://doi.org/10.1063/1.5092054
20. Radomska-Zalas, A., et al.: IT support for optimisation of abrasive water cutting process using the TOPSIS method. IOP Conf. Ser. Mater. Sci. Eng. **710**, 012008 (2019). https://doi.org/10.1088/1757-899X/710/1/012008
21. Senthilkumar, T.S., et al.: Surface morphology and parametric optimization of AWJM parameters using GRA on aluminum HMMC. Mater. Today Proc. (2019). https://doi.org/10.1016/j.matpr.2019.07.404
22. Spadło, S., Krajcarz, D.: Analysis of the surface quality for hardox 400 steel in abrasive water-jet cutting. Mechanik. 8–9, 726/431-726/441 (2015). https://doi.org/10.17814/mechanik.2015.8-9.454
23. Wojciechowski, S., et al.: Application of signal to noise ratio and grey relational analysis to minimize forces and vibrations during precise ball end milling. Precis. Eng. **51**, 582–596 (2018). https://doi.org/10.1016/j.precisioneng.2017.10.014
24. Zhang, D., et al.: Analysis of the relationships between waste cooking oil qualities and rejuvenated asphalt properties. Materials. **10**(5), 508 (2017). https://doi.org/10.3390/ma10050508
25. Jinhong Mining Catalog of Products. http://www.lygjhky.com/Hva_En/Catalog.pdf
26. SSAB. https://www.ssab.com/products/brands/hardox/products/hardox-500. Accessed 10 Oct 2019

Some Investigations into 1,000 MPa Pure Waterjet Cutting

Andrzej Perec[1,2](✉), Franz Trieb[3], and Frank Pude[1,4]

[1] inspire AG, Zurich, Switzerland
aperec@ajp.edu.pl
[2] Faculty of Technology, Jacob of Paradies University, Gorzów Wlkp., Poland
[3] BFT GmbH, Hönigsberg, Austria
[4] Steinbeis Consulting Center High-Pressure Waterjet Technology, Horgau, Germany

Abstract. This paper documents the application of the maximum available pressure for cutting operations using a special pump. This increase of the water pressure, and its conversion to an ultra-high-speed waterjet results in deeper and/or faster cutting. Theoretical aspects of using these ultra-high-speed waterjets are presented. The behavior of this waterjet when it hits the target material is shown. The thermal properties of the waterjet were analyzed. Research into the influence of selected pressure levels on the quality of the cutting edge of processed carbon fiber composite, and into the wear on the water nozzle subjected to ultra-high pressure (UHP) is presented.

Keywords: Ultra-high-pressure water jet · Cutting · Carbon fiber composite

1 Introduction

A more efficient processing by waterjet can be achieved by a higher amount of energy transferred into the workpiece. This can be obtained by increasing the hydraulic power in two ways: increasing the flow rate [7, 9] or increasing the working pressure [11, 16]. Working pressures of 400 MPa and a discharge of water of ca. 4 dm^3/min produced in a typical pressure intensifier are generally used with standard contemporary high-pressure abrasive jet equipment [6]. The new generation devices, which have recently been introduced into the market are able to reach 620 MPa pressure.

2 State of the Art

Higher pressure, of up to 900 MPa, has been reached in a few research centers. The possibility of processing materials by abrasive waterjet using a pressure level of 1,000 MPa was considered at the Koszalin University of Technology [8]. The movement of single grains in such jets was characterized, and their velocity and kinetic energy determined. Precise calculations of the tensions generated in the material treated were conducted.

D. Klichová et al. (Eds.): Water Jet 2019, LNME, pp. 155–163, 2021.
https://doi.org/10.1007/978-3-030-53491-2_17

The recommended length of the focusing tube and mesh size of the abrasive material for such treatment processes were determined as well.

Studies into the efficiency of abrasive water jet cutting materials using a pressure levels above 600 MPa were the subject of research at Hannover University [5]. This paper describes the cutting equipment, the way in which the consumption of abrasive material may be reduced, the capacity of standard abrasive waterjets for cutting curves of a smaller radius than pure waterjets are capable to cut, and finally offers a thorough discussion on the mechanisms of the material removal process when pure waterjets are used for cutting metals. Hannover University also conducted research into the temperature distribution in the tool and the target material hit by a pure waterjet [4], and presented the temperature generated by the cutting process using pure and also abrasive waterjets. The aim of that work was to scan the temperatures which occurred on the machined surface by using an infrared camera. The temperature distribution of the nozzle holder as influenced by the nozzle, and of the focusing tube, on the surface of the target material, from the top to the bottom of the cut, as well as of the abrasive waterjet as it leaves the kerf, is presented.

Generated cuts by pure and abrasive waterjets were also the subject of research conducted by the BHDT company [3, 13, 14]. These papers demonstrate the effect of adding abrasive material to UHP waterjets cutting at 800 MPa pressure level. Further, the effects on different materials were investigated, especially the interdependence of operating pressures, nozzle diameter, cutting speed, and cut surface quality. The pure waterjets generated at UHP are even able to cut thin layers of steel. Adding abrasive grains increases both cutting depth and speed tremendously. A comparison of UHP pure waterjet and abrasive waterjet shows the increased cutting performance of these processes. At Delft University preliminary studies on cutting with 700 MPa waterjets were conducted [12]. Uhde company was also working towards the design of a high-pressure intensifier working up to 700 MPa, with the aim of using it for waterjet cutting. Information associated with the construction of such devices has been presented by Koerner et al. [2].

This state of the art research presents a comprehensive picture of early developments as well as new fields of an application using this technology.

3 Flow Characteristics at UHP

The latest development is the use of high-pressure *autofrettage* pumps which generate pressures of up to 1,200 MPa with a water flow rate of 0.4 dm^3/min. To determine the maximum diameter of the nozzle which can be used in continuous operation see Fig. 1. For pressures of 1,200 MPa and flow rates of 0.4 dm^3/min, a nozzle with a typical discharge coefficient $C_d = 0.7$ has an inner diameter of 0.09 mm. Since the coefficient C_d at such high pressures is not known exactly, a nozzle with an internal diameter of 0.08 mm is selected. Selected sapphire nozzles are shown in Fig. 2.

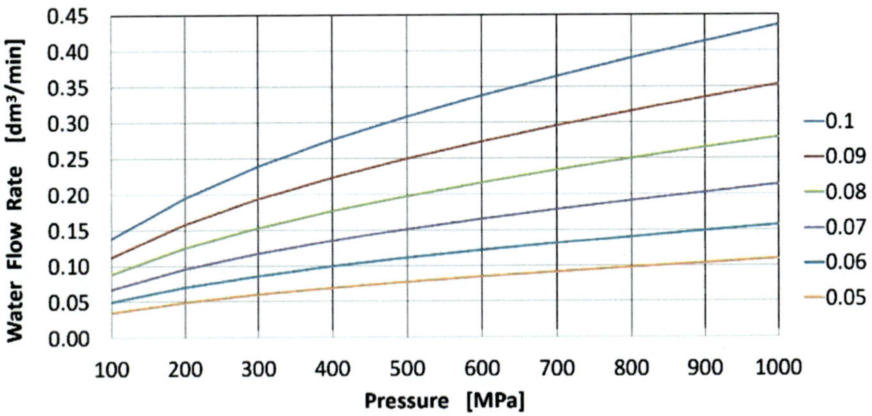

Fig. 1. Theoretical dependence of pressure and water flow rates on nozzle diameter. Discharge coefficient $C_d = 0.7$.

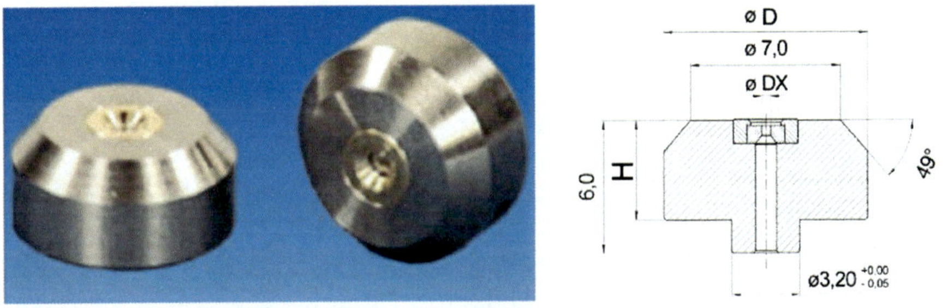

Fig. 2. Water nozzle type FLN-Y-1808, manufactured by Quick-Ohm [10].

4 Experimental Setup

4.1 The UHP Waterjet System

The source of the UHP was the *autofrettage* pump type BHDT1200.04, shown in Fig. 3. This apparatus was working in a system of two combined pressure intensifiers. To control the UHP pump an external dedicated microprocessor controller produced by BHDT is used.

The UHP pump used for the research generated discontinuous pressure amplitudes which could be identified by pulsations. This pulsation exists due to a non-installed pressure vessel (pressure accumulator) in the high-pressure line, which is normally used to smooth pulsations. This UHP pump was originally designed for use as an *autofrettage* pump. Pressure pulsations do not have a significant impact on the *autofrettage* process. A BHDT CNC machine (Fig. 4) with an internal control system and graphic user interface was used for moving the cutting head. The connection between the pump and the cutting head is realized by a special UHP (1,200 MPa) tube with a length of 6 m, outer diameter 3/8″ (9.5 mm) and inner diameter 1/16″ (1.5 mm).

Fig. 3. BHDT *autofrettage* 1200.04 pump.

Fig. 4. CNC waterjet machine.

4.2 High Speed and Thermographic Cameras

For high-speed imaging of the waterjet cutting process the high-speed camera Phantom v12.1 was used. The Phantom v12.1 camera can take more than 6,200 fps at full 1280 × 800 resolution, and up to 1,000,000 fps at reduced resolution [15]. For the thermographic imaging of the waterjet cutting process, the IR-TCM HD 1024 Thermographic Infrared camera was used. A pixel resolution of 2048 × 1536 is accurate. For visualizing or accurately measuring heat distributions the uncooled IR-TCM HD 1024 camera outputs detailed radiometric images of 2048 × 1536 pixels spatial resolution and a thermal resolution of 50 mK Noise Equivalent Temperature Difference (NETD). Operating at a frame rate of up to 30 Hz, the camera offers a real-time image resolution of 1024 × 768 pixels [1].

5 Results

5.1 Cutting Effects

Exemplary results of cut carbon fiber composites of total thickness equal to 8 mm are presented in Fig. 5.

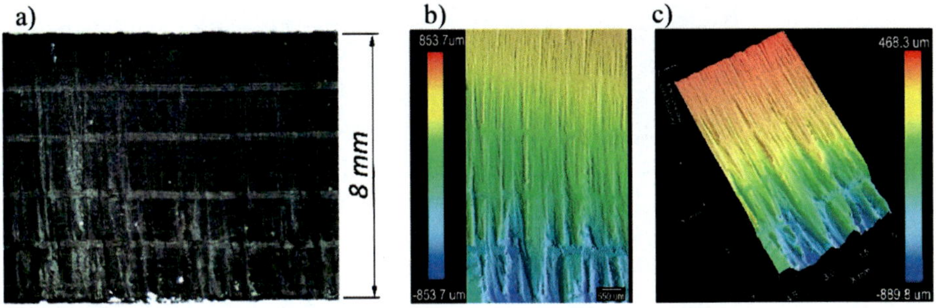

Fig. 5. Sample of CFC cut surface: a) macroscopic view, b) roughness map, c) 3D surface roughness. Pressure 1050 MPa, thickness 8 mm. None of the delamination effects which are typically observed for this type of composite material when treated by pure waterjets or waterjets with low abrasive mass flow rate occur during the experiments.

A typical structure of the surface after waterjet cutting with traverse speed of 100 mm/min, can be observed here. Starting from the middle position of the cut (middle between top and bottom of the cutting kerf) downwards, the roughness and waviness of the material surface increased dramatically. Sample surface of cut carbon fiber composites under EDX microscope are presented in Fig. 6.

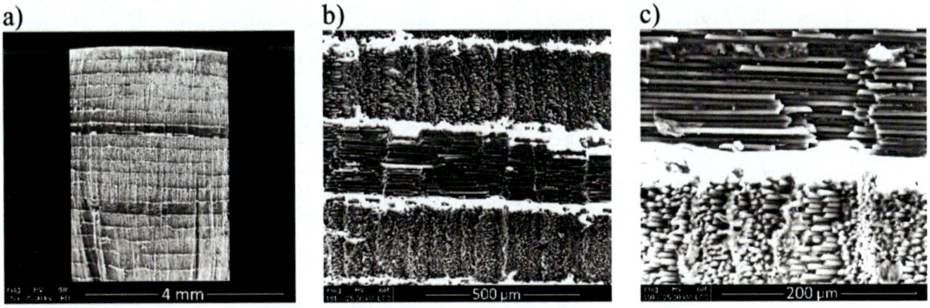

Fig. 6. Sample EDX microscope view of CFRP cut surface. Magnification: a) 25x, b) 160x, c) 500x.

5.2 High-Speed Imaging

In Fig. 7 high-speed camera images monitoring the whole tested range of 900 to 1,050 MPa are presented. Cylindrical, coherent sections of the water jet have a length of 20 mm for 1,050 MPa to 30 mm for 900 MPa. One can observe the gradually increasing dispersions of the water jet in the air below these sections.

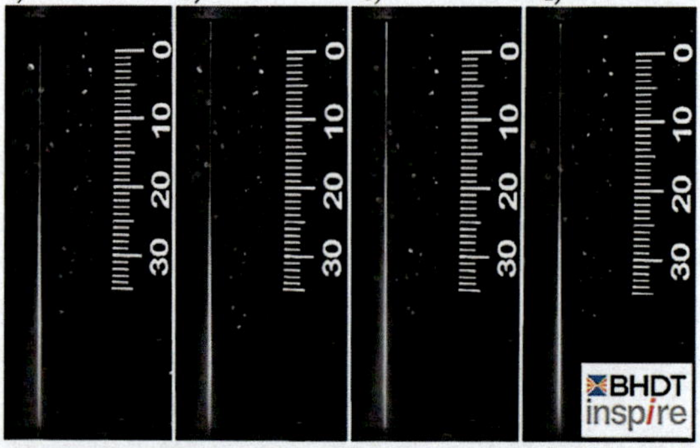

Fig. 7. View of UHP waterjets. The frame rate of exposition was 25,000 fps, exposure time was 40 μs. Nozzle ID = 0.08 mm, Pressure: a) 1,050 MPa, b) 1,000 MPa, c) 950 MPa, d) 900 MPa.

5.3 Temperatures of Sample Material and Used Components

A thermal analysis of different stages of the cutting process, using an infrared camera, is shown in Fig. 8.

At the time of first contact between the high-pressure waterjet and the target material (Fig. 8c), the highest temperature of $T = 62$ °C, is observed in the section of the jet reflected from the material closest to the point of impact. Beside this impact area the

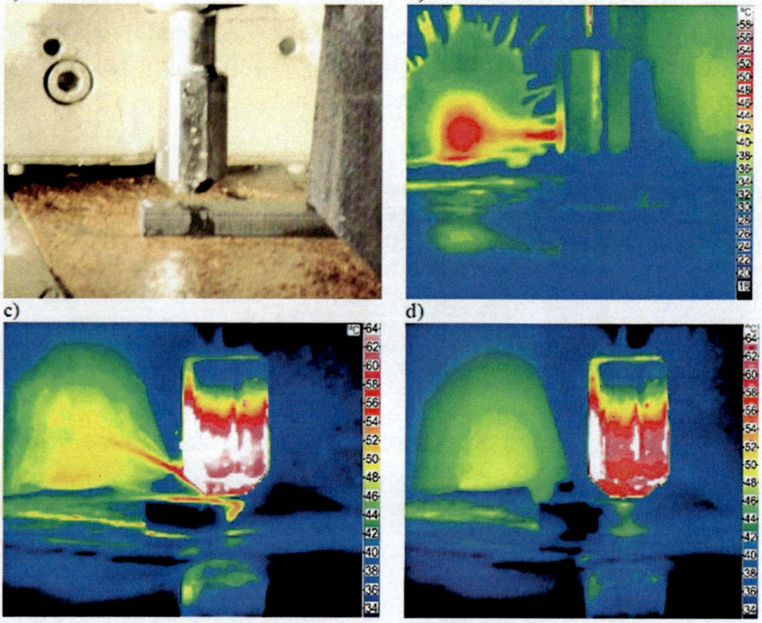

Fig. 8. Exemplary thermographic views on different stages of the cutting process at 1,050 MPa pressure level a) view in normal light conditions, b) thermographic view of water overtopping perpendicular to the outer wall of the nozzle holder, c) thermographic view of the start of the cutting process, d) thermographic view of the end of the cutting process.

temperature of the reflected jet decreases to levels of 52–50 °C. In the cutting zone and on the surface of the cut material, the temperature does not exceed 56 °C. At the bottom of the cutting zone (Fig. 8d) the jet temperature does not exceed 48 °C and the average temperature in the whole cutting area is appr. 46 °C. Only the nozzle holder reaches temperatures of appr. 65 °C. Due to an unwanted overtopping effect by a sealing with wrong torque the temperature of a generated spray jet at the side of the nozzle holder (Fig. 8b) was measured at a level of appr. 56 °C.

5.4 Technical Challenges Due to UHP Conditions

The challenges encountered during the research and the reasons for their arising, as well as their potential solutions are shown in Table 1.

Table 1. Technical challenges, reasons and solutions for UHP waterjet cutting at 1,000 MPa pressure level.

Technical challenges	Reasons	Solutions
Pulsation of pressure	Construction (design) of *autofrettage* pump	Addition of a UHP pressure vessel
Leaks between the UHP tube and the nozzle holder	Insufficient sealing forces	The use of bigger torque during assembly
Difficulties in building pressure over 1,050 MPa	Leaks	Use smaller nozzle (ID 0.07 mm)
	Too large nozzle internal diameter	Use more powerful UHP pump
Damage to UHP seals	Pulsation of pressure	Add UHP pressure vessel
Damage to water nozzle	Nozzle housing with insufficient strength	Change of nozzle housing material
	The sapphire with insufficient strength	Use of diamond nozzle
	Presence of particles from damaged seal which hit and destroy the top part of the nozzle	Use of UHP filter

One of the major challenges was the short lifetime of the high-pressure seals. Seals were charged by extreme forces, and in addition the generated pulsations created a lifetime reducing exposure overload. The pulsation effect could be reduced by the installation of a hydraulic accumulator in the high-pressure line. Figure 9 shows an exemplary view of a broken seal package.

Fig. 9. Wear on UHP sealing parts.

6 Conclusions

The following conclusions can be drawn based on the experiments conducted:

- The use of UHP pumps designed for the *autofrettage* process is possible for UHP waterjet cutting. This requires modification of the UHP line and the installation of an additional pressure vessel for smoothing pulsation effects.
- The use of sapphire nozzles because of limited lifetime under UHP conditions is not recommended.
- Diamond nozzles could serve as good substitute components.
- All unnecessary UHP fittings between the nozzle holder and the UHP pump should be eliminated.
- A special high strength steel housing for a diamond water nozzle needs to be designed for further testing of abrasive waterjet cutting at pressure levels of over 1,000 MPa.

References

1. Jenaoptik: Information materials about Jenaoptik IR-TCM HD 1024 Series Ther-mographic Infrared Cameras
2. Koerner, P., et al.: Design of reliable pressure intensifiers for water-jet cutting at 4 to 7 kbar. In: 16th International Conference on Jetting Technology, Aix en Provence, France (2002)
3. Kotchan, M., Trieb, F.: 800 MPa ultra high pressure for jetting applications. In: 15-th International Conference on Jetting Technology, Ronneby, Sweden (2000)
4. Louis, H., et al.: A thermographical map of the tool and workpiece during the cut-ting process by plain waterjet and abrasive waterjet up to 900 MPa. In: Water Jet-ting, 17th International Conference, Advances and Future Needs (2004)
5. Louis, H., Mohamed, M., Pude, F.: Cutting mechanism and cutting efficiency for water pressures above 600 MPA. In: WJTA American Waterjet Conference. Houston, Texas, USA, p. Paper 1–A (2003)

6. Perec, A.: Experimental research into alternative abrasive material for the abrasive water-jet cutting of titanium. Int. J. Adv. Manuf. Technol. **97**(1–4), 1529–1540 (2018). https://doi.org/10.1007/s00170-018-1957-2
7. Perec, A.: Investigation of limestone cutting efficiency by the abrasive water suspension jet. In: Hloch, S., et al. (eds.) Advances in Manufacturing Engineering and Materials, pp. 124–134. Springer, Cham (2019). https://doi.org/10.1007/978-3-319-99353-9_14
8. Perec, A.: Possibilities of materials machining with an abrasive waterjet under 1000 MPa pressure. In: Ruszaj, A. (ed.) The 3-rd International Conference on Machining and Measurements of Sculptured Surface, Kraków, Poland, pp. 435–444 (2003)
9. Perec, A., Radomska-Zalas, A.: Modeling of abrasive water suspension jet cutting process using response surface method. In: AIP Conference Proceedings, vol. 2078, pp. 0200511–0200518 (2019). https://doi.org/10.1063/1.5092054
10. Quick-Ohm: Sapphire Nozzles, Ultra High Pressure Nozzles, Long Life Nozzles. https://www.quick-keramik-shop.de/en/sapphire-waterjet-nozzles/. Accessed 01 Jan 2020
11. Sitek, L., et al.: Erosion capability of ultra-high-speed continuous waterjet and pulsating waterjet: the comparison. In: Water Jet 2017 – Research, Development, Applications. Ústav geoniky AV ČR, v. v. i. Ostrava 2017, Lednice, (2017)
12. Susuzlu, T., et al.: Initial research on the ultra-high pressure waterjet up to 700 MPa. J. Mater. Process. Technol. **149**, 30–36 (2004)
13. Trieb, F., Zamazal, K.: 800 MPa pure waterjet and abrasive waterjet cutting–what's next. In: American Waterjet Conference, Minneapolis, Minnesota, USA, p. Paper 61 (2001)
14. Trieb, F., Zamazal, K.: Waterjet cutting - 300 MPa high pressure to 800 MPa ultra high pressure. In: 6th Pacific Rim International Conference on Water Jetting Technology, Sydney, Australia, pp. 186–192 (2000)
15. VisionResearch: Phantom Camera v12.1 Data Sheet. http://www.adept.net.au/cameras/visionresearch/pdf/PhantomV121.pdf. Accessed 01 Jan 2020
16. Werth, H., et al.: Cutting performance and obtainable quality when applying 6000-bar abrasive water-jets. Strojniski Vestn.-J. Mech. Eng. **52**(7–8), 463–469 (2006)

Effect of Standoff Distance on the Erosion of Various Materials

Jakub Poloprudský[1], Alice Chlupová[1], Tomáš Kruml[2(✉)], Sergej Hloch[3], Petr Hlaváček[3], and Josef Foldyna[3]

[1] Institute of Physics of Materials, Czech Academy of Sciences, Brno, Czechia
[2] CEITEC IPM, Žižkova 22, Brno, Czechia
kruml@ipm.cz
[3] Institute of Geonics, Czech Academy of Sciences, Ostrava, Czechia

Abstract. The paper describes the interaction between pulsating water jet and surface of three different metals: 316L stainless steel, 6068 Al alloy and Ti6Al4V Ti alloy. The pulsating water jet is formed by using the ultrasonic sonotrode to the pressurized acoustic water chamber. It appears that the shape of the resulting water jet differs at various distances from the nozzle and the state of the surface of tested materials treated by the jet at various standoff distances differs very substantially. Moreover, the three materials are exposed to pulsating water jet at different standoff distances and results are discussed in terms of the capacity to absorb plastic deformation. As expected, at optimal standoff distance, the deepest eroded path appears in the Al alloy and the most resistant material is the Ti alloy.

Keywords: Pulsating water jet · 316L steel · Ti6al4v alloy · 6082 alloy · Surface topography · Fatigue

1 Introduction

Water jet (WJ) technology is used as an industrial tool applying pressure within range 400–600 MPa of water used for disintegration of a wide variety of materials [1–3]. When a mass of the abrasive substance is added to the stream of water then the method (abrasive water jet, AWJ) is even more efficient in material removal [4] due to increasing of the kinetic energy. A pulsating water jet (PWJ) is a modification to conventional continuous water jets (CWJ) which work by utilizing water pressures 7–100 MPa. In the PWJ process, pressure fluctuations in the form of standing waves are produced in the acoustic chamber using ultrasonic sonotrode. The orifice converts pressure fluctuation into the pulsating jet with various velocity within the standoff distance. Created clusters interact periodically with the material surface causing a water hammer effect, [1] inducing compressive stresses and creating surface asperities. This process is followed by inducing shear stresses by lateral flow in the material [2] leading to plastic deformation and thus material removal. PWJ provides higher efficiency of material removal with potential to practical applications [5]. The other proposed options are at use for medical applications

D. Klichová et al. (Eds.): Water Jet 2019, LNME, pp. 164–171, 2021.
https://doi.org/10.1007/978-3-030-53491-2_18

[6] or as mechanical surface strengthening method analogic to shot peening [7]. The specific feature of this technology is the standoff distance between the nozzle and the material's surface at which the PWJ erosion effect is highest [8].

2 Experimental Procedures

In our study, the effect of PWJ was used as a surface treatment method. PWJ was generated at technological conditions listed in Table 1 and applied on the polished surface of three types of materials: 316L austenitic stainless steel, Ti6Al4V titanium alloy and 6082 aluminium alloy.

Polishing was performed using emery paper with grit size up to 1200 and the direction of the last grinding was parallel to the surface of the specimen. In the case of 316L, the surface was also electrolytically polished at 0 °C for 15 s at 20 V using solution of ethanol, perchloric acid and nitric acid.

Initial values of the Vickers hardness of all materials under investigation are listed in Table 2.

In order to observe the different effect on the surface from hardening up to strong erosion, the standoff distance (SOD) was changed from 15 to 39 mm by 2 mm step as shown in Fig. 1.

Topography, i.e. appearance and roughness of the treated surface and cross-sectional cuts were studied using light microscopy (Leica S9i) and scanning electron microscopy (Tescan LYRA 3 XMU FEG/SEMxFIB).

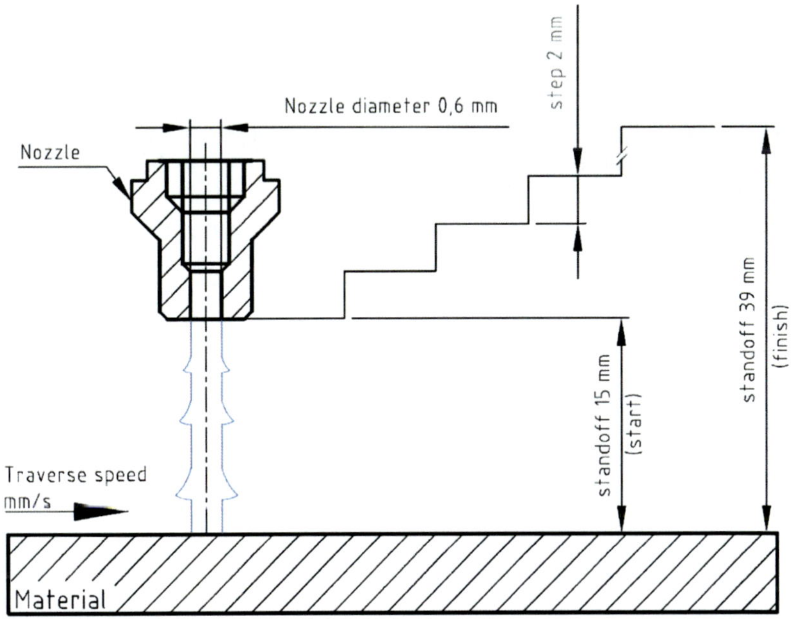

Fig. 1. Diagram showing the change of standoff distance in the experiment.

Table 1. Parameters of the PWJ process during the experiment.

Standoff distance [mm]	Pressure of water [MPa]	Frequency sonotrode [kHz]	Chamber length [mm]	Nozzle diameter [mm]	Flow speed [m/s]	Flow rate [l/min]	Traverse speed [mm/s]	Liquid density [kg/m^3]
15–39	40	20.1	12	0.6	254.81	4.32	0.25	998

Table 2. Vickers hardness of materials in the as-received state.

Material	316 L austenitic steel	Ti6Al4V titanium alloy	6082 aluminium alloy
Hardness	184 ± 10 HV0.2	372 ± 6 HV0.2	71 ± 1 HV0.2

3 Results

The surface of flat polished specimens of three different materials was treated with PWJ at different standoff distances and sub-sequentially observed by SEM. Cross-sections of treated samples were also observed. When very low (15 mm) and very high (39 mm) standoff distance was used, the surface of tested materials was plastically deformed or slightly damaged and the effect similar to cold-working process as a shot peening was observed. Under abrasive conditions and cyclic loading of PWJ at a particular standoff distance, the surface was strongly damaged or even partially removed.

Figure 2 and Fig. 3 show the comparison of three studied material surfaces at chosen standoff distances. 316L specimen shows the erosion groove even at the lowest applied standoff distance, where water pulses are not very well developed. At standoff distance of 17 and 19 mm, an area densely packed with persistent slip bands and cracks around the main groove can be seen. Between standoff distances of 25 and 33 mm, the PWJ path transforms from a single eroded groove into a number of small pits. And finally at a standoff distance of 39 mm mainly plastically deformed PWJ path with small amount of cracks appears. Al6068 specimen treated at a standoff distance of 17 mm is characterized by deformed PWJ path with a number of small cracks and voids. At standoff distance of 19 mm, all small cracks merged into a main erosion groove. With an increase in standoff distance, the eroded path becomes more prominent, before decreasing again at high standoff distances. The Al alloy specimen remained as a single eroded groove even for the highest studied standoff distance of 39 mm. For Ti6Al4V specimen a single eroded path appears at a standoff distance of 21 mm. At highest standoff distance of 39 mm, the PWJ path appears less distinct. In Fig. 3i there is a short interval in PWJ path without the presence of groove. This gap was caused by short loss of acoustic generator, i.e. the water jet was for a while only continuous without pulses.

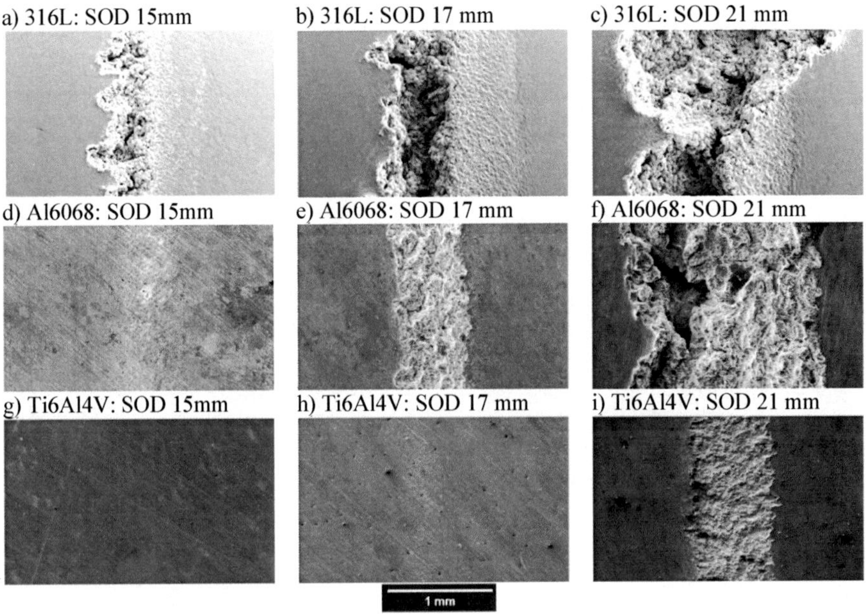

Fig. 2. SEM micrographs of PWJ paths for three different materials (316L, Al6068 and Ti6Al4V) at varying standoff distances (SOD 15 mm, 17 mm and 21 mm).

Fig. 3. SEM micrographs of PWJ paths for three different materials (316L, Al6068 and Ti6Al4V) at varying standoff distances (SOD 25 mm, 33 mm and 39 mm).

In Fig. 4 clear difference in cross-sections of three treated materials can be observed. For standoff distance from 25 to 33 mm, the depth of 316L eroded PWJ path decreased to almost zero. At the same distances, the depth of Al 6068 increased. Ti6Al4V has lowest erosion depth of all three materials at a standoff distance of 25 mm, however, the Ti alloy retains the single eroded PWJ path up to standoff distance of 33 mm.

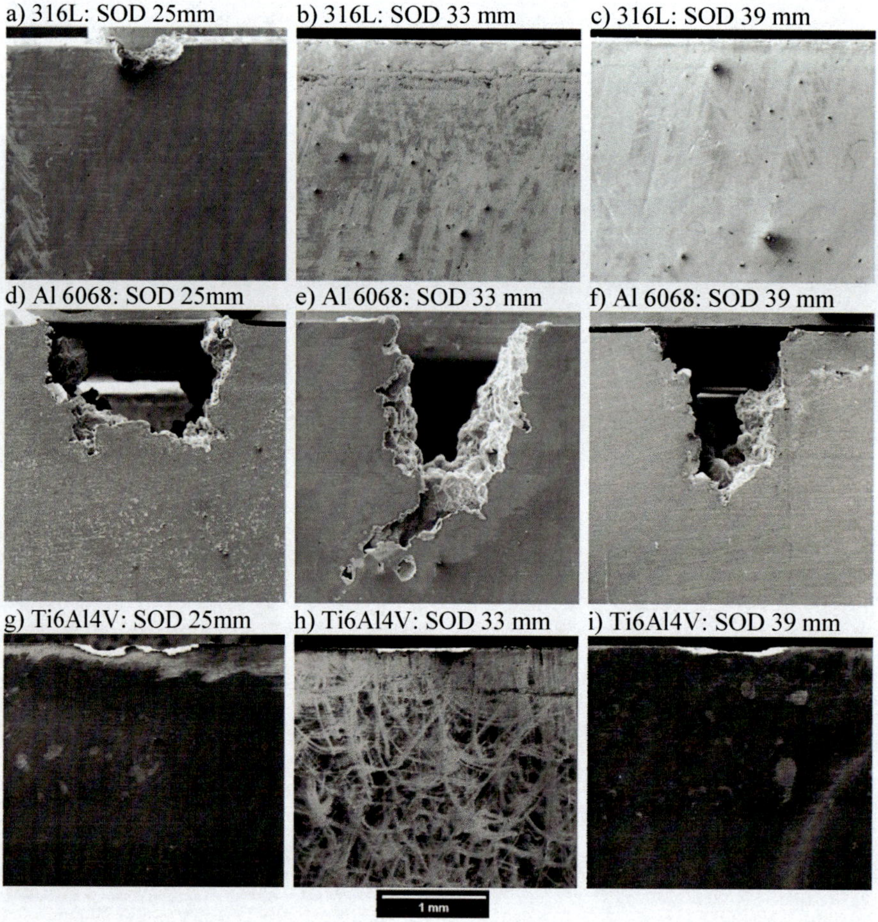

Fig. 4. SEM cross-sections of PWJ paths for three different materials (316L, Al6068 and Ti6Al4V) at varying standoff distances (SOD 25 mm, 33 mm and 39 mm).

At standoff distance of 27 mm (Fig. 5; Fig. 6), the single eroded path is evolved in all three materials. The clear distinction in the appearance of eroded surfaces of all materials is visible. In 316L specimen eroded surface is defined by deep cavities. The surface of 6082 treated alloy is characterised by a number of relatively flat areas. Ti6Al4V specimen is characterized by the surface with an absence of deep cavities and also with an absence of flat areas.

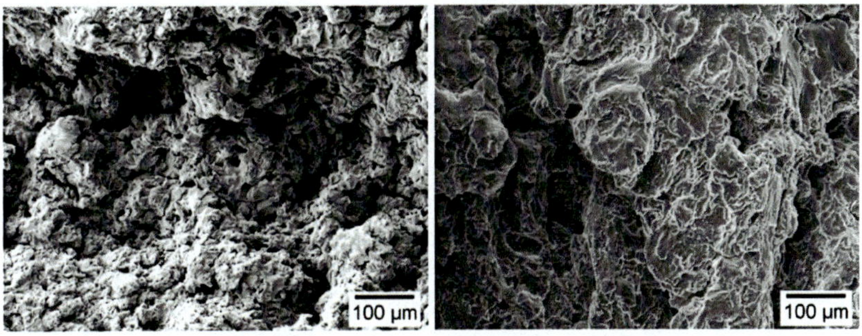

Fig. 5. SEM of eroded surfaces at a standoff distance of 27 mm a) 316L b) Al6082.

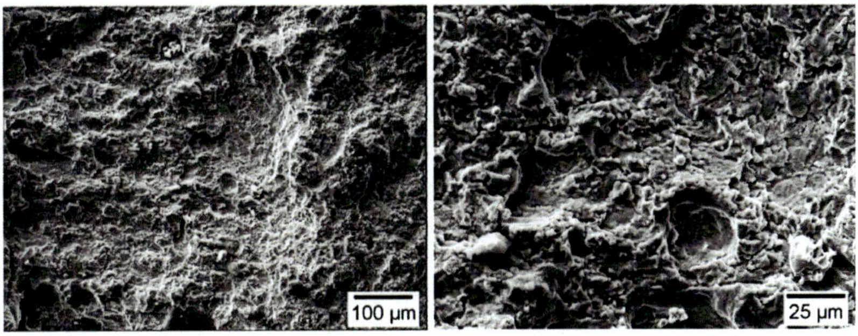

Fig. 6. Eroded surfaces at a standoff distance of 27 mm Ti6Al4V at different magnifications.

4 Discussion

Foldyna et al. [9] observed three stages of erosion of materials: 1) initial stage of brittle/plastic deformation, 2) creation and merging of erosion pits, 3) increasing of depth of main erosion groove. These three stages were indeed observed for Al and Ti alloys and stainless steel. The third erosion stage was observed at the shortest standoff distance of 15 mm for 316L stainless steel. Al-alloy exhibited the third stage at the standoff distance of 17 mm while Ti-alloy exhibited well-defined third erosion stage at the standoff distance of 21 mm. It was expected that Ti-alloy will be the most resistant material due to its high strength and that it would enter the third erosion stage close to ideal standoff distance when the water droplets are well-formed. The surprising fact is that 316L entered the third erosion stage before the Al-alloy having the lower strength. It means that the strength of the material, described by e.g. the yield stress or ultimate tensile stress, is not the decisive criterion for the difficulty of the erosion process. One of the possible explanations is the faster exhaustion of plasticity of material in the case of stainless steel. When the surface of the material cannot absorb more plasticity, cracks appear and the erosion process begins. With increasing standoff distance 316L steel is the only material that exhibits termination of the third stage before reaching the highest applied standoff distance of 39 mm.

The majority of materials studied in present literature show signs of ductile behaviour, such as aluminium alloy [4, 10], AISI 316 [9, 11], AISI 304 [12–14]. In accordance with this literature survey, a lot of persistent slip bands (PSB) was observed, both at the surface near the main groove and at cross-section near the main groove in the case of treated 316L specimens. However, during our observation, no visible persistent slip bands were found in Al alloys. The situation in the case of Ti alloys was similar, i.e. no PSB´s were present in the vicinity of the groove. The third erosion stage for 316L and Al alloys are also characterized by a number of subsurface voids and cavities which is in agreement with Hloch et al. [8]. No micro-voids or deep cavities were present in Ti6Al4V.

The resulting effect of PWJ process depends on two groups of parameters. The first group of process parameters including hydraulic properties, acoustic properties, nozzle geometry, and process configuration is already well studied in the present literature. The second group of sample parameters including sample geometry, surface roughness and material properties (hardness, toughness) is less developed.

The effect of the surface treatment i.e. surface roughness and properties of the hardened layer on subsequent PWJ treatment was observed in [8]. It was found that the less rough surface results in lower damage (more shallow groove) caused by subsequent PWJ treatment. In our case, samples of 316L were electrolytically polished, i.e. their surface was much finer and hardened layer was less prominent than in the case of Al alloy and Ti alloy samples, nevertheless, the 316L entered third erosion stage at the lowest standoff distance of tested materials. It seems, that in our case the opposite effect was reached, the further investigation will be necessary to compare this features in details and to avoid the interference of different roughness and hardness of surface layer using Ti and Al samples prepared by electrolytical polishing.

The author in [15] compared brass and bronze and concluded that hardness of material has a considerable effect on the depth of the groove but isn't the only effective parameter. Our results presented on the behaviour of 316L and Al alloy support this claim. In our case, early exhaustion of plasticity was considered to be the important factor, since PWJ damage is also consisting of fatigue parameters.

The erosion effect described in [8] or [16] shows that material toughness is the important parameter determining the shape of PWJ path. Our results concerning 316L and Al alloy are in agreement, nevertheless, in comparison with it the Ti6Al4V act differently, as shown.

5 Conclusions

- When the standoff distance is gradually increased, the austenitic stainless steel enters the third erosion stage at low standoff distance (with less defined pressure pulses) before the Al and Ti alloys.
- Persistent slip bands were observed both at cross-section and surface of treated 316L samples at distances larger than 50 μm from the groove.
- Al alloy shows the deepest erosion groove. Cracks parallel with the surface were found under the treated surface in the first/second stage of erosion damage, which could be negative when using this technology for peening.
- All three stages of erosion were observed for Ti alloy with a variation of standoff distance. Micro voids or cracks were not found in cross-section in any of those stages.
- The character of the eroded surface is different in all three eroded materials.

Acknowledgement. The authors are grateful for financial support of the project GACR No 19-000408S. This work was conducted in the frame of IPMinfra through the project LM2015069 and CEITEC – Central European Institute of Technology with research infrastructure supported by the project CZ.1.05/1.1.00/02.0068 financed from the European Regional Development Fund. Part of the work was carried out with the support of CEITEC Nano Research Infrastructure (ID LM2015041, MEYS CR, 2016–2019), CEITEC Brno University of Technology.

References

1. Fenoughty, K.A., Jawaid, A., Pashby, I.R.: Machining of advanced engineering materials using traditional and laser techniques. J. Mater. Process. Technol. **42**, 391–400 (1994)
2. Boud, F., Loo, L.F., Kinnell, P.K.: The impact of plain waterjet machining on the surface integrity of Aluminium 7475. In: 2nd CIRP Conference On Surface Integrity, Book Series: Procedia CIRP, vol. 13, pp. 382–386 (2014)
3. Paul, S., Hoogstrate, A.M., van Luttervelt, C.A., et al.: Analytical and experimental modelling of the abrasive water jet cutting of ductile materials. J. Mater. Process. Technol. **73**, 189–199 (1998)
4. Foldyna, J., Sitek, L., Ščučka, J., et al.: Effects of pulsating water jet impact on aluminium surface. J. Mater. Process. Technol. **209**, 6174–6180 (2009)
5. Klich, J., Foldyna, J., Hlavacek, P., et al.: Erosion of metals by pulsating water jet. Water Jet **2011**, 165–174 (2011)
6. Hloch, S., Foldyna, J., Pude, F., et al.: Experimental in-vitro bone cements disintegration with ultrasonic pulsating water jet for revision arthroplasty. Tehnicki Vjesnik-Technical Gazette **22**(6), 1609–1615 (2015)
7. Meguid, S.A., Shagal, G., Stranart, J.C.: Finite element modelling of shot-peening residual stresses. J. Mater. Process. Technol. **92–93**, 401–404 (1999)
8. Hloch, S., Adamčik, P., Nag, A., et al.: Hydrodynamic ductile erosion of aluminium by a pulsed water jet moving in an inclined trajectory. Wear **428–429**, 178–192 (2019)
9. Foldyna, J., Klich, J., Hlavacek, P., et al.: Erosion of metals by pulsating water jet. Tehnicki Vjesnik-Technical Gazette **19**(2), 381–386 (2012)
10. Foldyna, J.: Recent developments in pulsating water jets. In: Hloch, S., Klichová, D., Krolczyk, G.M., Chattopadhyaya, S., Ruppenthalová, L. (eds.) Advances in Manufacturing Engineering and Materials. LNME, pp. 54–62. Springer, Cham (2019). https://doi.org/10.1007/978-3-319-99353-9_7
11. Hlaváček, P., Kruml, T., Foldyna, J., et al.: Effect of pulsating water jet peening on stainless steel. In: Water Jet 2015, Ústav geoniky AV ČR, v. v. i. (2015). ISBN 978-80-86407-56-2
12. Srivastava, M., Hloch, S., Tripathi, R., et al.: Ultrasonically generated pulsed water jet peening of austenitic stainless-steel surfaces. J. Manuf. Process. **32**, 455–468 (2018). https://doi.org/10.1016/j.jmapro.2018.03.016
13. Srivastava, M., Hloch, S., Krejci, L., et al.: Residual stress and surface properties of stainless steel welded joints induced by ultrasonic pulsed water jet peening. Measurement **127**, 453–462 (2018). https://doi.org/10.1016/j.measurement.2018.06.012
14. Hloch, S., Srivastava, M., Krolczyk, J.B., et al.: Strengthening effect after disintegration of stainless steel using pulsating water jet. Tehnicki Vjesnik - Technical Gazette **25**(4), 1075–1079 (2018). https://doi.org/10.17559/TV-20170327134630
15. Lehocka, D., Klich, J., Foldyna, J., et al.: Copper alloys disintegration using pulsating water jet. Measurement **82**, 375–383 (2016). https://doi.org/10.1016/j.measurement.2016.01.014
16. Lehocká, D., Klich, J., Botko, F., et al.: Pulsating water jet erosion effect on a brass flat solid surface. Int. J. Adv. Manuf. Technol. **97**, 1099–1112 (2018). http://doi.org/10.1007/s00170-018-1882-4

Check for updates

Development of an Assistance and Control System for Waterjet Cutting of Free-Form Workpieces

Nermin Redžić(✉), Felix Pfeiffer, Marco Witt, and Philipp Klimant

Chemnitz University of Technology, Institute for Machine Tools and Production Processes (IWP), Chemnitz, Germany
nermin.redzic@mb.tu-chemnitz.de

Abstract. The process accuracy when cutting free-form parts using waterjet technology depends on several factors, including various preparatory tasks, such as defining the workpiece coordinate system. Some manual procedures for determining it are already used. Those methods are based on the usage of laser pointers or dial gauges, which are fixed on the machine head. As a computer-aided alternative, measuring probes can be used to define the zero point offset of a workpiece, or to compare its dimensions to the ones predefined in the CAD data. When processing flat parts, optical systems can be used for layout alignment on the blank material. In order to redefine how the workpiece coordinate system is measured on a waterjet cutting machine, a new assistance and control system is being developed. It features an optical system that scans marked clamping and positioning devices whose relative positions to the machined part are precisely defined. By returning data about their location to the control unit, the workpiece coordinate system can be calculated. This paper describes the steps necessary for developing such a system, as well as the design and work principle of a fixturing system that supports the clamping and positioning of such free-form workpieces.

Keywords: Waterjet cutting · Clamping devices · Workpiece coordinate system · Assistance and control system

1 Introduction

The advance in aerospace and automotive industry requires the use of lightweight engineering materials that exhibit good thermal and mechanical properties [1]. These materials are usually difficult to cut using conventional machining technologies. As a technology that is able to cut almost any material [2], waterjet cutting becomes more and more important for their processing. At the same time, these industries tend to the manufacturing of parts with complex free-form shapes. The limitations in processing of such parts can be overcome by using CAD/CAM and CNC technology [3]. For machining of free-form workpieces, made of difficult to cut materials, abrasive waterjet machining can be used. The versatility of this technology for machining complex 3D parts was

D. Klichová et al. (Eds.): Water Jet 2019, LNME, pp. 172–185, 2021.
https://doi.org/10.1007/978-3-030-53491-2_19

demonstrated in [2, 4, 5] and [6]. In this paper, a novel assistance and control system that supports waterjet machining of free-form workpieces is presented. The basis of this system is an optical scanning process that gathers location and orientation data of fixtures that are used for clamping and positioning free-form parts. Since the position of each fixture relative to the machined workpiece can be defined with a sufficient accuracy, this data can be used to calculate the workpiece coordinate system.

2 State of the Art

When machining free-form workpieces using abrasive waterjet machining, multiple strategies can be applied. In their research, Rennau and Dix [5] defined 2.5D and 3D as two different cutting approaches for machining free-form parts. The 2.5D presents a cutting strategy where the cutting head follows complex spatial trajectories, but remains perpendicular to the XY-plane (see Fig. 1a.) of the machine in every point of the trajectory. The 3D strategy implies that the cutting head stays perpendicular to the surface of the machined workpiece (see Fig. 1b.) in every point of its trajectory. The 2.5D approach can be considered as a one that utilizes three axes of a waterjet cutting machine, while the 3D requires benefits of the 5-axis motion.

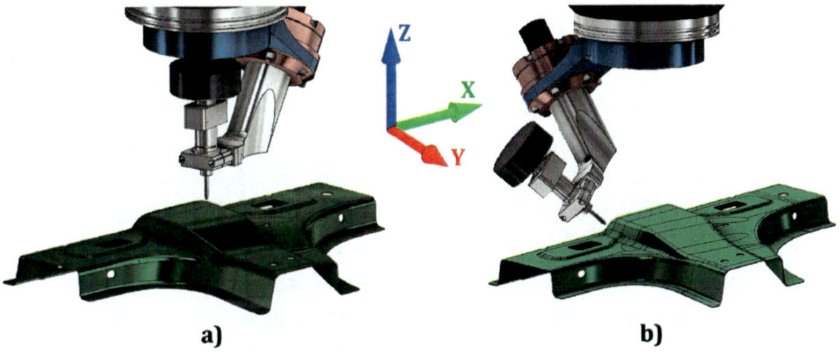

Fig. 1. Cutting strategies for processing free-form parts on a waterjet cutting machine: a) 3-axis motion; b) 5-axis motion.

Cutting contours on free-form parts using the 3-axis motion of a waterjet cutting machine has limitations considering quality of cuts and resulting part tolerances. The properties of this motion result in a variable nozzle stand-off distance, cutting depth and inclination angle of the waterjet relative to the machined surface. These parameter variations influence the resulting kerf characteristics and surface quality [7] and limit the use of this approach on cutting simple free-form parts.

The main advantage of utilizing a 5-axis approach when machining free-form parts is that, due to keeping a constant cutting depth at every point, there is no necessary feedrate reduction, which improves the uniformity of cuts [5]. This cutting strategy can also be applied using robotic abrasive waterjet systems [6]. The properties of this machining approach make it suitable for processing complex free-form workpieces.

2.1 Waterjet Cutting of Free-Form Workpieces

Although there are commercially available 5-axis abrasive waterjet cutting machines, their capabilities are limited on machining relatively simple 3D parts [4]. One challenge presents damage prevention due to the residual energy of the waterjet after piercing or cutting the workpiece. This decaying jet is stopped in a water-filled tank, or by using appropriate catching accessories [4].

Another challenge when processing free-form parts is the general non-alignment of the free-form part with the CAD data. This occurs because the CAD model cannot predict process disturbance and possible material thickness variations, wrinkles after forming, adjustment of tools and molds previous to part forming and similar factors [3]. Because of having complex features and irregular shapes, such workpieces are also difficult to position and clamp on grates or slats on a waterjet cutting machine. This procedure influences the accuracy of defining the workpiece coordinate system, which furthermore influences the processing accuracy of the finished part. Since free-form parts have features that need to be maintained after cutting and that define the relative position of cuts within a certain tolerance level, it is of high importance to achieve precise part positioning and defining of a workpiece coordinate system.

There are several methods for determining the workpiece coordinate system on a waterjet cutting machine. In general, this zero point offset can be measured manually or with the use of a computer-aided system. Manual methods include the use of a dial gauge or a machine-integrated laser pointer. These methods usually suffice when cutting parts out of blanks that have no or only simple features to which the relative position of cuts needs to be maintained within a certain tolerance level. Dial gauges are used for manual alignment of a certain part feature to an axis of a waterjet machine. The laser pointer is fixed on the machine cutting head and, when active, shows the current position of the waterjet cutting nozzle relative to the workpiece. It is utilized by manually moving the machine axes to a point where the workpiece coordinate system has to be defined. However, using this approach does not result in a precise zero point offset relative to the real part that has to be machined. This is why it is often used when machining flat blanks with no important features that need to stay within a certain tolerance range to the cuts.

The accuracy of manual procedures for determining the zero point offset is often influenced by the human factor and can introduce a certain error in the measurement. To facilitate this preparatory task, some industrial computer-aided solutions were developed. A computer-aided optical feature scanning system can be used for finding the position of preprocessed blank materials. Such a system uses a camera that scans predefined features of a workpiece, and calculates the workpiece coordinate system, using acquired location data for each part feature [8].

When machining free-form parts, advanced measurement and control systems need to be used. In this case, the workpiece coordinate system not only influences the cutting accuracy, but also affects the nozzle standoff distance in each point of the programmed toolpath. This distance needs to be defined precisely and maintained within a certain tolerance range during the whole processing time, so that possible collision between the tool and the workpiece are avoided. One approach for measuring the zero point offset of a free-form part is by using a measuring probe [9]. The probe is either mounted on the machine cutting head, or installed on a separate tool holding head that runs independently in a programmable measuring cycle. These advanced systems can be used for probing the zero point offset of a workpiece, or to compensate the non-alignment of a real part with the CAD data, in order to achieve a constant nozzle standoff distance in the cutting process later on. [10]. To keep a constant nozzle standoff distance from the workpiece during the cutting process without previously having to measure the non-alignment, a height control system can be used when utilizing the 3-axis motion of the cutting head [11]. There are also some systems that actively compensate the tool center point in order to keep a constant nozzle standoff distance [12].

2.2 Optical Position Determination with Fiducial Markers

For the system to work, it is necessary to use a reliable and robust method for position and pose estimation. There are a lot of industrial tasks that have already been supported by the usage of optical scanning systems, for example in self-driving automobiles [13–15], in autonomous robotics [16–19] or in virtual assisting systems [20–23]. A camera is in general a comparatively cheap variant of a sensor, which can be scaled up to reach higher resolutions and therefore higher precision and/or higher frame rates, when investing in industrial systems.

If circumstances could be assumed as ideal, natural features like whole textures or significant keypoints would be used to lay out the scanned environment [24–26]. Since the environment of the described system is mostly artificial and therefore lacks, at least in general, the needed rich distribution of distinct and redundant visual features, those techniques are usually not applicable. At least this creates the opportunity to avoid some of the still remaining challenges of the mentioned natural features method: instability problems and usually significant computational efforts.

A common alternative are artificial landmarks. The usage of fiducial markers is a popular approach. Those markers usually offer precise, robust and often fast tracking. The most basic variants include simple dots printed in a plane, on a planar surface or in three dimensions: LEDs or retroreflective elements.

Though already useful for pose and position estimation itself, the targeted application requires the markers to be distinguishable and therefore identifiable. Some steps have been taken to integrate identification procedures into artificial landmark detection techniques [27, 28], e.g. using barcodes [29, 30]. A very popular approach is the usage of square planar fiducial markers. They are a construct of an external high contrast (often black) border and an internal binary coding. This gives the marker not only an ID, but, if distributed correctly, the possibility to compute its pose.

One of the earliest implementations was ARToolKit [31] which was extended to ARToolKit+ [32] and is now privatized as The Studierstube AR Project [33]. ARToolKit itself and its derivate ARTag [34] are no longer actively developed. AprilTags [35] is a variant of a visual fiducial system that integrates error detection and correction.

A very accessible and popular fiducial marker detection library is ArUco [36], which is also the library chosen to implement the optical detection system, described in this paper. It is implemented as a contribution module inside the OpenCV framework. It features a fairly reliable detection algorithm, which is robust to non-uniform illumination and does implement error detection and correction of the binary codes. It also implements a subpixel corner refinement algorithm, which is critical for precise localization.

Technically, the detection is employed by thresholding the image to extract all objects that have black corners, which essentially creates a binary image. This is then fed into a contour extraction pipeline, which reduces it to edges of polygons. This set of polygons is further condensed by finding only those which form a four sided convex polygon. The next step is to remove the projection by computing and applying the homography matrix. To extract the binary code, the resulting square picture is divided into a regular grid and the average pixel count in each grid cell is used for binarization. After successfully identifying the marker as valid, the edges that form the corners are subjected to linear regression to refine their position inside the image with subpixel accuracy. That way the corners and the whole marker can be located with more precision than just the pixels of the image would suggest.

To enhance this refinement algorithm even more, enclosed markers have been proposed (see Fig. 2) [37]. They essentially expand the edges of the marker in the opposite direction, making more pixels around a corner accessible for a linear regression.

Fig. 2. Example of an ArUco/OpenCV enclosed marker that enhances subpixel accuracy of pose detection by being enclosed with 4 squares and providing longer edges for orientation detection.

3 Design of the Fixturing System

The overall processing quality on a waterjet cutting machine depends on several factors, including the accuracy of positioning and clamping of the workpiece. The fact that the waterjet penetrates through the machined part and can damage the support underneath it is one of the reasons why grates and slats are used as workpiece support on these machines. Generally, slats are considered as more economical, even though there is a drawback of providing limited clamping options. Although grates are characterized by better sturdiness and offer more clamping possibilities than slats, their industrial application is limited due to economic reasons.

Manual clamping procedures with non-standardized clamping devices often increase error possibilities and do not enable precise workpiece positioning. To achieve high processing accuracy and eliminate the possible clamping error, a new workpiece support system has been designed, as well as flexible clamping and positioning devices that can be applied for a wide range of free-form parts.

3.1 Design of a Workpiece Support System

In order to improve the way free-form parts are being positioned and clamped on a waterjet cutting machine and fulfill the requirements necessary for the development of the assistance and control system described in Sect. 3, a novel fixturing system has been developed. The system consists of rails that are fixed on the machine water tank and slides that carry beams on which clamping and positioning devices can be placed. The developed system is presented in Fig. 3.

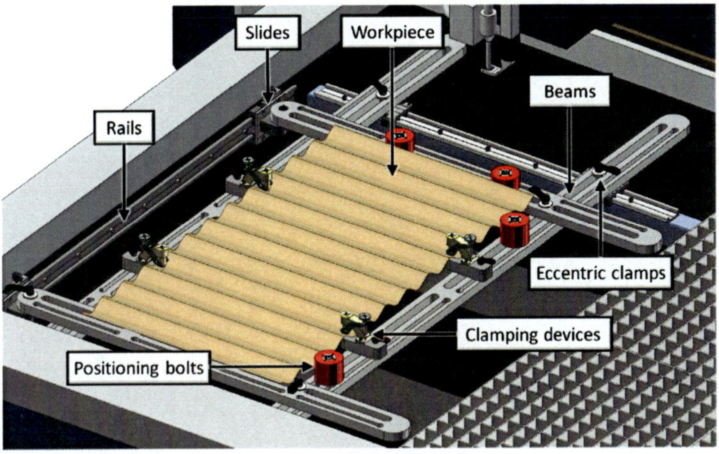

Fig. 3. Design of the flexible workpiece support system (Developed in cooperation with STM Waterjet Germany GmbH).

Prior to the clamping of the workpiece, the machine operator has to adjust the position of each beam in order to create a support for it. After the adjustment of the beams has been completed, eccentric clamps are used to fix their position. These fixed beams serve as a basis for the following clamping procedure. Such a support system has several advantages compared to current state of the art systems, including:

1. Flexible adjustment for a wide range of machined workpieces;
2. The penetrating waterjet does not damage the workpiece support;
3. All elements can be replaced in case of wear or damage;
4. All elements, except beams, are standardized;
5. Beams can be manufactured on a waterjet cutting machine;
6. The system is sturdy and waterproof.

Disadvantages of such a system include low stiffness when machining heavy-duty parts and limitations for clamping and supporting the workpiece only on its edges. These issues can be solved by utilizing grates as an additional workpiece support system. Since the flexible rail system with slides does not require the complete workspace of a waterjet cutting machine, grates can also be set up on the rest work area of the machine on which the rail system has been built. This widens the diversity of workpieces that can be processed on a single waterjet cutting machine.

3.2 Design of Flexible Clamping Devices

In comparison with other production technologies, waterjet machining results in relatively low cutting forces. Generally, the impact force has values between 10 and 15 N [38], while cutting forces for continuous cuts have values in the range between 15 and 40 N [39], depending on process parameters and jet diameter. The clamping devices need to assure that cutting forces, as well as forces due to jet reflection or an upward float force, cannot move the workpiece. Forces that clamping devices have to exert on workpieces are relatively low, so a clamping device that utilizes the force of a compression spring has been designed, which is presented in Fig. 4.

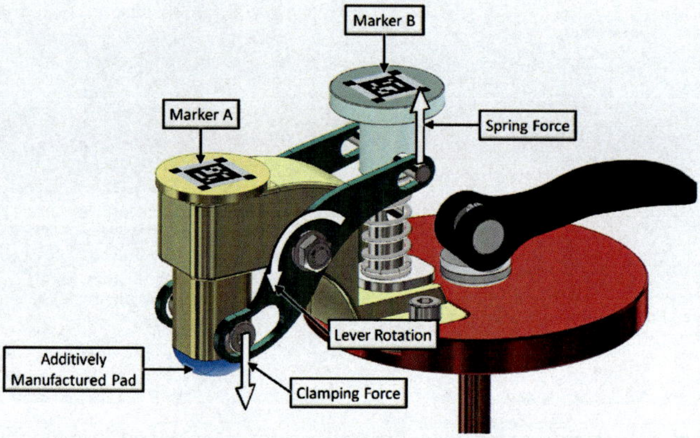

Fig. 4. Features and work principle of the clamping device.

The spring has to be preloaded and it is chosen so that its spring coefficient enables force variation within the range of 40 to 75 N, depending on the change of the spring length. The force is transferred on the clamping pad using a lever with a transfer rate 1:1, which means that the clamping force approximately equals the exerted spring force.

Free-form workpieces often require special fixtures that suite their spatial geometry. To enable the usage of a single clamping device for a wide range of workpieces, an additively manufactured pad is inserted at the contact zone of the clamping device. Such a customized pad shape results in tight fitting with the part that has to be machined. The higher the complexity of the clamping zone on the free-form workpiece, the more there is an advantage in using additive technologies for the production of customized fixtures [3]. Examples of customized pads are presented in Fig. 5.

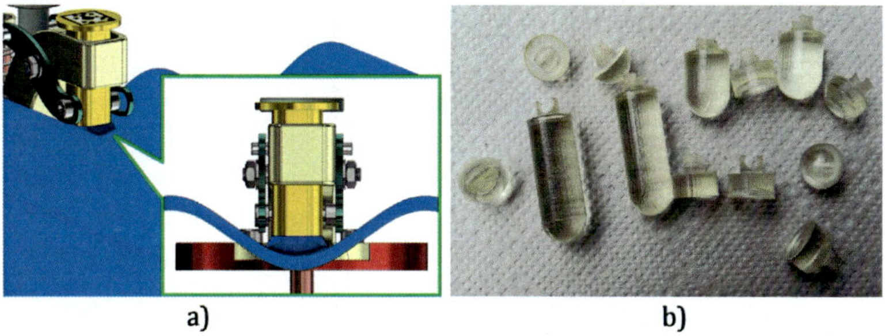

Fig. 5. Customized clamping pads: a) shape optimization for clamping a corrugated sheet;. b) example of additively manufactured pads.

3.3 Utilization of Positioning Bolts

As shown in Fig. 3, positioning bolts are used to define a clear workpiece position and orientation in the machine work area. The part is then fixed using clamping devices. After fixturing the part for the machining process, the scanning process is conducted and results in location data for positioning bolts. The gathered information is used to calculate the workpiece coordinate system. A detailed description of this process is presented in Sect. 4.

When machining free-form parts that cannot be placed and fixed on a single planar surface, the height of clamping devices has to be adjusted in order to achieve mid-air fixturing, which limits the use of positioning bolts. In such a case, the workpiece coordinate system has to be defined by scanning location data for clamping devices and using the gathered information to calculate it. Such a fixturing case is presented in Fig. 6.

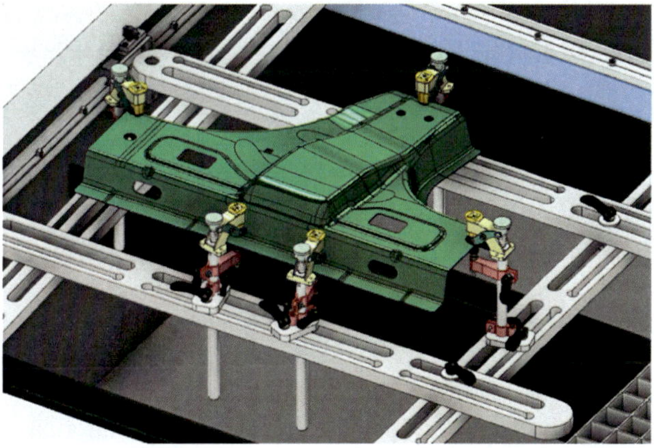

Fig. 6. Clamping a free-form workpiece mid-air using flexible clamping devices (Developed in cooperation with STM Waterjet Germany GmbH).

4 Principles of the Assistance and Control System

Defining the workpiece coordinate system precisely is a necessary preparatory task that influences the outcome of the waterjet cutting process. An improperly defined coordinate system can lead to a variety of processing issues due to deviations in the resulting tool trajectory relative to the workpiece. This error can result in a too small or too big nozzle standoff distance, collision of the tool with the part or with a clamping device, irregular cuts, etc. Some of these machining errors are illustrated in Fig. 7.

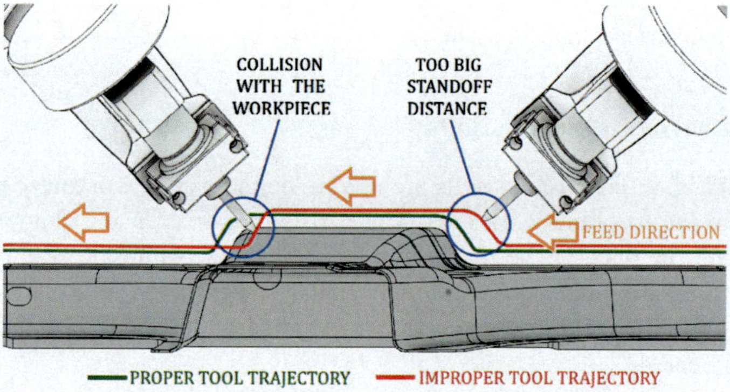

Fig. 7. Processing errors when using an improperly defined workpiece coordinate system for waterjet cutting of free-form workpieces.

Even though different approaches for defining a workpiece coordinate system already exist (see Sect. 2.1) and machines can exhibit the necessary 5-axis motion to follow complex spatial tool paths, there still are some challenges that have to be overcome in order to enable a fully automatized approach to 5-axis waterjet machining. In order to support this task and raise the overall process automation level, a novel assistance and control system is being developed.

The work principle of this system implies defining the workpiece coordinate system by optically scanning the location and orientation of marked clamping and positioning devices, whose position relative to the workpiece are predefined. After gathering relevant data about positions of these fixtures, the workpiece coordinate system can be calculated. This data is also used to predict and prevent possible tool collision with the machined part during the cutting process. Figure 8 presents the schematic view of the work principle of this system.

The first step required for utilizing the assistance and control system is defining all relevant workpiece, tool and process data in CAM. Additionally, information about fixture location and orientation has to be specified. Fixtures include clamping and positioning devices, as well as all other workpiece specific marked elements that are necessary for the determination of its position in the workpiece coordinate system. Such an approach avoids a direct coordinate system measurement procedure that is usually conducted on the part and utilizes the fact that each marked element's position relative to it is precisely defined. This precise position definition of each clamping and positioning device is a prerequisite in order to calculate the workpiece coordinate system accurately. To achieve this goal, marked elements have to be placed on exact positions relative to the machined part. However, considering the fact that markers are to be placed on clamping and positioning devices, then these devices would have to fix more than one degree of freedom of the part, which represents a challenging design task when fixturing free-form parts. This is why a different calculation method has been developed.

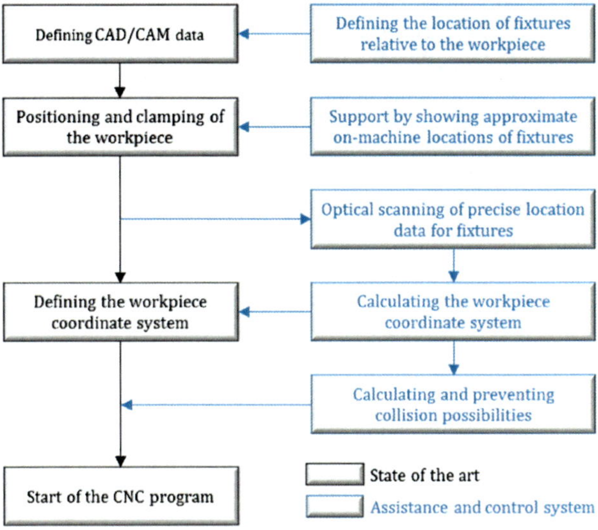

Fig. 8. Tasks of the assistance and control system.

The setup for calculating the workpiece coordinate system is presented in Fig. 9. Considering that the clamping and positioning devices are placed next to an edge of the workpiece, the position of each marked element has to be specified relative to these edges. When looking at a single positioning device, the distance to the reference edge (presented as "a" in Fig. 9) can be defined precisely, unlike the distance from an adjacent edge (presented as "b" in Fig. 9). To disregard this undefined value, the workpiece coordinate system has to be calculated using only the accurately defined distance of the marker to its reference edge. When placing multiple positioning devices on a single straight edge, an auxiliary line that includes the markers can be defined. For two markers this line can be defined as a straight line that includes these two points, while for three or more, a linear least square interpolation can be used to define the line equation. Such an approach can be utilized on multiple edges of the workpiece, and the cross point of auxiliary lines can be defined as zero point offset. The auxiliary lines show the directions of the X and Y coordinate in the workpiece coordinate system. For parts with curved or multiple edges, auxiliary curves or points can be defined, as necessary, for defining the workpiece coordinate system. These structures are then defined with multiple markers and calculated using least square interpolation for scanned marker locations.

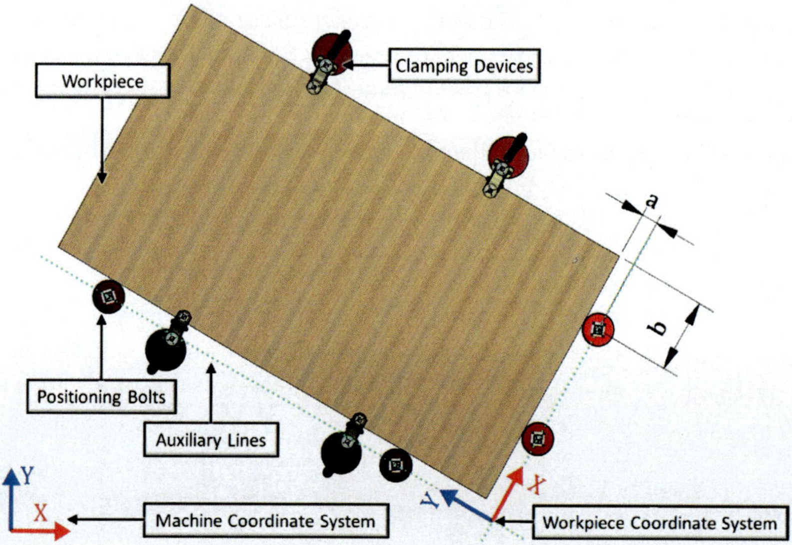

Fig. 9. Setup for calculating the workpiece coordinate system.

After defining process relevant data in CAM, the workpiece can be placed on the machine table. The fixturing process is supported by the assistance and control system, which shows approximate positions for each clamping and positioning device on the machine table, using an integrated laser pointer that is fixed on the machine cutting head. Due to safety reasons, each movement of the cutting head needs to be approved on the control unit, which allows the placement of only one clamping or positioning device at a time. The next step includes defining the workpiece coordinate system. In order to gather precise location and orientation data of the workpiece, an optical scanning process needs to be executed. The scanning process collects this information for each clamping and positioning device.

To find correct positions of clamping devices, they have to be detected, identified and localized. Also their poses have to be estimated. Finding a single clamping device will be done by reading two high contrast enclosed ArUco markers, which have to be laser engraved on its upper side, as shown in Fig. 4 (markers A and B). Since the coarse position and orientation of the clamp is already defined inside the CAD setup, the machine head can be moved to the position of marker A within a defined tolerance range, considering the cameras positional offset. This way the marker detection system should always have a full view of the marker inside the image of the camera.

A stepwise correction of the camera position is done by minimizing the correction vector between the marker and the camera until the marker is centered. The marker detection system is also able to compute the orientation of the marker. This is needed to subtract the camera offset vector to find the coordinates of the clamping device in machine terms. Unfortunately the determination process for orientations of one marker lacks the required precision. While it is not usable to compute the actual pose of the clamping device from it, it can be used to estimate the position of a second marker B. Once the machine head is positioned above this secondary marker, considering again the camera offset, the correction process applied to the first marker is repeated in respect of marker B. As a result the precise locations of two principal points of the clamping device are now known in machine coordinates and are then used to compute the planar orientation of it.

The same scanning principle is used for locating the positioning bolts. Since their form is symmetric, their planar orientation is irrelevant. Therefore one marker suffices in order to define their location. After gathering relevant location data, the workpiece coordinate system can be calculated by determining auxiliary lines and their cross point. Additionally, a simulation can be run in order to analyze and prevent collision possibilities.

5 Conclusion

The presented state of the art shows that there are many ways of determining a workpiece coordinate system on a waterjet cutting machine. Most of currently available systems measure the zero point offset directly on the workpiece, using either an optical or mechanical measuring principle. The novel optical system that is currently in the development phase and that was presented in this research paper considers a different approach, based on indirectly calculating the workpiece coordinate system by gathering location data from clamping and positioning devices. Based on this information, the workpiece zero point offset can be defined.

Acknowledgements. This work has been realized in affiliation with STM Waterjet Germany GmbH and funded by the Federal Ministry for Economic Affairs and Energy on behalf of the German Bundestag. The authors are grateful for their support.

References

1. Hassan, A.I., Kosmol, J.: An overview of abrasive waterjet machining (AWJM), Prace Naukowe Katedry Budowy Maszyn/ Politechnika Śląska, pp. 181–205 (1997)
2. Hashish, M.: WaterJet Machining Process, in Handbook of Manufacturing Engineering and Technology, pp. 1651–1686. London, Springer (2015)
3. Violante, M.G., Iuliano, L., Minetola, P.: Design and production of fixtures for free-form components using selective laser sintering. Rapid Prototyp. J. **13**(1), 30–37 (2007)
4. Liu, H.T., Olsen, J.H.: Application of Abrasive-Waterjet for 3D machining, in 2013 WJTA-IMCA Conference and Expo. Houston, Texas (2013)
5. Rennau, A., Dix, M.: Präzisionsbeschnitt von CFK-Metall-Verbunden. ZWF - Zeitschrift für wirtschaftlichen Fabrikbetrieb **112**, 680–683 (2017)
6. Snider, D.: 6 axis robotic abrasivejet advancements in accuracy for quality and production. In: American WJTA Conference and Expo, Houston, Texas (2007)
7. Duflou, J.R., Kruth, J.-P., Bohez, E.L.: Contour cutting of pre-formed parts with abrasive waterjet using 3-axis nozzle control. J. Mater. Process. Technol. **115**, 38–43 (2001)
8. OMAX: Precision Optical Locator (2019). https://www.omax.com/accessories/precision-optical-locator. Accessed 05 Sept 2019
9. Water Jet Sweden: FiveX - Front line technology for advanced free form cutting (2019). https://www.waterjetsweden.com/products/fivex. Accessed 05 Sept 2019

10. Ridder, H.G.: Automatisierungs-GmbH, 3D-Abstandssysteme - Automatisches 3D Abstandssystem für das Wasserstrahlschneiden (2019). https://www.ridder.de/waricut_zubehoer_3dabstandssysteme.php. [Accessed 02 09 2019]
11. WARDJet: Save Money and Increase Part Accuracy with a Waterjet Height Sensor (2019). https://wardjet.com/news/save-money-and-increase-part-accuracy-with-a-waterjet-height-sensor. Accessed 05 Sept 2019
12. Flow Inc.: Dynamic Waterjet XD. (2019). https://www.flowwaterjet.de/Schneidkopfe/Dynamic-Waterjet-XD. Accessed 02 Sept 2019
13. Sim, R., Little, J.J.: Autonomous vision-based robotic exploration and mapping using hybrid maps and particle filters, Image and Vision Computing **27** (1), Bd. Canadian Robotic Vision 2005 and 2006, pp. 167–177 (2009)
14. Pichler, A., Akkaladevi, S.C., Ikeda, M., Hofmann, M., Plasch, M., Wögerer, C., Fritz, G.: Towards shared autonomy for robotic tasks in manufacturing. Procedia Manuf. **11**, 72–82 (2017)
15. Valencia-Garcia, R., Martinez-Béjar, R.: An intelligent framework for simulating robot-assisted surgical operations. Expert Syst. Appl. **28**(3), 425–433 (2005)
16. Broggi, A., Dickmanns, E.: Applications of computer vision to intelligent vehicles. Image Vis. Comput. **18**(5), 365–366 (2000)
17. Patterson, T., McClean, S., Morrow, P., Parr, G., Luo, C.: Timely autonomous identification of UAV safe landing zones. Image Vis. Comput. **32**(9), 568–578 (2014)
18. Sanchez-Lopez, J.L., Pestana, J., de la Puente, P., Campoy, P.: A reliable open-source system architecture for the fast designing and prototyping of autonomous multi-UAV systems: simulation and experimentation. J. Intell. Robot. Syst. **84**, 1–19 (2015)
19. Olivares-Mendez, M., Kannan, S., Voos, H.: Vision based fuzzy control autonomous landing with UAVs: from v-rep to real experiments. Control and Automation (MED), pp. 14–21 (2015)
20. Pflugi, S., Vasireddy, S., Lerch, T., Lerch, T., Ecker, T.M., Tannast, M., Boemke, N., Siebenrock, K., Zheng, G.: Augmented marker tracking for periacetabular osteotomy surgery. Int. J. Comput. Assisted Radiol. Surgery **13**(2), 291–304 (2018)
21. Lima, J.P., Roberto, R., Simoes, F., Almeida, M., Figueiredo, L., Teixeira, J.M., Teichrieb, V.: Markerless tracking system for augmented reality in the automotive industry. Expert Syst. Appl. **82**, 100–114 (2017)
22. Chen, P., Peng, Z., Li, D., Yang, L.: An improved augmented reality system based on AndAR. J. Vis. Commun. Image Represent. **37**, 63–69 (2016)
23. Khattak, S., Cowan, B., Chepurna, I., Hogue, A: A real-time reconstructed 3d environment augmented with virtual objects rendered with correct occlusion, Games Media Entertainment, pp. 1–8 (2014)
24. Engel, J., Schöps, T., Cremers, D.: LSD-SLAM: large-scale direct monocular SLAM. In: European Conference on Computer Vision, Zürich, Switzerland (2014)
25. Mur-Artal, R., Montiel J., M.M., Tardós, J.D.: Orb-slam: a versatile and accurate monocular slam system. IEEE Trans. Robot. **31**(5), 1147–1163 (2015)
26. Zhong, S., Liu Y., Chen, Q.: Visual orientation inhomogeneity based scale-invariant feature transform. Expert Syst. Appl. **42**(13), 5658–5667 (2015)
27. Knyaz, V.A., Sibiryakov, R.V.: The development of new coded targets for automated point identification and non-contact surface measurements. 3D Surface Measurements, International Archives of Photogrammetry and Remote Sensing, Vol. XXXII, No. 5, pp. 80–85 (1998)
28. Naimark, L., Foxlin, E.: Circular data matrix fiducial system and robust image processing for a wearable vision-inertial self-tracker. In: Proceedings of the 1st International Symposium on Mixed and Augmented Reality (ISMAR), pp. 27–36 (2002)

29. Rekimoto, J., Ayatsuka, Y.: CyberCode: designing augmented reality environments with visual tags. In: Proceedings of DARE 2000 - Designing Augmented Reality Environments, New York (2000)
30. Rohs, M., Gfeller, B.: Using camera-equipped mobile phones for interacting with real-world objects. In: Advances in Pervasive Computing, pp. 265–271 (2004)
31. Kato, H., Billinghurst, M.: Marker tracking and HMD calibration for a video-based augmented reality conferencing system. In: The 2nd International Workshop on Augmented Reality (IWAR 1999), pp. 85–94 (1999)
32. Wagner, D., Schmalstieg, D.: ARToolKitPlus for pose tracking on mobile devices. Computer Vision Winter Workshop, pp. 139–146 (2007)
33. Schmalstieg, D., Fuhrmann, A., Hesina, G., Szalavári, Z., Encarnacäo, L.M., Gervautz, M., Purgathofer, W.: The studierstube augmented reality project. Presence: Teleoper. Virtual Environ. **11**(1), 33–54 (2002)
34. Fiala, M.: Designing highly reliable fiducial markers. IEEE Trans. Pattern Anal. Mach. Intell. **32**(7), 1317–1324 (2010)
35. Olson, E.: AprilTag: a robust and flexible visual fiducial system. In: Proceedings - IEEE International Conference on Robotics and Automation, pp. 3400–3407 (2011)
36. Garrido-Jurado, S., Munoz Salinas, R., Madrid-Cuevas, F.J., Marín-Jímenez, M.J.: Automatic generation and detection of highly reliable fiducial markers under occlusion, Pattern Recogn. **47**(3), 2280–2292 (2014)
37. Garrido-Jurado, S., Munoz Salinas, R., Madrid-Cuevas, F.J., Medina-Carnicer, R.: Generation of fiducial marker dictionaries using mixed integer linear programming. Pattern Recogn. **51**, 481–491 (2016)
38. Momber, A.W.: Energy transfer during the mixing of air and solid particles into a high-speed waterjet: an impact-force study. Experimental Thermal Fluid Sci. **25**, 31–41 (2001)
39. Hassan, A.I., Chen, C., Kovacevic, R.: On-line monitoring of depth of cut in AWJ cutting. Int. J. Mach. Tools Manuf **44**, 595–605 (2004)

Pulsating Abrasive Water Jet Cutting Using a Standard Abrasive Injection Cutting Head – Preliminary Tests

Libor Sitek[1(✉)], Petr Hlaváček[1], Josef Foldyna[1], Michael Jarchau[2], and Vladimír Foldyna[1]

[1] Institute of Geonics, Czech Academy of Sciences, Ostrava, Czechia
libor.sitek@ugn.cas.cz

[2] Hammelmann GmbH, Carl-Zeiss-Straße 6-8, Oelde, Germany

Abstract. The paper focuses on research into disintegration effects of a pulsating abrasive water jet using a standard abrasive injection cutting head. Forced pulses of a frequency of 20 kHz were generated by a piezoelectric pulse generator using a sonotrode placed in a high pressure system in front of a water jet nozzle. Several variants of the water jet generation in the cutting head were investigated - a continuous and pulsating water jets without the presence of the focusing tube to verify the pulse formation in the mixing chamber. Consequently, an abrasive water jet and a pulsating abrasive water jet with an installed focusing tube were studied. At a selected water pressure of 70 MPa, an increase in cutting effects of the pulsating abrasive water jet on average by 15–20% compared to the continuous abrasive water jet of the same parameters was observed. The authors consider that the increase in efficiency is mainly caused by better distribution of abrasives over the focusing tube cross-section, largely due to separated slugs of water forming the pulsating water jet. In this regard, the authors see the potential of further increase in the efficiency of abrasive water jets. Modeling and simulation of the water flow using an appropriate computational system can contribute to achieving an optimal configuration of the pulsating abrasive water jet.

Keywords: Abrasive water jet · Pulsating abrasive water jet · Depth of cut · Abrasive cutting head

Abbreviations

WJ	Water jet
PWJ	Pulsating water jet
AWJ	Abrasive water jet
PAWJ	Pulsating abrasive water jet

D. Klichová et al. (Eds.): Water Jet 2019, LNME, pp. 186–196, 2021.
https://doi.org/10.1007/978-3-030-53491-2_20

1 Introduction

An increase in the efficiency of high-speed water jets is one of traditional efforts of particular science and research centers around the world. Besides the addition of abrasive particles to the water jet in the form of widespread abrasive water jets (AWJ), a cyclically varying impact velocity of the water jet applied in the pulsating water jet (PWJ) technology has been used in recent decades. Effects of the both aforementioned water jet techniques of material disintegration are well-known among professionals and have been often published in scientific literature. The both water jets are widely applicable in various fields of human activity due to their significant increase in efficiency compared to the conventional continuous water jets. However, the combination of the two principles of the increase in the water jet effects is rather neglected and less attention is paid to this issue in the literature.

2 Background

Ren et al. [1] and Bai et al. [2] experimentally investigated the use of forced pulsating abrasive water jets for cutting and surface preparation at pressures of approximately 35 MPa. They designed a special nozzle configuration that allows the addition of abrasive particles into the pulsating water jet. Sizes of adaptors and mixing tubes were designed to find out, if the jets would effectively scour large areas per pass. The most important condition in their design was to preserve the characteristics of the water pulses exiting the mixing chamber. The main findings described in the above-mentioned publications were as follows: (i) pulses in the chamber increase suction, leading to an increased airflow rate, which, in turn, increases the flow rate of abrasive powders; (ii) in cutting tests, the efficacy of the garnet-laden pulsating abrasive water jet (PAWJ) was almost three times that of a conventional abrasive water jet; (iii) pulsing an abrasive water jet increases its effectiveness in eroding layers of various types of coating; (iv) the fine size abrasive particles remove the paint more effectively and uniformly than coarser abrasives.

A self-excited oscillation pulsed abrasive water jet which uses a Helmholtz resonator when cutting metal materials at pressures of about 25 MPa was published by Xue [3]. Unlike in previous publications, where abrasive particles were added to the water jet, Xue uses an abrasive suspension which passes through the Helmholtz resonator. The author presents results after cutting steel blocks. Unfortunately, a comparison with a conventional AWJ is not provided.

Our method of PAWJ generation was based on a rather simple configuration. An ultrasonic generator of pressure pulses was installed in front of the injection abrasive cutting head. After the pressurized water passes through the water nozzle, the pressure pulsations change into speed pulsations. At a certain distance from the nozzle outlet, the initially continuous water flow breaks into individual slugs of water and a pulsating water jet is formed. The promising disintegration potential of impacting slugs of water on material or the increase in efficiency of a continuous water jet by its breakage on slugs of water has been widely published and documented in appropriate studies (e.g. [4–9], etc.). Similar to the continuous water jet, abrasives are mixed with the pulsating water jet in the mixing chamber downstream the water nozzle [10]. Whether the cutting capabilities of the PAWJ with a standard cutting head will be improved has been investigated and this issue has become the subject of this paper.

3 Experiments

3.1 Equipment and Setting of Experimental Parameters

In the experimental research of cutting capabilities of the PAWJ, a plunger crankshaft high-pressure pump Hammelmann HDP 253 was used. This pump with a maximum output of 200 kW is capable of delivering up to 67 l·min^{-1} at a water pressure of 160 MPa. The ABB IRB 6640-180//2.55 Master Robot manipulated the cutting head. A key element of the experimental chain was the injection abrasive cutting head Hammelmann Type B 1500 for higher water flow rates (25–40 l·min^{-1}) and higher abrasive mass flow rates. A patented ultrasonic generator of pulses with a sonotrode was placed in front of the water nozzle in the high-pressure system [11, 12].

Operating parameters of individual water jet variants were as follows: water pressure of 70 MPa, water nozzle diameter of 1.5 mm, focusing tube inner diameter of 3 mm, focusing tube length of 76.4 mm and cutting rate from 1 to 10 mm·s^{-1}. The standoff distance was changing from 2 mm to 52 mm according to expected significant changes in disintegration efficiency of the PWJ and PAWJ depending on greater standoff distances. Slugs of water of the pulsating water jet were generated at the frequency of 20.2 kHz, acoustic power consumed was 120 W. For abrasive water jets, the Australian garnet concentrate GMA 80 mesh was used at the abrasive mass flow rates of 1 and 1.5 kg·min^{-1}.

The metal test specimens were made of AW 5754 aluminum alloy (AlMg3) of 1.5 mm thickness (for tests with the continuous and pulsating water jets) and AISI 304 stainless steel of 15 mm thickness (for tests with abrasive water and pulsating abrasive water jets). During experiments, kerfs were created in the specimens by means of individual water jets. The achieved depth of cut at defined water jet parameters was chosen as a measure of the cutting ability assessment of individual jets. The depth of cut was determined using a contactless optical profilometer MicroProf FRT.

The obtained results are presented below predominantly in form of graphs for various water jet parameters. All graphs are displayed on the same scale for better comparison of individual results.

3.2 Methods and Results

In the first phase of the research, it was investigated whether the water jet breaks into slugs of water typical for a pulsating water jet in the used Hammelmann cutting head (i) without and (ii) with the focusing tube installed. Therefore, kerfs were created in an aluminum sheet by means of the pulsating water jet formed in the cutting head mentioned-above at a gradually increasing standoff distance. The obtained results clearly indicate the formation of pulses (slugs of water) and thus an increase in the water jet efficiency. In the case of absence of the focusing tube, material disintegration significantly increased from a distance of about 40 mm from the cutting head outlet due to the generation of pulses and forced water jet breakage [9, 13]. It has been found that the inner geometry of the cutting head does not affect the successful generation and subsequent development of water pulses. A comparison with effects of the identical water jet without pulses is well illustrated in Fig. 1 (kerfs A and B). When the focusing tube is installed below the mixing tube, disintegration effects of the PWJ are noticeably worse with earlier breakage of slugs of water which are, in addition, probably not sufficiently developed, as shown in Fig. 2 (kerf C). Nevertheless, disintegration effects of the PWJ are considerably better than those of the continuous water jet generated by the cutting head with the focusing tube (Fig. 2 - kerf D). In this case, the standoff distance was measured from the end of the focusing tube which extends about 50 mm from the body of the cutting head (compare Fig. 1 and Fig. 2).

Next experiments focused on determining disintegration capabilities of the abrasive pulsating water jet generated in the complete cutting head with the installed focusing tube. As we expected greater cutting effects when using abrasives, the test kerfs were made on thicker stainless steel specimens. Tests done under identical operating conditions proved better performance of the pulsating abrasive water jet in comparison to the conventional abrasive water jet (approximately of 15 to 20% higher) – Figs. 3, 4, 5 and 6. However, the difference was much lower than that reported by Ren et al. [1] and Bai et al. [2]. The special design of their mixing chamber and used focusing tubes most likely allowed good preservation of characteristics of the water pulses; however, it was inappropriate for the conventional abrasive water jet generation.

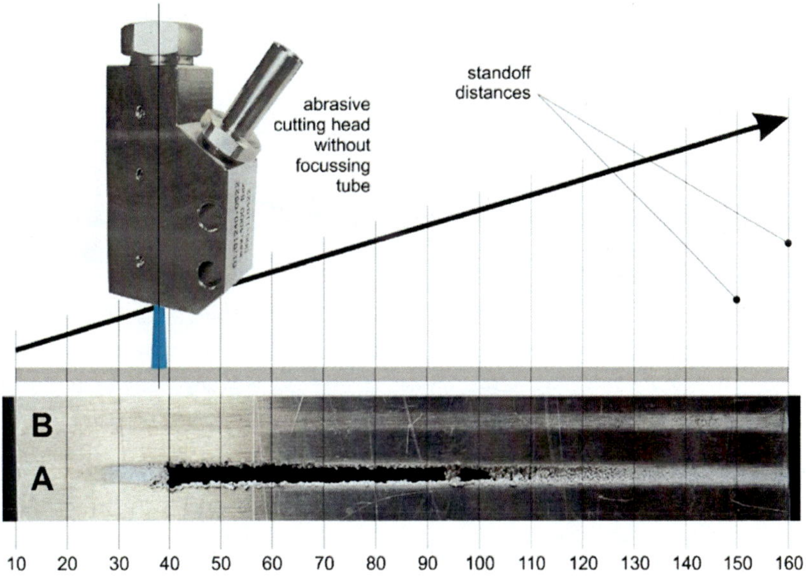

Fig. 1. Kerfs made in an aluminum sheet by A – pulsating water jet and B – continuous water jet (abrasive cutting head without the focusing tube, gradually increasing standoff distance in mm, cutting rate: 10 mm·s^{-1}).

4 Discussion

Design of a standard abrasive cutting head allows trouble-free PWJ generation and development of individual slugs of water just in case of the absence of the focusing tube. In the focusing tube, pulsations in the water jet are likely to be damped and the water jet is not sufficiently broken down into individual slugs of water. Nevertheless, the PAWJ effects are demonstrably higher compared to the conventional AWJ. By simple comparison of depth values in the graphs, we obtain values of cutting ability improvement of about 15–20% (see Figs. 3, 4, 5 and 6). However, if we study in detail the increase in cutting ability in relation to the disintegration effects of the PAWJ and the AWJ, we will find that the above-mentioned percentage improvement is only an average value. The improvement strongly depends on the standoff distance and probably on the abrasive mass flow rate (Fig. 7). This phenomenon should be examined in more detail in further research.

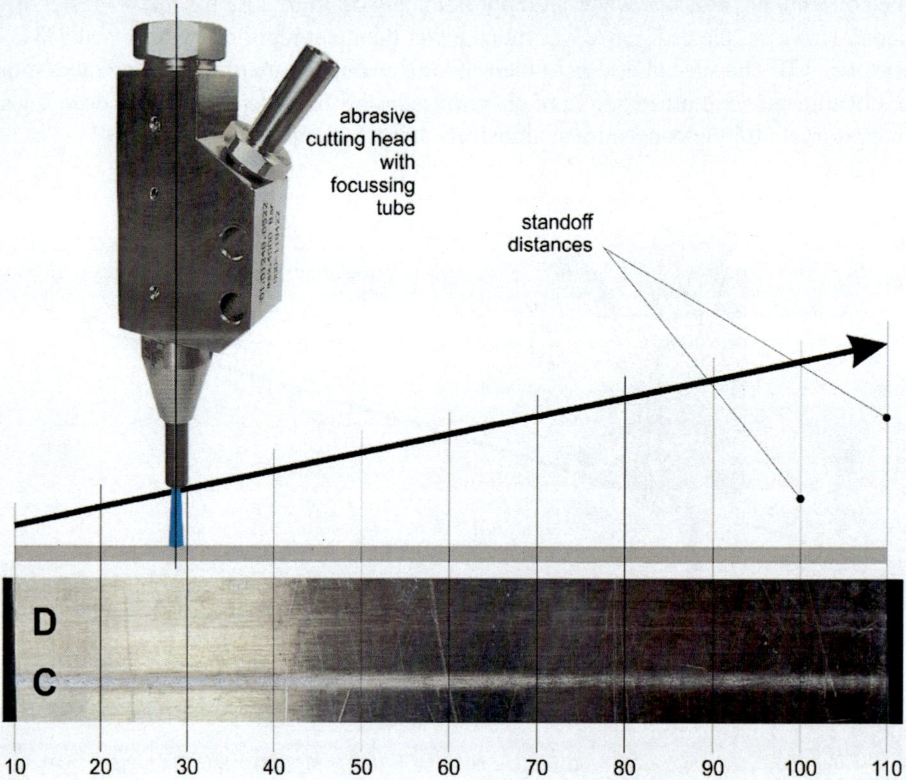

Fig. 2. Kerfs made in an aluminum sheet by C – pulsating water jet and D – continuous water jet (abrasive cutting head with the focusing tube, gradually increasing standoff distance in mm, cutting rate: 10 mm·s^{-1}).

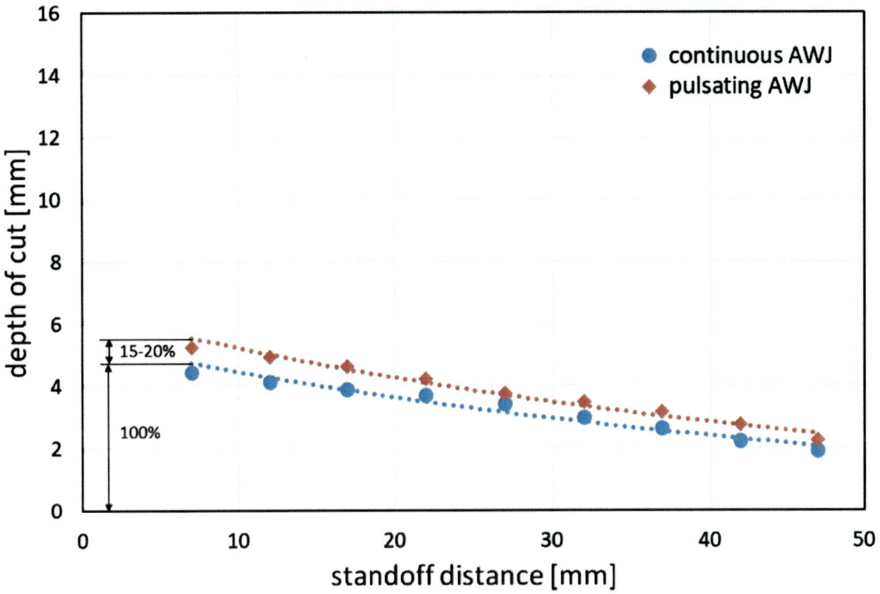

Fig. 3. Effect of standoff distance of the continuous AWJ and the pulsating AWJ on depth of cut (cutting rate: 4 mm·s^{-1}, abrasive mass flow rate: 1 kg·min^{-1}).

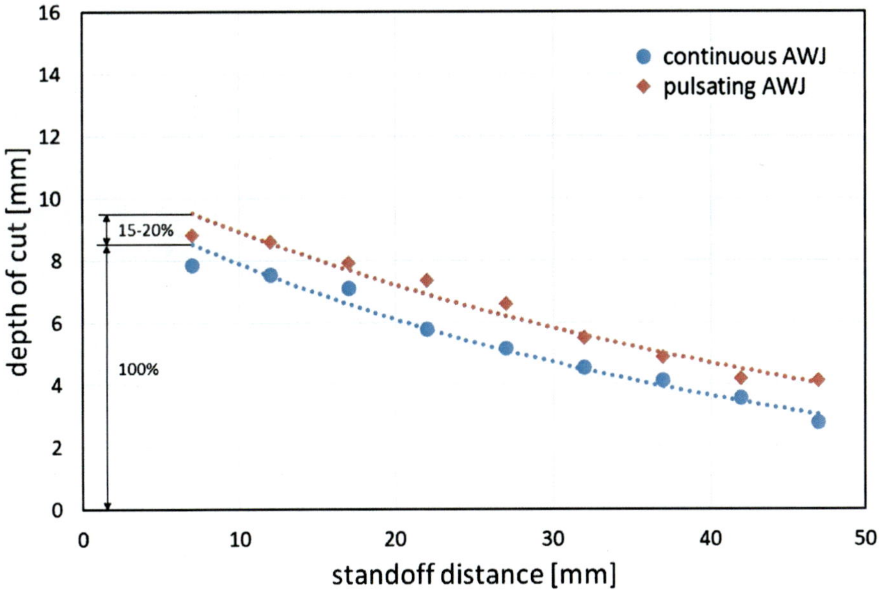

Fig. 4. Effect of standoff distance of the continuous AWJ and the pulsating AWJ on depth of cut (cutting rate: 3 mm·s^{-1}, abrasive mass flow rate: 1 kg·min^{-1}).

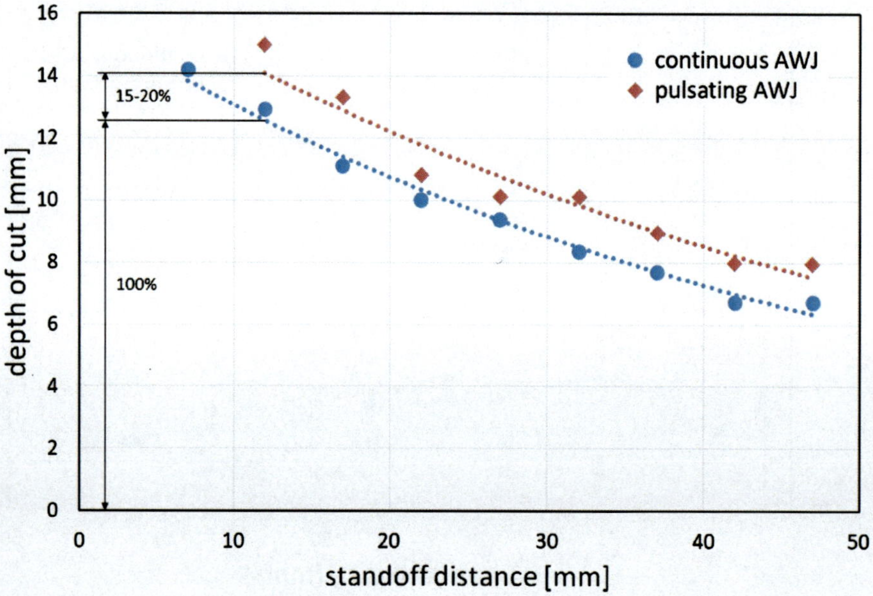

Fig. 5. Effect of standoff distance of the continuous AWJ and the pulsating AWJ on depth of cut (cutting rate: 2 mm·s^{-1}, abrasive mass flow rate: 1 kg·min^{-1}).

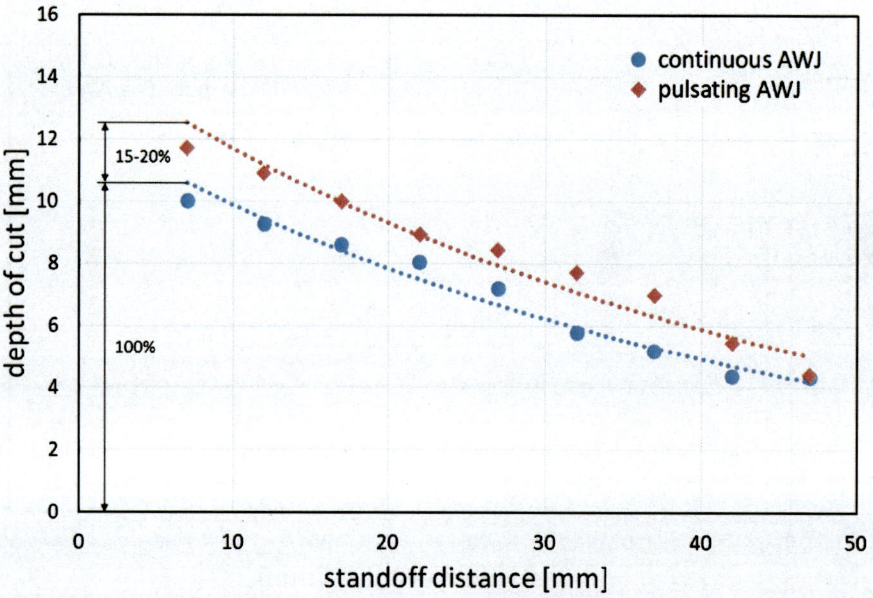

Fig. 6. Effect of standoff distance of the continuous AWJ and the pulsating AWJ on depth of cut (cutting rate: 3 mm·s^{-1}, abrasive mass flow rate: 1.5 kg·min^{-1}).

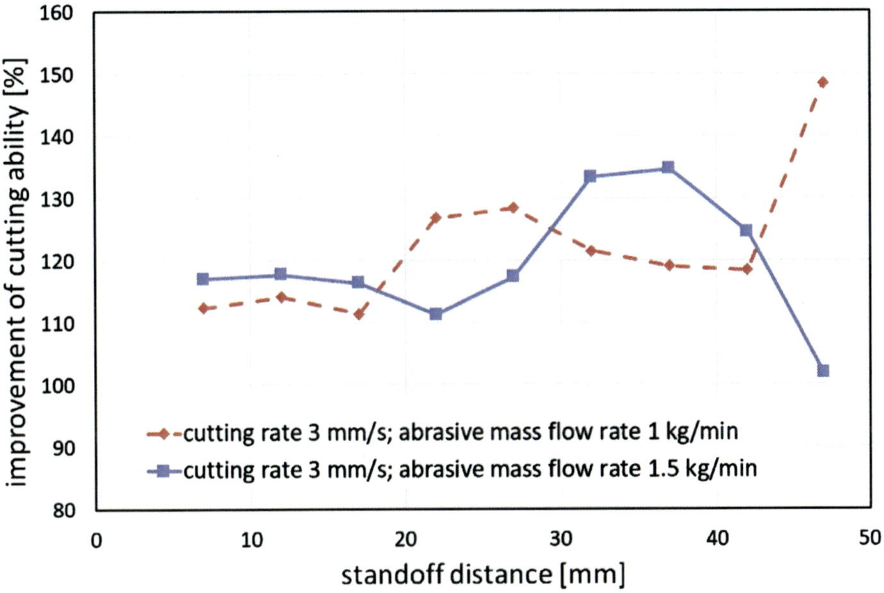

Fig. 7. Effect of standoff distance and abrasive mass flow rate on the improvement of cutting ability of the pulsating AWJ compared to the continuous AWJ (improvement of cutting ability was calculated as the ratio of the depth of cut achieved by the PAWJ to the depth achieved by the AWJ under the same conditions).

The mentioned improvement of PAWJ cutting effects compared to the AWJ can be explained as follows. Due to the lateral velocity distribution in high-speed water jets, the velocity decreases towards the water jet edges due to friction with the surrounding stationary environment [14–21]. In water jets with abrasive particles, the particles are concentrated mainly in the central part of the focusing tube due to corresponding pressure changes in the direction perpendicular to the water flow direction (see Fig. 8) [22]. Similar accumulation of abrasives in the axis of the water jet was also observed during abrasive water jet simulations as a part of our earlier research activities into the behavior of water jets after the exit from the abrasive cutting head (Fig. 9).

In the PAWJ, on the other hand, the abrasive particles are distributed differently (Fig. 8). The authors assume that better performance of the pulsating abrasive water jet compared to the conventional AWJ can be explained by the structure of the pulsating water jet downstream of the nozzle exit. The jet consists of equally spaced mushroom-like slugs of water which allow better mixing and acceleration of abrasive particles in the mixing chamber and the focusing tube. The kinetic energy of abrasives is thus better used when disintegrating a material being cut. However, this area has not yet been fully investigated. Using suitable methods of computer simulation of the high-speed water flow, design of some parts of the cutting head can be modified in a way to further improve the above-mentioned increase in the cutting capability of the cutting head.

Such a tool can be used in cutting applications, treatment of material surfaces and modification their properties at low pressures and lower water flow rates (food industry, medical sciences). In addition, it can be applied for accelerated mechanical simulations of the abrasive or cavitation wear caused by, for example, fast flowing fluids (resistance of parts of concrete structures, etc.) [23].

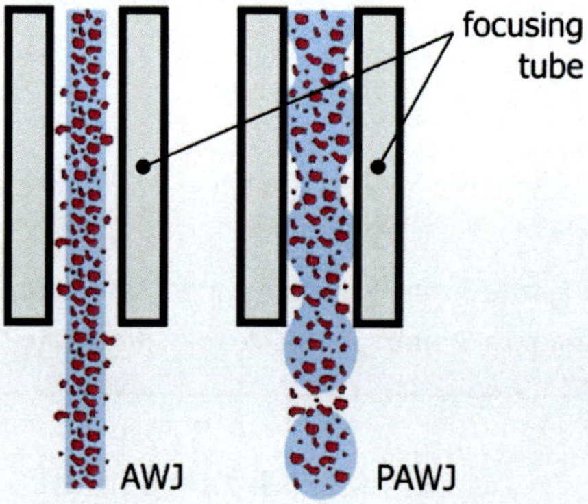

Fig. 8. Idea of the distribution of abrasive grains in the focusing tube using the AWJ and PAWJ.

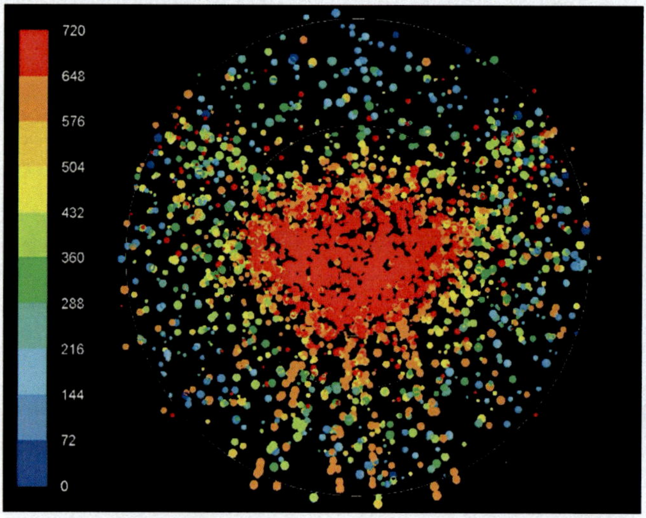

Fig. 9. Simulation of the distribution of abrasive grains in the AWJ. The color scale shows the velocity of particles v in $m{\cdot}s^{-1}$ (water pressure: 415 MPa, abrasive mass flow rate: 100 $g{\cdot}min^{-1}$, diameter of abrasive grains: 0.22–0.65 mm, water nozzle diameter: 0.33 mm, inner diameter of the focusing tube: 1.02 mm).

5 Conclusions

The pulsating water jet increases the efficiency of the abrasive water jet during its generation. Rather than by changing the velocity of the cutting medium, as in the case of the pulsating water jets, the effectiveness of the pulsating abrasive water jet is increased by better mixing of abrasive particles in the mixing chamber and their distribution in the focusing tube. Based on our own experience, we assume that modification of the standard abrasive head using methods of the numerical flow simulation will enable to further improve cutting effects of the pulsating abrasive water jet. It can be concluded that the pulsating abrasive water jets have a great potential especially for applications where lower pressures and lower water consumption are required.

Acknowledgement. This work was supported by a grant of the Czech Science Foundation (18-25035S), by a project of the Technology Agency of the Czech Republic (TH04020218) and by a project for the long-term conceptual development of research organisations RVO: 68145535. The authors are very thankful for the support.

References

1. Ren, B., Daniels, B., Yan, W., Tieu, A., Vijay, M.: Abrasive-entrained forced pulsed waterjet technique – basic study. In: Proceedings of 2005 WJTA American Waterjet Conference, Paper 2B-3 (2005)
2. Bai, C., Chandra, S., Daniels, B., Ren, B., Yan, W., Tieu, A., Vijay, M.: Abrasive-entrained high-frequency pulsed waterjet: basic study and applications. In: Proceedings of the 18th International Conference on Water Jetting, pp. 325–336 (2006)
3. Xue, F.: A novel metal-cutting technology for high-risk environments using self-excited oscillation pulsed abrasive water jet. In: 3rd International Forum on Energy, Environment Science and Materials Advances in Engineering Research, vol. 120, 2017 (IFEESM 2017) pp. 714–718 (2018)
4. Hancox, N.L., Brunton, J.H.: The erosion of solids by the repeated impact of liquid drops. Philosophical Trans. Royal Soc. London. Series A, Math. Phys. Sci. vol. 260, no. 1110, pp. 121–139 (1966)
5. Heymann, F.J.: High-speed impact between a liquid drop and a solid surface. J. Appl. Phys. **40**(13), 5113–5122 (1969)
6. Thomas, G.P., Brunton, J.H.: Drop impingement erosion of metals. In: Proceedings of the Royal Society of London. Series A, Mathematical and Physical Sciences, vol. 314, no. 1519, pp. 549–565 (1970)
7. Vijay, M.M., Foldyna, J.: Ultrasonically modulated pulsed jets: basic study. In: 12th International Conference on Jet Cutting Technology, N. G. Allen, Editor, BHR Group Conference Series, Publication No. 13, Mechanical Engineering Publications Limited, London, pp. 15–35 (1994)
8. Foldyna, J., Sitek, L., Švehla, B., Švehla, Š.: Utilization of ultrasound to enhance high-speed water jet effects. Ultrasonic Sonochemistry **11**(3–4), 131–137 (2004)
9. Foldyna, J., Sitek, L., Ščučka, J., Martinec, P., Valíček, J., Páleníková, K.: Effects of pulsating water jet impact on aluminium surface. J. Mater. Process. Technol. vol. 209, no. 20, pp. 6174–6180 (2009). Elsevier, 2009. ISSN 0924-0136

10. Jarchau, M.: Method for manufacturing water abrasive jet, involves supplying abrasive agent under high pressure water jet according to injector principle, and interrupting water jet before supplying abrasive agent. German patent application. DE 102008008 A1 (2008)
11. Foldyna, J., Švehla, B.: Method of generation of liquid jet pulsations and apparatus for implementation of this method. European Patent EP 1 863 601 B1, Date of Patent: 5 January (2011)
12. Foldyna, J., Švehla, B.: Method of generation of pressure pulsations and apparatus for implementation of this method. 2011, US Patent US 7,934,666 B2, Date of Patent: 3 May (2011)
13. Vijay, M.M.: Main lecture: properties and parameters of water jets. In: Rakowski, Z. (ed.) Proceedings of International Conference Geomechanics, Balkema, Rotterdam, vol. 91, pp. 207–222 (1992)
14. Leach, S.J., Walker, G.J.: Some aspect of rock cutting by high speed water jets. Phil. Trans. R. Soc., A **260**, 395–310 (1966)
15. Yanaida, K.: Flow characteristics of water jets. In: Proceedings of the 2nd International Symposium on Jet Cutting Technology, BHRA, Cambridge (1974)
16. Yanaida, K., Ohashi, A.: Flow characteristics of water jets in air. In: Proceedings of the 5th International Symposium on Jet Cutting Technology, BHRA, Hannover (1980)
17. Przyklenk, K., Schlatter, M.: Simulation of the cutting process in water jetting with the finite element method. In: Proceedings of the 8th International Symposium on Jet Cutting Technology, BHRA, Durham (1986)
18. Hlaváč, L.M., Sochor, T.: A contribution to the physics of a high velocity abrasive particle interaction with brittle non-homogeneous materials. In: Allen, N.G. (ed.) Proceedings of the 12th International Conference Jet Cutting Technology, Mech. Eng. Pub. Ltd., London, p. 117–126 (1994)
19. König, W., Wulf, Ch.: The influence of the cutting parameters on jet forces and the geometry of the kerf. In: Proceedings of the 7th International Symposium on Jet Cutting Technology, BHRA, Ottava (1984)
20. Zakin, J.L., Summers, D.A.: The effect of viscoelastic additives on jet structures. In: Proceedings of the 3rd International Symposium on Jet Cutting Technology. Chicago, USA, Paper A4 (1976)
21. Summers, D.A.: Waterjetting Technology. E & FN SPON, London (1995)
22. Sitek, L., Foldyna, J., Martinec, P., Klich, J., Mašláň, M.: On the preparation of precursors and carriers of nanoparticles by water jet technology. Tech. Gazette **19**(3), 465–474 (2012)
23. Sitek, L., Hlaváček, P., Bodnárová, L.: Use of high-speed water flows for accelerated mechanical modelling of erosive wear of concrete surfaces. In: Stančeková, D., Vaško, M., Rudawska, A., Čuboňová, N., Sapietová, A., Mrázik, J., Szabelski, J., Tlach, V. (eds.) Innovative Technologies in Engineering Production (ITEP 2018), Bojnice, Slovak RepublicMATEC Web of Conferences, vol. 244, p. 02007 (2018), 11–13 September (2018). https://doi.org/10.1051/matecconf/201824402007

Mechanical Strengthening of Anti-Corrosive Surface Layers by Water Jet

Dana Stancekova[1(✉)], Sergej Hloch[2], Dominik Cuha[2], and Michal Sajgalik[1]

[1] University of Žilina, Žilina, Slovakia
dana.stancekova@fstroj.uniza.sk
[2] Technical University of Kosice, Presov, Slovakia

Abstract. A brief definition of progressive technologies is difficult to implement due to the widely different processes that fall into this category. There is a consensus in the literature that this group includes processes that use conventional forms of energy in a new way. One of such technologies includes strengthening of surface layers with a pulsating water jet, of which we do not yet have enough verified information. Mechanical strengthening of materials is widely used in the automotive and aerospace industries, as well as in the manufacturing of production machines and equipment. The beneficial effects of mechanical strengthening of surface layers, especially for dynamically stressed components, have been known for decades. Evidence of the beneficial effects of mechanical strengthening of surface layers in terms of increasing the service life of components processed in this way is already available. The aim of the present experimental tests is to use a pulsating water jet to solidify the surface layer of materials, specifically chromium-nickel steel, to determine suitable parameters so that the solidified surface has the required properties and is not damaged by the water jet. Evaluation of the experimental verifications was done by measuring microhardness and residual stresses in the surface layer.

Keywords: Strengthening of surface · Pulsating water jet · Microhardness · Chromium-nickel steel · Residual stresses

1 Introduction

Advanced manufacturing technologies and their application in various fields of material treatment represent a fast-growing trend these days. These technologies are mostly applied to increase production rates, reduce costs and improve treatment processes and machining of materials. One of such advanced manufacturing technologies is water jet (WJ) and its technological modifications such as abrasive water jet (AWJ) and pulsating water jet (PWJ). This machining technology is commonly used to disintegrate and machine various kinds of materials. Besides that, it is also possible to strengthen the material surface under certain conditions [1, 2]. The technology of water jet peening is now in the state of testing and research. By a proper selection of technological parameters, a required strengthened surface can be achieved [3]. By the evaluation of

D. Klichová et al. (Eds.): Water Jet 2019, LNME, pp. 197–207, 2021.
https://doi.org/10.1007/978-3-030-53491-2_21

researched parameters of surface integrity, such as residual stress in the subsurface layer and microhardness of the strengthened area, we can conclude that it may be used in further research and development of the mechanisms for the surface strengthening of the material by water jet [3–5].

Beneficial effects of mechanical strengthening of surface layers, especially in the case of dynamically strained components, have been known for decades. Methods of mechanical strengthening of surface layers expanded, especially in the period from 1940 to 1970 majorly in the area of the automobile industry, in aviation as well as in the area of specialised production and, to a lesser extent, in the sphere of manufacturing machines and apparatuses [6, 7].

The tests of surface strengthening using various parameters are performed on AISI 316 Ti. It is non-magnetic material with a tendency towards cold drawing when it obtains the property of a poorly machinable material [8, 9]. Austenitic stainless steels cannot be hardened by heat treatment because of low carbon content, and the amount of plastic strain induced strongly depends on the stress required to further deform the material [10, 11].

The aim of the experiment is to select suitable parameters for the strengthening of surface layers of samples by water jet (Fig. 1). Moreover, the evaluation of results of hardness, residual stresses and a change of surface structure of samples by the application of various parameters on the strengthening of surface layers of the material by PWJ were studied. Pulsating water jet was used for the strengthening due to the fact that it does not contain any abrasives and thus does not cause any erosion on the surface of samples. On the contrary, compressive strength is induced by pulses and thus strengthening is caused. Erosion may take place as a result of too high operating pressure or a shorter distance of the nozzle from the surface. Due to this fact, the selection of appropriate technological parameters is a key factor in the process of strengthening by water jet. The purity of an applied medium (water) should also be considered, because disruption of material surface integrity may take place due by the presence of small impurities in water.

Fig. 1. The process of surface strengthening by pulsating water jet.

2 Characteristics of the Process of Strengthening by Water Jet

2.1 Equipment Used For Surface Strengthening

The equipment for surface strengthening by PWJ consists of the high-pressure pump Hammelmann HDP 253 with the maximum operating pressure 160 MPa and the maximum flow rate 67 $l.min^{-1}$, the industrial robot ABB IRB 6640-180/2.55 which controls the motion of the PWJ head and the ultrasonic generator ECOSON WJ-UG-630-40 to generate pulsations (Fig. 2) Pressurised water was fed into the acoustic chamber from where the pulsations were induced in the water, and the pulsating water exited through the nozzle. A circular nozzle (StoneAge) with diameter $d = 0.6$ mm was used.

Fig. 2. Robot ABB with the ultrasound generator ECOSON.

2.2 Strengthened Material

The sample material on which tests of surface strengthening using various parameters were performed is AISI 316 Ti. It is an austenitic chromium-nickel stainless steel, stabilised with titanium. It has good corrosion and chemical resistance to the low content of hydrochloric and organic acids. Its chemical composition is stated in Table 1.

Table 1. Chemical composition of the material AiSi 316 Ti in %.

C	Si	Mn	P	S	Cr	Mo	Ni	Ti
0.08	1	2	0.045	0.015	16.5–18.8	2–2.5	10.5–13.5	5xC

AISI 316 Ti is a non-magnetic and non-hardenable material. It tends to be cold worked, and it has satisfactory machinability. It may contain residual martensite from high temperatures that causes partial magnetisation. These modifications affect the corrosion resistance of the material.

Mechanical properties:

- Tensile strength Rm 520–690 $N.mm^{-2}$,
- Yield strength Rp 0.2 min. 220 $N.mm^{-2}$,
- Elongation A80mm min. 40%,
- 264 HV10.

Application: construction components, machines and apparatuses in chemical industries with required halogen resistance and non-oxidising acids. Higher yield strength provides advantages in mechanically stressed constructions. It is applied in photographic, pharmaceutical and petrochemical industries.

2.3 Parameters of Surface Strengthening of Samples

Due to the novelty of this technology for surface strengthening, there do not exist any specific table values of pressure, flow rate, feed rate, standoff distance, power or frequency of the pulsating sonotrode. The appropriate technological parameters were found through experiments.

3 Experimental Definition of Proper Parameters of Surface Strengthening

Pilot experiments for determination of technological parameters were carried out to achieve the required strengthened area. These parameters define the nature of the strengthened area formed on the sample. For the pilot run, the nozzle followed a stepped trajectory with a step length of 5 mm and a height of 2 mm. The pressure of the water jet, power or frequency of the pulsating apparatus was not varied due to the fact that these parameters were most appropriate for the strengthened surface.

The initial programmed standoff distance was 18 mm, and from this distance, the water jet worked on the material. The final distance from the tested material was 32 mm. To form a gradually strengthened surface, a circular nozzle with a diameter of 0.6 mm was used.

A gradually strengthened area with a width of 2 mm across the whole tested material (Fig. 3) was formed. After the evaluation of the area, optimal standoff distance of 24 mm, i.e., a spot in the centre of the sample where ideal conditions for strengthening took place, was selected. At this standoff distance, the surface was not deformed by erosion, and it was simultaneously affected by the jet of the high-pressure medium (water). In areas in edges of the sample in which the standoff distance was too short or too long, on the contrary, the material did not manifest favourable characteristics for strengthening.

3.1 Parameters of Surface Strengthening of Samples

For strengthening, the optimal feed rate of 2 $mm.s^{-1}$ and the nozzle standoff distance of 24 mm with a nozzle diameter of 0.6 mm were selected after the pilot test. The power and frequency of the sonotrode were 200 W and 20.18 kHz, respectively.

Fig. 3. A gradually strengthened area across the tested sample for a definition of a proper distance.

After the experimental verification of the optimal parameters for the strengthening of the sample surface layer, further experiments were carried out with various levels of parameters such as feed rate and standoff distance on mechanical properties of the surface.

4 Experimental Strengthening of Sample Surfaces

Strengthening of the material surface took place with different parametric settings of feed rate and standoff distance.

For comparison, six various strengthened areas were formed. In each of these areas, different process strengthening parameters were applied (Table 2). This provides a difference between surfaces affected by the pulsating water jet, and it also provides various characteristics of the surfaces and their mechanical properties.

Table 2. Input parameters for experimental strengthening.

	Pressure [MPa]	Frequency [kHz]	Nozzle [mm]	Power [W]	Distance [mm]	Feed rate [mm.s^{-1}]
1	35	20.18	0.6	200	22	2
2	35	20.18	0.6	200	24	2
3	35	20.18	0.6	200	26	2
4	35	20.18	0.6	200	22	0.5
5	35	20.18	0.6	200	22	2
6	35	20.18	0.6	200	22	8

An individual numbering of affected areas on the samples corresponds with a numeric designation of the input parameters in Table 2 and Fig. 4.

After the formation of strengthened surface areas, evaluation of the surfaces was carried out. They were evaluated from the standpoint of their visual and physical modifications.

4.1 Measurement of Residual Stresses

Residual stresses are side effects of the strengthening process, and it is important to know about their magnitude and directions.

Fig. 4. Strengthened areas by different parameters.

After the measurement of residual stresses of individual affected areas, Fig. 4, final values were obtained, which were consequently compared, Table 3. A sample protocol for the measurement of residual stresses is given in the appendix to this article.

Table 3. Final values of measured residual stresses in the strengthened areas.

Measurement	σ [MPa]	τ [MPa]
1	−527.5 ± 6.2	−13.7 ± 3.2
2	−531.5 ± 7.7	−0.8 ± 4
3	−346.7 ± 13.9	−27.7 ± 7.2
4	−493.7 ± 8.1	−13.5 ± 4.2
5	−420.8 ± 8.5	−6.2 ± 4.4
6	−458.6 ± 11	−17.4 ± 5.7

After the evaluation of measurement of residual stresses of the individual strengthened, the lowest values of residual stresses occurred in the measurement points 2 and 5. These measurement areas show the best values from the standpoint of residual stresses.

4.2 Microscopic Evaluation of Sample Surfaces

For the evaluation of strengthened sample surfaces, a surface integrity study of the strengthened surface is also important. Machining accuracy is critical for the quality of a mechanical product and is an important consideration for any manufacturer [12–16]. After the experimental strengthening of sample surfaces by the pulsating water jet with different parameters, we zoomed in the individual surfaces of strengthened layers by a microscope. This visual evaluation showed which process parameters of the strengthened surface have the most erosive impact on the surface integrity of the strengthened material.

The microscopic evaluation of the samples was performed on the digital stereo microscope Zeiss Stemi DV4 with 10× magnification. Results of the evaluation are stated in Fig. 5.

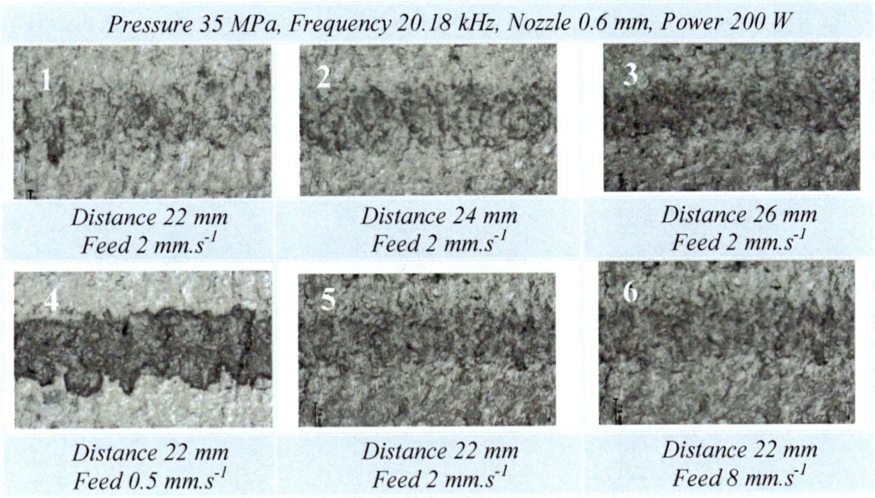

Fig. 5. Results of erosive defects of the material by different strengthening parameters.

From the comparison with the stereo microscope, it is evident that a decrease of the feed rate of the nozzle to the value 0. 5 $mm.s^{-1}$ (Test 4) has an impact on the integrity of the strengthened surface. At this feed rate, distinctive erosion occurs which makes it inadmissible in the process of strengthening of the material surface.

A visible disintegration of the surface can be also observed in Test 1. This disintegration of the surface is caused due to the standoff distance. At lower standoff distance, partial surface defects occur.

In other parametric settings, no distinctive erosive defects occurred, and these parameters can be evaluated as good from the standpoint of the surface integrity.

4.3 Measurement of the Hardness of Strengthened Layers

To measure the hardness of individual strengthened surface layers, Vickers hardness was measured by the hardness testing machine HPO 250/AQ, Fig. 6. This method proved to be the most suitable.

Fig. 6. Hardness measurement.

The hardness of the surface layer and stereo-microscopic photograph of the impression corresponding to individual parameters of the strengthening process were recorded (Fig. 7).

Pressure 35 MPa, Frequency 20.18 kHz, Nozzle 0.6 mm, Power 200 W		

Distance 22 mm, Feed 2 mm.s^{-1}	*Distance 24 mm, Feed 2 mm.s*$^{-1}$	*Distance 26 mm, Feed 2 mm.s*$^{-1}$

Distance 22 mm, Feed 0.5 mm.s^{-1}	*Distance 22 mm, Feed 2 mm.s*$^{-1}$	*Distance 22 mm, Feed 8 mm.s*$^{-1}$

Fig. 7. Values of the hardness of the strengthened surfaces regarding strengthening parameters.

Based on the values in Fig. 7, it can be concluded that the highest value of hardness against the basic non-strengthened material (264 HV10) is measured in the surfaces 2 and 5. In these strengthened surfaces, optimal strengthening parameters were kept and were experimentally verified. In case of the strengthened surface 4, a substantial surface disintegration occurs due to a low feed rate during the strengthening process (see the stereo-microscopic sample evaluation). Due to disintegration, the material hardness was not reliably measured, and thus the results of hardness were inaccurate. This inaccuracy, however, was not important regarding the fact that the defined parameters of surface strengthening were not suitable due to distinctive material surface disintegration. The final values of hardness for individual parameters of strengthening were low, but even despite that we may define the most optimal parameters of strengthening regarding the hardness of the final strengthened layer.

5 Conclusion

For AISI 316 Ti material, the optimal parameters of the process were verified by experimental tests of material surface strengthening. These parameters were defined by experimental tests by means of gradual material strengthening with the change in standoff distance. Quality of strengthened materials was evaluated from the standpoint of residual stresses, surface integrity deterioration and hardness. After the measurement of values of residual stresses, hardness and surface integrity, optimal technological parameters found are stated in Table 4. In these areas, the lowest values of residual stresses were measured as well as the highest surface hardness and suitable surface integrity of the strengthened layer were obtained.

Table 4. The most suitable strengthening parameters regarding the final surface hardness.

Pressure (MPa)	Frequency (kHz)	Nozzle (mm)	Power (W)	Height (mm)	Feed (mm.s^{-1})
35	20.18	0.6	200	24	2

Acknowledgements. The article was made under the support of grant project KEGA 022ŽU-4/2017 Implementation of on-line education in the area of precise technologies with an impact on the educational process to increase skills and flexibility of students of engineering fields of study and partial support by supported by the Slovak Research and Development Agency under Contract No. APVV-17-0490.

Appendix

Residual Voltage Measurement Protocol

STRESS REPORT

File name:	15APR19_0012
Gain file:	D:\044_Austenit\19APR18_0004.Gain
User	Ing. Michal Šajgalík, PhD.
Date:	April, 15, 2019, 22:35:42
Description	1,2,3,4,5,6-1ab|2ab|3ab|...

Beta Angle [8]: ; X = 10.00; Y = 0.00; Z = 0.02

Stress = - 493.7 ± 8.1 MPa

Shear Stress = - 13.5 ± 4.2 MPa

Intensity Ratio:	1.81	1.31
Average Peak Breadth:	4.09 ± 0.34	3.53 ± 0.20
Average Peak FWHM:	3.422 ± 0.13	3.086 ± 0.17
Average Peak FWHM fit:	3.894 ± 0.16	3.494 ± 0.16

Detector: 1

Beta Angles	*Psi*	*Sin^2psi*	*DSpacing*	*2Theta*	*Strain*E3*	*FWHM*	*FWHM fit*	*Breadth*	*Intensity*
35.00	21.40	0.1331	1.081269	153.08	-0.589	3.653	4.316	4.67	595.00
30.38	16.78	0.0834	1.081532	152.96	-0.346	3.568	4.124	4.40	585.06
25.74	12.14	0.0442	1.081838	152.83	-0.063	3.424	3.971	4.43	572.79
20.94	7.34	0.0163	1.081914	152.80	0.008	3.378	3.811	4.31	650.17
13.60	0.00	0.0000	1.081905	152.80	-0.000	3.244	3.822	4.08	660.10
10.04	-3.56	0.0039	1.081875	152.81	-0.028	3.344	3.840	4.09	627.07
2.87	-10.73	0.0347	1.081771	152.86	-0.124	3.352	3.819	4.31	539.88
0.00	-13.60	0.0553	1.081669	152.90	-0.218	3.492	3.868	4.24	534.19
-2.87	-16.47	0.0804	1.081648	152.91	-0.238	3.400	3.847	4.21	536.91
-10.04	-23.64	0.1608	1.081396	153.02	-0.471	3.404	3.785	4.07	589.27
-13.60	-27.20	0.2089	1.081187	153.12	-0.664	3.319	3.806	4.04	592.42
-20.94	-34.54	0.3215	1.080775	153.30	-1.045	3.329	3.688	3.67	533.47
-25.74	-39.34	0.4019	1.080386	153.47	-1.404	3.591	4.045	3.70	444.80
-30.38	-43.98	0.4823	1.080196	153.56	-1.580	3.574	3.868	3.69	375.09
-35.00	-48.60	0.5627	1.079805	153.74	-1.941	3.254	3.803	3.45	364.51

Detector : 2

Beta Angles	*Psi*	*Sin^2psi*	*DSpacing*	*2Theta*	*Strain*E3*	*FWHM*	*FWHM fit:*	*Breadth*	*Intensity*
35.00	48.60	0.5627	1.079772	153.75	-1.972	2.902	3.539	3.26	479.68
30.38	43.98	0.4823	1.079892	153.70	-1.861	3.201	3.530	3.57	482.62
25.74	39.34	0.4019	1.080180	153.57	-1.595	3.092	3.413	3.56	532.33
20.94	34.54	0.3215	1.080546	153.40	-1.257	2.965	3.471	3.39	628.05
13.60	27.20	0.2089	1.080938	153.23	-0.895	3.107	3.316	3.52	566.53
10.04	23.64	0.1608	1.081236	153.09	-0.619	3.113	3.394	3.40	571.69
2.87	16.47	0.0804	1.081433	153.01	-0.437	2.922	3.301	3.34	515.12
0.00	13.60	0.0553	1.081538	152.96	-0.340	2.785	3.286	3.30	497.93
-2.87	10.73	0.0347	1.081664	152.91	-0.224	2.995	3.415	3.38	538.80
-10.04	3.56	0.0039	1.081919	152.79	0.013	3.059	3.442	3.54	596.20
-13.60	0.00	0.0000	1.081905	152.80	-0.000	3.012	3.333	3.46	600.32
-20.94	-7.34	0.0163	1.081867	152.82	-0.035	3.215	3.631	3.75	529.81
-25.74	-12.14	0.0442	1.081772	152.86	-0.123	3.193	3.700	3.73	502.46
-30.38	-16.78	0.0834	1.081578	152.94	-0.303	3.285	3.733	3.84	509.20
-35.00	-21.40	0.1331	1.081454	153.00	-0.417	3.449	3.911	3.93	549.39

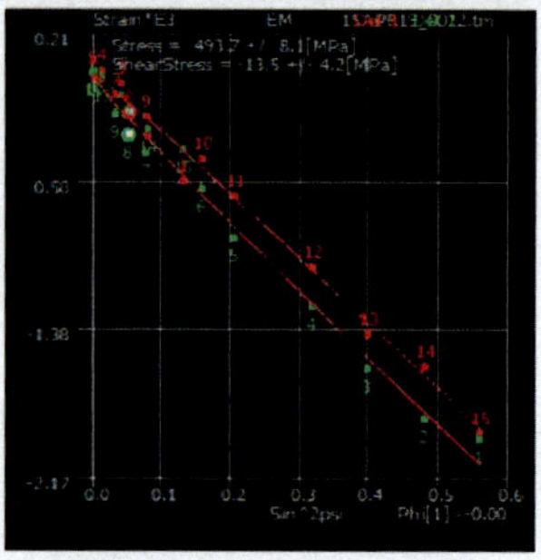

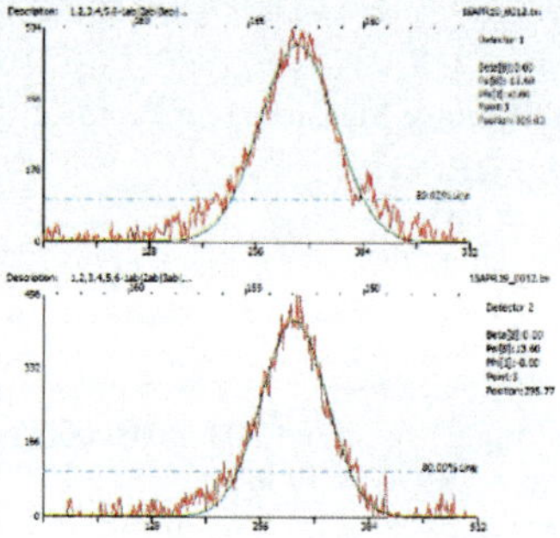

PARAMETERS

Material: ST.STEEL - AUST. (Mn_Ka) (FCC, hkl-311)
Tube: Mn_K-Alpha
Wavelength: 2.103
KV: 13.00; *mA*: 4.00
Beta Osc: 2.00; *Phi Osc*: 0.00; *X Osc*: 0.00; *Y Osc*: 0.00;
Exp. time: 2.00 2.00

Nr of exp: 20 20
Aperture: 1
Filters: 1

Units: MPa
Bragg Angle: 152.80
Peak Location: Gaussian80%
Gain Correction: P/G(s)
PeakShift: Absolute Peak
(1/2)S2: 7.18 E-6 [1/MPa]
-S1: 1.20 E-6 [1/MPa]
DSpacing (Angstroms): 1.0819055
LPA Correction: YES
Psi Zero Assignment (Pixel): 307.91 300.26

Background static fit:

LeftStartPixel: 64 66
RightStartPixel: 480 470
ROI Left: 108 108

LeftStopPixel: 94 96
RightStopPixel: 510 500
ROI Right: 475 461

Comments:

B# 0, F#: 016BFEEF, S#: 170, B# 0, F#: 016BFEEF, S#: 170,

References

1. Srivastava, M., Tripathi, R., Hloch, S., Rajput, A., Khublani, D., Chattopadhyaya, S., Dixit, A.R., Foldyna, J., Adamčík, P., Klich, J., Zelenak, M., Klichová, D.: Surface treatment of AISI 304 using pulsating water jet peening. In: Singh, M.K., Kushvah, B.S., Seth, G.S., Prakash, J. (eds.) Applications of Fluid Dynamics. LNME, pp. 535–548. Springer, Singapore (2018). https://doi.org/10.1007/978-981-10-5329-0_40
2. Srivastava, M., Hloch, S., Muller, M., Hromasová, M., Cais, J., Chattopadhyaya, S., Dixit, A.R., Klich, J.: Effect of frequency change during pulsed waterjet interaction with stainless steel. In: Hloch, S., Klichová, D., Krolczyk, G.M., Chattopadhyaya, S., Ruppenthalová, L. (eds.) Advances in Manufacturing Engineering and Materials. LNME, pp. 85–96. Springer, Cham (2019). https://doi.org/10.1007/978-3-319-99353-9_10
3. Srivastava, M., Hloch, S., Krejci, L., et al.: Residual stress and surface properties of stainless steel welded joints induced by ultrasonic pulsed water jet peening. Measurement **143**, 81–92 (2019). https://doi.org/10.1016/j.measurement.2018.06.012
4. Sitek, L., Foldyna, J., Martinec, P., Scucka, J., Bodnárová, L., Hela, R.: Use of pulsatingwater jet technology for removal of concrete in repair of concrete structures. Balt. J. Road Bridge Eng. **6**(4), 235–242 (2011). https://doi.org/10.3846/bjrbe.2011.30
5. Hloch, S., Adamčík, P., Nag, A., Srivastava, M., Čuha, D., Müller, M., Hromasová, M., Klich, J.: Hydrodynamic ductile erosion of aluminium by a pulsed water jet moving in an inclined trajectory. Wear **428–429**, 178–192 (2019). https://doi.org/10.1016/j.wear.2019.03.015
6. Bilík, J., Baca, J.: Mechanical reinforcement of surface layers by dynamic throwing bullets. MtF STU (2004) http://konference.tanger.cz/data/metal2004/sbornik/papersú23.pdf
7. Ťavodová, M., Náprstková, N.: Evaluation of material surface quality after cutting AWJ. Strojírenská Technol. **17**(3), 186–192 (2012)
8. Skočovský, P., Vaško, A.: Materials and Technologies. EDIS, Žilina (2004)
9. Wang, H., Song, G., Tang. G.: Effect of electropulsing on surface mechanical properties and microstructure of AISI304 stainless steel during ultrasonic surface rolling process. Mater. Sci. Eng. A **662**, 456–467 (2016). https://doi.org/10.1016/j.msea.2016.03.097
10. Bagherifard, S., Slawik, S., Fernández-Pariente, I., Pauly, C., Mucklich, F., Guaglianoa, M.: Nanoscale surface modification of AISI 316L stainless steel by severe shot peening. Mater. Des. **102**, 68–77 (2016). https://doi.org/10.1016/j.matdes.2016.03.162
11. Arola, D., Ramulu, M.: Material removal in abrasive waterjet machin-ing of metals surface integrity and texture. Wear **210**, 50–58 (1997). https://doi.org/10.1016/S0043-1648(97)00087-2
12. Sedlák, J., Tropp, p., Chladil, J., Kouřil, K. Polzer, A. Osička, K.: Analysis of selected aspects of turned bearing rings regarding required workpiece quality. Manuf. Technol. **16**(3), 612–622 (2016)
13. Sadílek, M., Fojtík, F., Sadílková, Z., Kolařík, K., Petrů, J.: A study of effects of changing the position of the tool axis against the machined surface. Trans. FAMENA **39**(2), 33–46 (2015)
14. Tripathi, R., Srivastava, M., Hloch, S., Adamčík, P., Chattopadhyaya, S. Das AK.: Monitoring of acoustic emission during the disintegration of rock. Procedia Eng. **149**, 481–488 (2016). https://doi.org/10.1016/j.proeng.2016.06.695
15. Nag, A., Hloch, S., Čuha, D., Dixit, A.R., Tozan, H., Petrů, J., Hromasová, M., Müller, M.: Acoustic chamber length performance analysis in ultrasonic pulsating water jet erosion of ductile material. J. Manuf. Process. **47**, 347–356 (2019). https://doi.org/10.1016/j.jmapro.2019.10.008
16. Náprstková, N., Cais, J., Stančeková, D., Kraus, P.: Analysis of the composition of selected inserts using electron microscopy. Manuf. Technol. **16**(5), 1063–1067 (2016)

Remarks to Abrasive Waterjet (AWJ) Forces Measurements

Adam Štefek[1], Libor M. Hlaváč[1](✉), Martin Tyč[1], Pavel Barták[2], and Jiří Kozelský[1]

[1] VŠB – Technical University of Ostrava, Ostrava, Czechia
libor.hlavac@vsb.cz
[2] Kámen Ostroměř s.r.o., Ostroměř, Czechia

Abstract. Measurement of forces during cutting process has a potential to be a method applicable for online monitoring of the AWJ cutting process. Force signal can help with understanding the interaction between jet and material, namely with determination of the cutting to wear mode ratio. The force measurements also detect some undesired effects, such as not fully penetrated material and/or other product distortions. Therefore, such measurements could be efficient tool in the online process monitoring and/or control, when appropriate device for industrial applicability is constructed. Some weak points are discussed in the contribution as well.

Keywords: Abrasive waterjet · Measurement · Force · Cutting · Quality

1 Introduction

Since its introduction into practice in the early nineties of the last century, the abrasive waterjet became an efficient tool for cutting plane parallel plates [1] and drilling [2]. However, utilization of this efficient tool for more complex and difficult operations is in progress up to date. It is caused by rapid progress in material sciences. Preparation of new, difficult to machine materials, has triggered the need to find adequate machine tools for these materials. Therefore, the investigation of AWJ tool utilization for other machining processes, not just cutting and drilling, has been intensified in recent years. AWJ is applicable for machining of almost every type of material, independently on its thickness, provided that sufficient energy supply is used. However, using of the AWJ for milling [3], turning [4], 3D machining [5, 6], grinding [7] or polishing [8] increased the necessity of prediction of the process progress, process monitoring and control. The prediction can be performed using some analytical [9], semi-analytical [10] and empirical [11] models, or applying the own experimental research. Sometimes it is possible to use some statistical (regression) models, like [12–14], provided that the scope of machining variables and machined material match the ones used for models' preparation. Otherwise, the results of such models could be substantially different causing so damage of the product, as it was referred in [15]. Better understanding of the erosion process seems to be essential for improvement of the contemporary models and/or creation of new and more complex

D. Klichová et al. (Eds.): Water Jet 2019, LNME, pp. 208–218, 2021.
https://doi.org/10.1007/978-3-030-53491-2_22

ones. Therefore, the monitoring and control of the AWJ machining processes are the important challenges of the contemporary industrial applications.

However, online monitoring of AWJ interaction with material is very hard task, because the jet damages all objects in its path and, sometimes, also in its vicinity (by a residual jet reflected on some parts of holding and fixing system). This is the reason for investigation of the contactless measurement methods (analyses of sound [16, 17], acoustic emissions [18] and ultrasonic emissions [19]), or contact methods with possibility of sensor mounting far from the interaction zone (monitoring of vibrations [20]). Some results aimed at 3D machining [21, 22] or studying of cutting head vibrations regarding the abrasive mass flow rate and impact of AWJ on various materials [23] show that information is inconclusive. Moreover, according to Mikler [24], more than 25% of the information from acoustic emissions or vibrations may be incorrect. Therefore, the device measuring forces in three orthogonal axes x-y-z during AWJ cutting has been designed and constructed and patented [25]. It is applied for studying forces regarding cutting and deformation wear that can help with determination of cutting quality regarding some anomalies as described in previous research results aimed at AWJ precision [26]. Results showing some specific features were noticed and they are presented and discussed in this contribution.

2 Measuring Device

The patented force sensor [25] is used for measurement of the AWJ cutting forces. The design of the sensors is shown in Fig. 1. The black boxes contain amplifiers of the electric signals generated in piezoresistive sensors mounted on the deformation elements, which are in contact with sample holder.

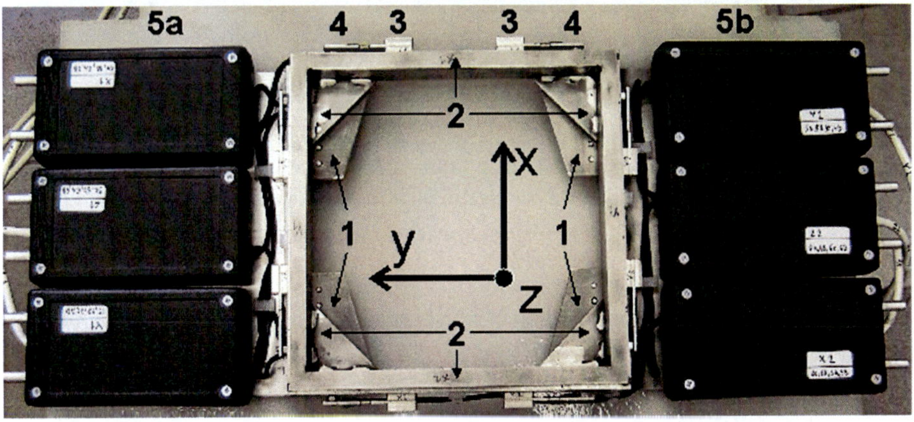

Fig. 1. Force sensor overall view: 1 – main measuring frame, 2 – sample holder frame, 3 – fixing of x^+ axis deformation elements, 4 – x^+ axis deformation elements with strain gauges, 5 – boxes containing amplifiers for strain gauges bridges (5a – x^+, y^+ and one part of the z^- direction; 5b – x^-, y^- and second part of the z^- direction).

The main measuring square frame is prepared from 4 mm thick stainless steel that is toughen in corners by trapezoidal elements used simultaneously as holders for deformation elements in z axis. The sample holder frame, toughen in corners by triangle elements, is also prepared from 4 mm thick stainless steel. This holder is put on four thorns mounted on deformation elements of the z axis. Side walls are in contact with thorns mounted on deformation elements for both x axis and y axis.

Signals from all axes are recorded to computer via the Signal Express program. The voltage signals are processed in LabVIEW and converted to forces in Newtons. Measurements were performed in the firm Kámen Ostroměř s.r.o. (Fig. 2) on the cutting table MACH 2 M403 (203lb) with pump Flow HYPLEX® PRIME M-416. Experimental variables are summarized in Table 1.

Fig. 2. Set-up of the force measuring system during measurement in the firm Kámen Ostroměř (Czech Republic) – sensor is laid down on the grid of the cutting table and fixed by two big stones put on the electronic boxes.

Table 1. Summary of factors and parameters in experiments.

Variable (unit)	Value
Pump pressure (MPa)	350
Water orifice diameter (mm)	0.305
Focusing tube diameter (mm)	0.889

(*continued*)

Table 1. (*continued*)

Variable (unit)	Value
Focusing tube length (mm)	101.6
Abrasive mass flow rate (g/min)	300
Indian abrasive material average grain size (mm)	0.200
China abrasive material average grain size (mm)	0.150
Abrasive material type in experiments 2017	Indian garnet
Abrasive material type in experiments 2019	China garnet
Stand-off distance (mm)	2

The firm Kámen Ostroměř s.r.o. is one of the firms with close cooperation with VŠB – Technical University of Ostrava. Therefore, it was chosen for testing of measuring device under industrial conditions. However, between measurements in 2017 and 2019 the firm changed the type of used abrasive material and so it was impossible to repeat measurements with identical abrasive material.

Granite samples were cut at three traverse speeds (200, 300 and 400 mm/min) oriented towards x^+ axis in 2017 (jet was flowing into z direction). An example of the signal for traverse speed 400 mm/min cutting the medium hard granite Azul Noche is presented in Fig. 3.

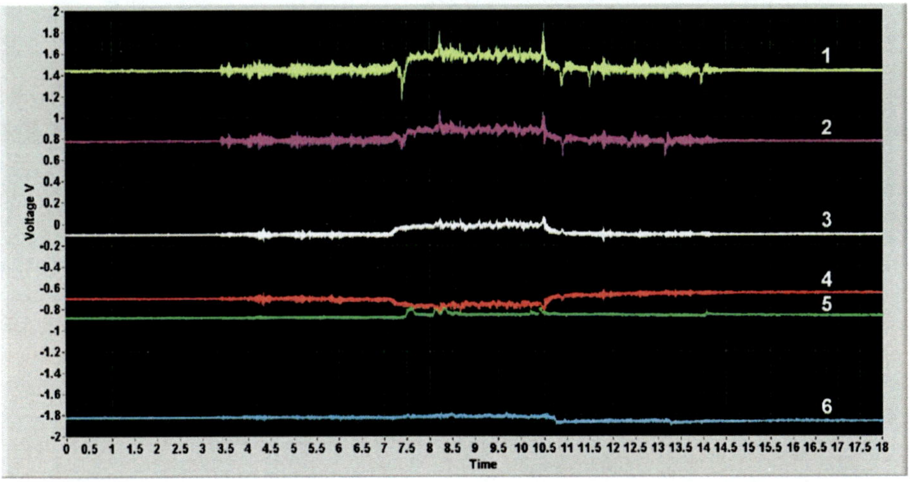

Fig. 3. Measured signal on medium hard granite Azul Noche: 1, 2 – both parts of z^- axis, 3 – x^+ axis, 4 – x^- axis, 5 – y^+ axis, 6 – y^- axis.

The repeated experiments in 2019 needed to be performed using traverse speeds 100, 150 and 200 mm/min to cause the same cutting results on the same rock samples. Direction of traverse speeds and jet flowing were identical as in 2017. The most probable reason of traverse speed decrease is much less efficiency of Chinese garnet regarding the Indian one. It can be the result of the much smaller average grain size.

3 Experimental Results

Typical experimental results – time signals – can be described as the one presented in Fig. 4. Signal recorded during cutting is processed in program prepared in LabVIEW software. This program enables determination of differences of selected parts of the signal. Typical selection of the signal parts and respective determination of the normal force (NF) value is evident from signal description presented in Fig. 5.

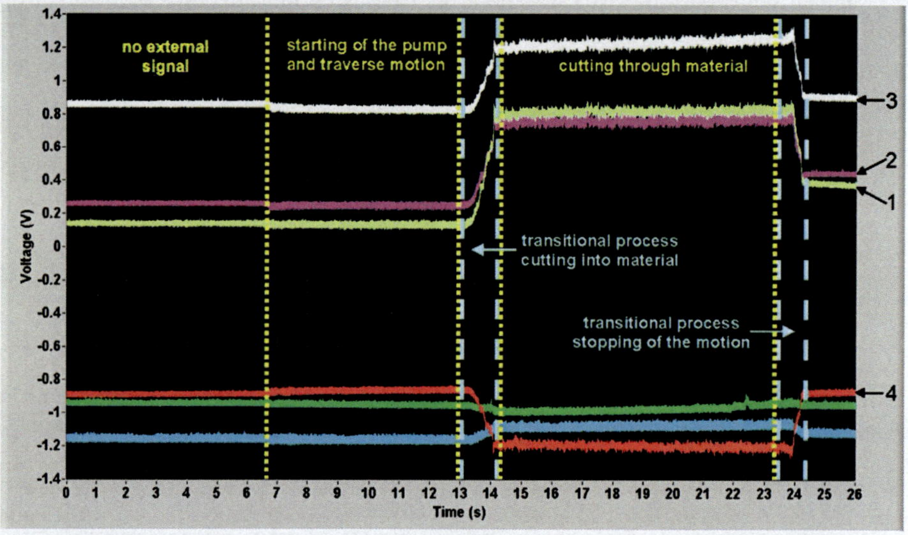

Fig. 4. Description of the typical good time signal recorded from force sensor: 1 – one portion of normal force signal (yellow line), 2 – second portion of normal force signal (magenta line), 3 – tangential force signal positive (white line is in the x^+ direction), 4 – tangential force signal negative (red line is in the x^- direction), green and blue lines are y^+ and y^- directions, i.e. lateral force signals (not investigated in this article – very small forces).

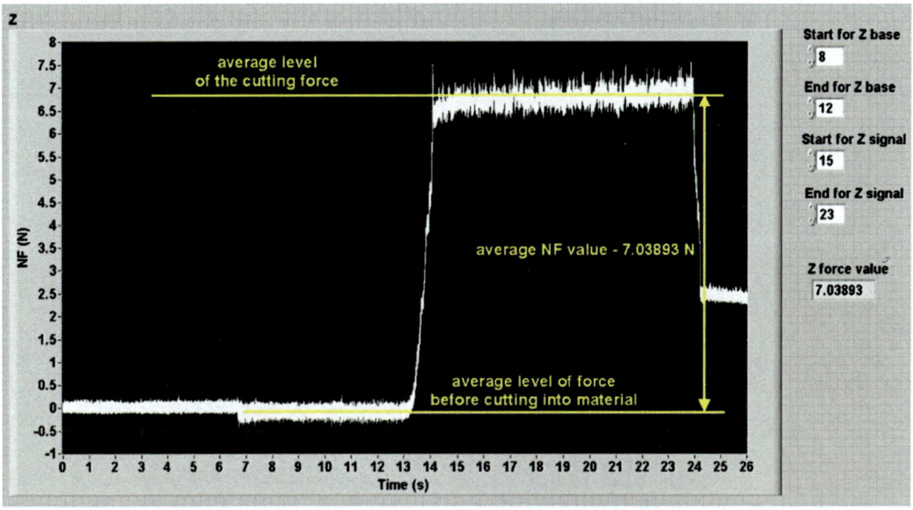

Fig. 5. Determination of the average force value in z-axis; settings on the right side of the graph enable selection of definite part of the signal and respective base level before cutting (base level of the signal is selected from 8 to 12 s, stable cutting signal is selected from 15 to 23 s).

Beside the nice records, like those presented in Fig. 4 and Fig. 5, sometimes some other types of signal occurred. One of those atypical signal recordings is presented in Fig. 6 – sudden shift of the measuring level.

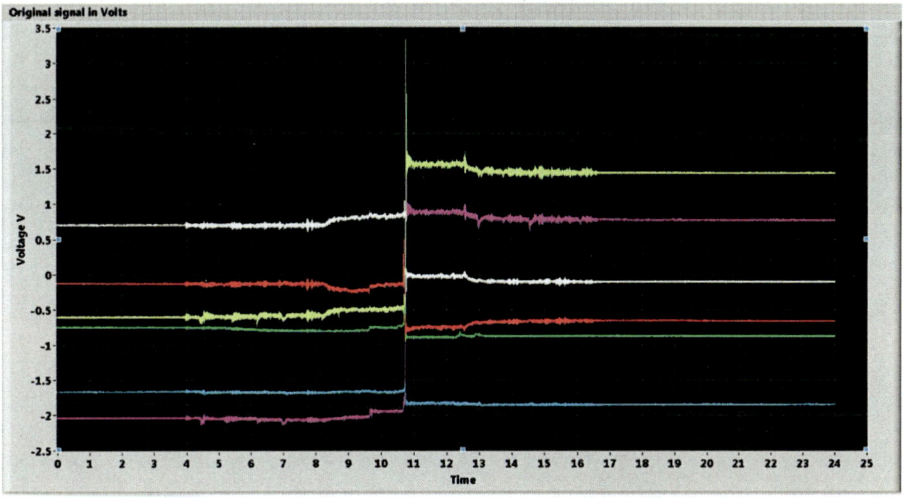

Fig. 6. Unusual signal recorded in firm Kámen Ostroměř on granite sample.

Another type of spoiled signal is presented in Fig. 7. It was recorded between two normal signals on the identical sample and with the same sample mounting. There were performed three parallel cuts in distance of 5 mm. The first and the third signal records were normal, but the middle one is extremely noisy and two drop downs in the baseline occurred (at 4.2 s and 6.8 s) before the cutting started at the 9th second.

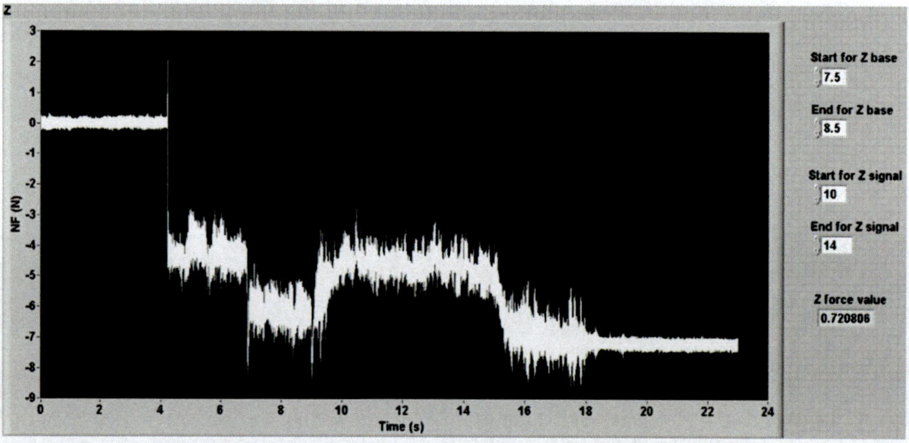

Fig. 7. Noisy signal with strong shift in the base level.

Similar problems were registered during repeated experiments in firm Kámen Ostroměř in 2019. This time all three cuts were recorded into one file. However, as it could be seen in Fig. 8, the unusual shift of the base level occurred again (in some records). The shift occurred several times even in the case when no sample was inserted into the force measuring device (Fig. 9).

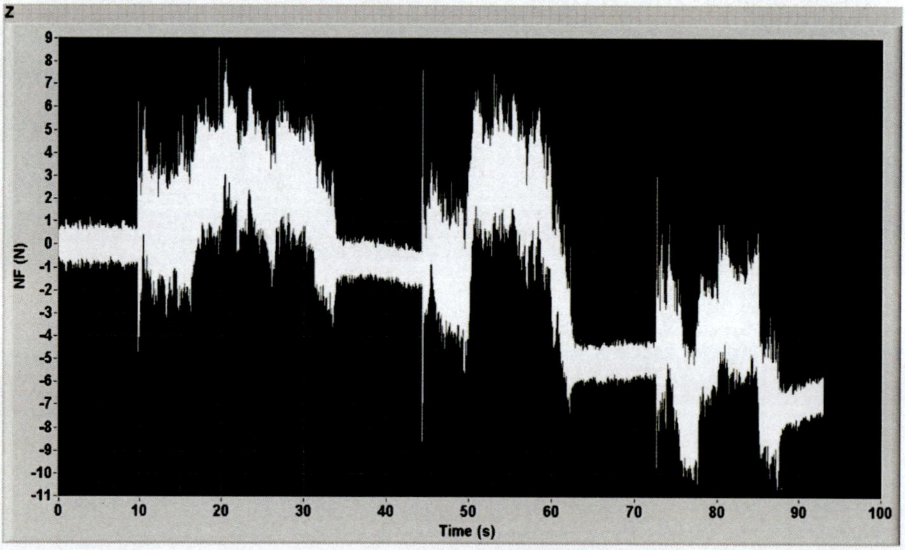

Fig. 8. Noisy signal with shifting base signal level recorded on one of the granite samples.

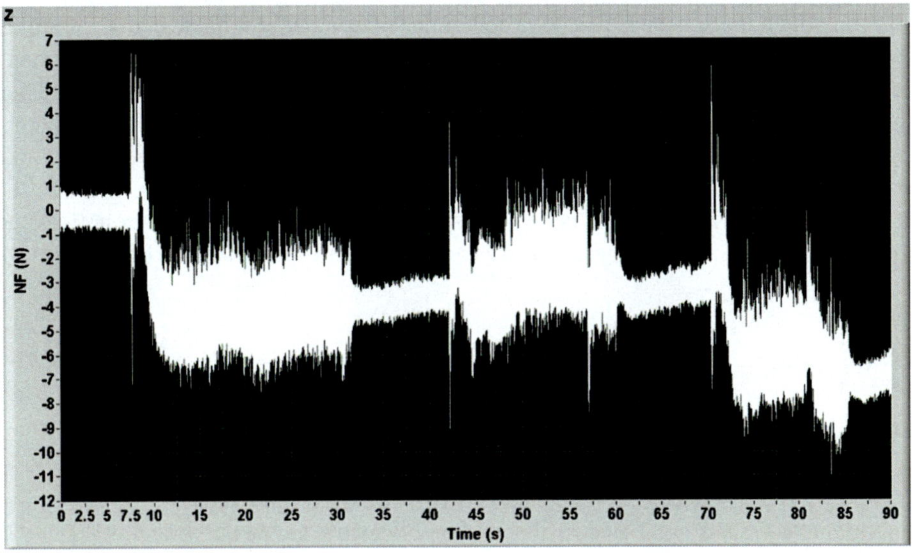

Fig. 9. Noisy signal with shifting base signal level without sample input to the force sensor.

4 Discussion

Because the force sensor is very sensitive on stochastic phenomena, it is very hard to arrange the same conditions for all measurements. However, it was observed that shifts in absolute values (caused by different fixing of measuring device or sample) are of the corresponding size in all axes. Therefore, the cutting to deformation force ratio (ratio of average values in axes x and z) could be analysed.

It can be seen in Fig. 3 (and others) that signal in both parts of y direction is negligible. It is also evident that signals in x and z directions are quite low. It is caused by high cutting ability of AWJ on rocks. The high level of dynamics is caused by these factors: the sample is relatively light and the water level in the collecting vessel dissipating residual AWJ energy is too high – water waves sometimes impact the sensor; AWJ dynamics is transmitted to the water vibration in the collecting vessel being not stiff enough – vibrations are transmitted to the sensor. Impact of the waves can be observed as peaks in z axis at 8.2 s and 10.5 s in Fig. 3. The gap at 7.4 s is caused by partial material breakage during cutting into material – a small piece was broken off.

Some of the signal distortions or shifts cannot be explained by evident causes. However, knowing the fact that cutting centre of the firm is a part of railway station sheds and, moreover, many more processes are in progress during tests (cutting on the diamond saws of great diameter, transportation of heavy blocks of stones, their laying down and picking up by forklifts), some suggestions can be submitted. The intensive impulse evident in Fig. 6 and subsequent shift of signal in all axes could be caused by floor shaking due to fall of the heavy object, but electrical signal disturbance cannot be excluded. Contrary, noise evident in Figs. 7, 8 and 9 is caused by construction of waste collecting vessel – even when the water level is lowered, the waves are strongly influencing sensor placement, namely the grid. The most probable reason for high noise is vibration of the slats on which the sensor is placed. The similar reason could influence also the initial signal shift evident namely in Fig. 7. This shift could be also caused by insufficient precision during placing of the pad with mounted sample into the sensor frame. Then the initial vibration may cause the sample pad settling in the sensor holder. Another explanation is that the zero potential level may shift during measurement. Finally, the electromagnetic interference caused by other machines in the factory hall can be the source of the high signal noise in certain measurements. These factors are not easy to investigate during normal operation, so special capacities will need to be allocated to their research.

5 Conclusions

Final conclusions can be divided into two groups. The first one includes suggestions regarding the measuring device. The most important one is strengthening of the base frame of the force sensor. Use of piezoelectric instead of piezo-resistive chips is also important, as well as good electrical shielding of electric signal amplifiers.

The second group of suggestions is aimed at sensor positioning. To obtain the best results, it is essential that all sources of vibration and electrical disturbances are reduced, in particular, the influence of waves in the residual energy absorbing vessel. Therefore, the sensor should be placed on some holder that is not in a contact with the waste collecting vessel (container).

Acknowledgements. The work was supported by the project SP 2019/26 of the Ministry of Education, Youth and Sports of the Czech Republic.

References

1. Hlaváč, L.M., Hlaváčová, I.M., Plančár, Š., Krenický, T., Geryk, V.: Deformation of products cut on AWJ x-y tables and its suppression. In: IOP Conference Series: Materials Science and Engineering, **307**(1), Article number 0120152017, IOP Publishing Ltd, Bristol (2018)
2. Phapale, K., Singh, R., Patil, S., Singh, R.K.P.: Delamination characterization and comparative assessment of delamination control techniques in abrasive water jet drilling of CFRP. In: 44th North American Manufacturing Research Conference, pp. 521–535, Procedia Manufacturing, Blacksburg VA (2016)
3. Hejjaji, A., Zitoune, R., Toubal, L., Crouzeix, L., Collombet, F.: Influence of controlled depth abrasive water jet milling on the fatigue behavior of carbon/epoxy composites. Composites Part A-Appl. Sci. Manuf. **121**, 397–410 (2019)
4. Hou, R.G., Huang, C.Z., Zhu, H.T., Wang, J.: Study on experiment device of abrasive water jet micro-turning. 10th Asia-Pacific Conference on Materials Processing. Advanced Materials Research, vol. 500, pp. 339–344. Trans Tech Publications Ltd., Switzerland (2012)
5. Kovacevic, R., Yong, Z.: Modelling of 3D abrasive waterjet machining: Part 1—theoretical basis. In: Gee, C. (ed.) Jetting Technology, vol. 21, pp. 73–82. Mech. Eng. Publication Ltd., Bury St Edmunds (1996)
6. Yong, Z., Kovacevic, R.: Modelling of 3D abrasive waterjet machining: Part 2—simulation of machining. In: Gee, C. (ed.) Jetting Technology, vol. 21, pp. 83–89. Mech. Eng. Publication Ltd., Bury St Edmunds (1996)

7. Hlaváč, L.M., Hlaváčová, L.M., Vašek, J., Jandačka, P., Mádr, V.: Material grinding by water jets—Feasibility and limits. J. Sci. Conf. Proc. **2**(1), 8–14 (2010)
8. Chen, Fengjun, Miao, Xiangliang, Tang, Yu., Yin, Shaohui: A review on recent advances in machining methods based on abrasive jet polishing (AJP). Int. J. Adv. Manuf. Technol. **90**(1–4), 785–799 (2016). https://doi.org/10.1007/s00170-016-9405-7
9. Hlaváč, L.M.: JETCUT - software for prediction of high-energy waterjet efficiency. In: Louis, H. (ed.) Jetting Technology, vol. 32, pp. 25–37, Prof. Eng. Pub. Ltd., Bury St Edmunds (1998)
10. Pahuja, R., Ramulu, M.: Abrasive water jet machining of titanium (Ti6Al4V)-CFRP stacks - a semi-analytical modeling approach in the prediction of kerf geometry. J. Manuf. Processes **39**, 327–337 (2019)
11. Zhang, S.J., Wu, Y.Q., Wang, Y.L.: An empirical surface roughness model of titanium alloys cut with abrasive water jet. In: Jiang, Z.Y., Li, S.Q., Zeng, J.M., Liao, X.P., Yang, D.G. (eds.) Advanced Materials Research, vol. 189–193, pp. 4231–4244, Trans. Tech. Publications Ltd., Switzerland (2011)
12. Chaturvedi, V., Singh, D.: Multi response optimization of process parameters of abrasive water jet machining for stainless steel AISI 304 using VIKOR approach coupled with signal to noise ratio methodology. J. Adv. Manuf. Syst. **14**(2), 107–121 (2015)
13. Singh, D., Chaturvedi, V.: Investigation of optimal processing condition for abrasive water jet machining for stainless steel AISI 304 using grey relational analysis coupled with S/N ratio. In: Balasubramanian, K.R., Sivapirakasam, S.P., Anand, R. (eds.) 16th International Mechanical Engineering Congress (IMEC), Applied Mechanics and Materials, vol. 592–594, pp. 438–443, Trans. Tech. Publications Ltd., Switzerland (2014)
14. Kumaran, S.T., Ko, T.J., Uthayakumar, M., Islam, M.M.: Prediction of surface roughness in abrasive water jet machining of CFRP composites using regression analysis. J. Alloys Compounds **724**, 1037–1045 (2017)
15. Hlaváč, L.M., Krajcarz, D., Hlaváčová, I.M., Spadło, S.: Precision comparison of analytical and statistical-regression models for AWJ cutting. Precision Eng. **50**, 148–159 (2017)
16. Merchant, H.C., Chalupnik, J.D.: Case-history - abrasive water jet sound power measurement. Noise Control Eng. J. **29**(3), 85–89 (1987)
17. Jurisevic, B., Brissaud, D., Junkar, M.: Monitoring of abrasive water jet (AWJ) cutting using sound detection. Int. J. Adv. Manuf. Technol. **24**(9–10), 733–737 (2004). https://doi.org/10.1007/s00170-003-1752-5
18. Hreha, P., Hloch, S., Perzel, V.: Analysis of acoustic emission recorded during monitoring of abrasive waterjet cutting of stainless steel AISI 309. Tehnicki Vjesnik-Technical Gazette **19**(2), 355–359 (2012)
19. Zhao, W., Guo, C., Wang, F.-C., Zhou, D.-X.: Online monitoring of abrasive amount in high-pressure vessel of ASJ cutting system using ultrasonic detection. Int. J. Adv. Manuf. Technol. **85**(5–8), 1107–1114 (2015). https://doi.org/10.1007/s00170-015-7454-y
20. Miron, A., Arghir, M., Balc, N., Popan, A., Miron-Borzan, C.: Determination of cutting head vibrations during abrasive water jet cutting process. Acta Technica Napocensis Ser. Appl. Math. Mech. Eng. **58**(3), 431–434 (2015)
21. Rabani, A., Madariaga, J., Bouvier, C., Axinte, D.: An approach for using iterative learning for controlling the jet penetration depth on abrasive waterjet milling. J. Manuf. Processes **22**, 99–107 (2006)
22. Rabani, A., Marinescu, I., Axinte, D.: Acoustic emission energy transfer rate: a method for monitoring abrasive water jet milling. Int. J. Mach. Tools Manuf. **61**, 80–89 (2012)
23. Salokyová, Š.: Measurement and analysis of technological head vibrations in hydroabrasive cutting technology. Acad. J. Manuf. Eng. **12**, 90–95 (2014)
24. Mikler, J.: On use of acoustic emission in monitoring of under and over abrasion during a water jet milling process. J. Mach. Eng. **142**, 104–115 (2014)

25. Mádr, V., Lupták, M., Hlaváč, L.: Force sensor and method of force sensing in the process of abrasive water jet cutting. Protective document CZ 303189. Bureau of the industrial property, Prague, Czech Republic. Patent owner: VSB – Technical University of Ostrava, Ostrava-Poruba, CZ“, registration of the application: 15.2.2011, number of patent application: 2011-82, granting of patent: 5.4.2012, patent publishing: 16.5.2012, patent publication: 16.5.2012 (2012)
26. Hlaváč, L.M., Hlaváčová, I.M., Arleo, F., Viganò, F., Annoni, M.P.G., Geryk, V.: Shape distortion reduction method for abrasive water jet (AWJ) cutting. Precision Eng. **53**, 194–202 (2018)

Evolution of Microstructure of Silicon Steel After Pulsating Water Jet Treatment

Eva Švábenská[1](✉), Alice Chlupová[1], Josef Foldyna[2], and Oldřich Schneeweiss[1,3]

[1] Institute of Physics of Materials of the Czech Academy of Sciences, Brno, Czechia
svabenska@ipm.cz
[2] Institute of Geonics of the Czech Academy of Sciences, Ostrava, Czechia
[3] CEITEC IPM, Brno, Czechia

Abstract. The present study is focused on the evolution of microstructure of silicon steel (Fe-Si 6 wt%) after treatment by pulsating water jet (PWJ). The various stages of erosion by pulsating water jet have been studied. The initial damage shows formation of depressions and isolates erosion pits. Repeated passing of water initialized further pit formation and coalescence of pits to craters. During this stages small cracks formation was observed. The repeated passing of water induced deepening of the craters and material removal. The PWJ caused plastic deformation and increase of values of hardness of the subsurface region. The cleavage fracture was present on contained eroded surface.

Keywords: Pulsating water jet · Silicon steel · Microstructure

1 Introduction

The technology of water jet is exploitable for cleaning, cutting, milling, surface layer removal, etc. [1]. Two basic methods - water jet (WJ) and abrasive water jet (AWJ) are commonly used.

Impact of AWJ method on the material is higher because this method profit from the improvement of efficiency due to presence of abrasive particles in water jet. The high-speed liquid stream together with input mass of abrasives represent a tremendous kinetic energy that is converted impact energy causing erosion of materials. The interaction of abrasive particles with targets material can be under way by different form. It possible observed micro-cutting, lateral cracking, ploughing or cutting deformation, erosion, etc. Obviously, these mechanisms usually do not act separately, but in combination. Their impact in erosion process depends on several factors, such as properties of abrasives (e.g. type, diameter) and target materials, working condition (i.e., pump pressure, flow rate, traverse speed, or impact angle of particles).

Another methods for machining is method pulsating water jet (PWJ). The PWJ use only pure water; the continuous stream of water is modulated by ultrasonic waves. The PWJ method uses piezoelectric transducer for generation of ultrasonic vibration. This

D. Klichová et al. (Eds.): Water Jet 2019, LNME, pp. 219–228, 2021.
https://doi.org/10.1007/978-3-030-53491-2_23

ultrasonic vibration converts water stream from continuous jet into discontinuous pulsed jet [2–4]. The impact of high-velocity liquid stream to solid material can cause various damage of material surface but also could improve some material properties. Water jets can be used as alternative to common peening method (shot peening, glass beads peening) for strengthening of the metal surface [5, 6].

The liquid impact on a solid surface occurs in two stages. In the first stage, causes the effect of the so-called water hammer effect where the major damage is caused by the impact of fluid. The fluid behaves in a compressible manner and generates water hammer pressure. The second stage commences when this pressure releases. It is characteristic by liquid moving away from the point of impact. In this process the liquid flow is tangential and velocity of liquid flow may be five times higher than during the impact [2, 3].

In the literature it is possible to find various model of mechanism of damage during liquid flow impact on the surface of sample. A little differently, however, all models describe the erosion mechanisms that follow the impact of water on a given material. The Kong [7] described two levels of damage mechanism (initial and evolved) after fractographic investigation on the sample of gamma titanium aluminide treated by plain waterjet milling. The schematic illustration of material removal mechanism during Liquid Impingement Erosion test mentioned Ma [8] for titanium alloy. Also for titanium alloy (rolled Ti–6Al–4V) has been suggested mechanism of droplet erosion by Kamkar [9]. The microscopic irregularities or asperities take into consideration the model described by Heymann [10] for aluminum alloy.

The iron-silicon steel has been extensively used in the electrical industry as soft-magnetic material in several applications where the high permeability and low losses are the most important characteristics [11]. The magnetic properties improve with increases silicon content in the steel, nevertheless increasing content of silicon decreasing strongly the ductility of the silicon steel. Steel with concentration of silicon about 7 at.% (~3.5 wt%) becomes extremely brittle to be rolled without cracking during conventional rolling method for production of a thin sheet. The alternative ways for preparing steel with high silicon content has been studied [12–14].

The various behavior of silicon steel has been studied [15]. The study dealing with effect of shock wave on microstructure induced the other question about mechanism of erosion of silicon steel. It was the main reason for this paper, where the main aim was to observe and evaluate the evolution of microstructure of silicon steel after treatment by PWJ process.

2 Experimental Procedures

2.1 Materials

Sample for PWJ experiments was cut using spark erosion from sheet of hot-rolled Fe-6 wt% Si alloy. This silicon steel is composed from 93.97 wt% Fe 5.97 wt% Si (~11 at%). The small amount of impurities (Al, Mn, S) was detected by Glow discharge optical emission spectrometry (GD-Profiler 2 instrument with the Quantum™ XP software; Horiba Jobin Yvon, France) and confirmed by EDX analysis.

The surface of the sample before PWJ experiment were ground, polished and shortly washed by 5% Nital etchant. This final chemical etching was used to remove a thin surface layer damaged by mechanical polishing.

2.2 PWJ Experiments

The surface of sample was treated by pulsating water jet at Institute of Geonics of CAS in Ostrava, using parameters mentioned in Table 1.

Table 1. Parameters of sample preparation - pulsating water jet (abbreviation used in table: *d* - nozzle diameter [mm], α - spraying angle [°], *f* - frequency [kHz], *z* - stand of distance [mm], *P* - water pressure [MPa], *v* - traverse speed [mm/s], *n* - number of passes).

Water jet typed	*d* [mm]	α [°]	*f* [kHz]	*P* [MPa]	*z* [mm]	*v* [mm/s]	*n*
Hammelmann flat	2	10°	20.4	40	55	2	10x

2.3 Scanning Electron Microscopy Observation

The samples were investigated by scanning electron microscope Tescan LYRA 3 XMU FEG/SEMxFIB equipped with X-Max80 EDS detector for X-ray microanalysis and EBSD detector Nordlys by Oxford Instruments with the Aztec control system.

3 Results and Discussions

3.1 Surface

The evolution of water erosion on the surface of silicon steel has been observed. The initial damage of surface by one drop of water resulted in formation of depressions and isolated erosion pits of different size (Fig. 1A). The creation of erosion pits has not shown any preferences for the placement of pits. The similar amount of pits was observed both at grain boundaries and inside the grains.

The repeated passing of water jet initialized further pits formation and pits coalescence (Fig. 1B) to local craters. The materials on the edge of pits was little bit elevated under the original surface of specimen (further described as asperity) and this effect was possible to observed partially also at the some grain boundaries as is documented in Fig. 1B (mark with white arrow). The asperity on the edge of pits and grain boundaries was present with increasing quantity in dependence on number of passes of water jet and the location of asperity was influenced by direction of water jet stream, i.e. by the direction of stress wave propagation.

Figure 1C shows the later stage of repetitive falls of water droplets on surface. The surface damage evolution continues as subsequent combination of various erosion stages:

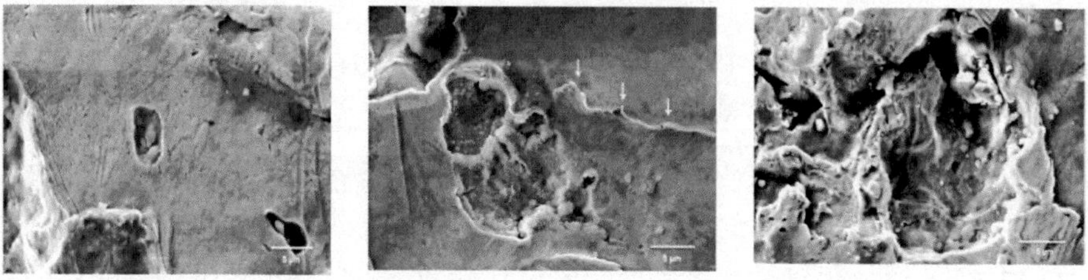

A B C

Fig. 1. The evolution of damage surface of silicon steel caused by increasing number of impacts (scale bar = 5 μm): A - initial damages by one impact, B - repeated impacts, C - pits coalescence.

merging of craters and pits to form larger craters, cracking and fracture. The process of material detachment to fragments starts. Potential formation of lateral sub-tunnels with rough surface occurs. In this stage, due to accumulated impact of water jet the speed of material removal increases. The signs of cleavage fracture on the eroded surface of grain have been reported.

The initiation of small micro-cracks at the bottom of craters has been recorded (Fig. 2). The combination of cracks propagation and pits coalescence at the next impact of water droplets leads to beginning of material removal. The repeated passing of water induced new pits as well as deepened the existing craters.

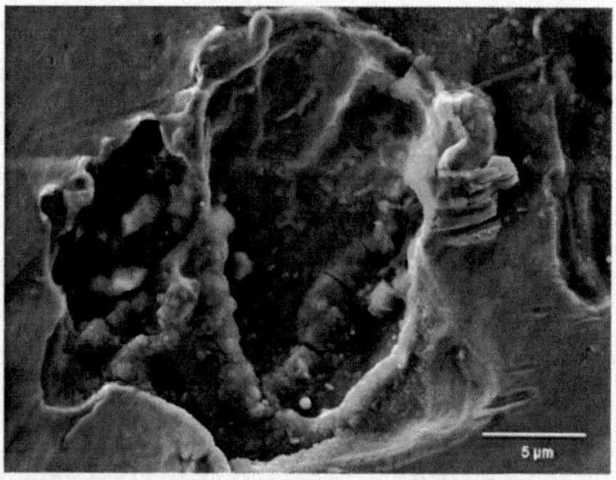

Fig. 2. Example of asperity on the edge of craters and small micro-cracks at the bottom of the craters.

The comparison of Fig. 3 shows structure of surface after impact of several water drop (Fig. 3A) and after ten times passing of water jet (Fig. 3B). On the left side of Fig. 3A could be observed erosion craters but also grain boundary. On the right side could be observed evolution of cleavage facets and propagation of material removing.

Only cleavage facets accompany in some cases with deeper cavity have been observed on Fig. 3B. The comparison of these images shows propagation or enlargement of the initial damages by higher intensity or frequency of impacts. The value of water-hammer pressures and the number of passes through of sample surface corresponds with the damage of material.

Basically, the erosion of materials is result of repeated impact of water drops. The mechanism includes several phases. Plastic deformation is the initial damage due to water hammer and stress wave. Other damage especially at surface with some irregularities is caused by lateral water jets traveling across surface at velocities up to five times speed higher than of water droplet impact [3, 10]. Later the crack formation starts. The subsequent water droplet impact opened/merged the cracks to cavities or come into existence of sub-tunnels. The other crack formation continued together with detachments the fragments of materials. The following damage is combination of each of erosion stages. As a results of initial damage in some materials can be observed triple split, intergranular cracking, translamellar of interlamellar fracture. In general the damage depends on the value of water-hammer pressure, number of passes and also type of water jet [7].

A B

Fig. 3. Comparison of the initial stage after impact of several water drops and surface after ten passes with high roughness and presence of cleavage fracture.

3.2 Cross-Section

In the right side on Fig. 4A (mark with white arrow) there are two narrow trenches separated by each other by peace of material. Structure of these narrow trenches corresponds with sub-tunnels documented on cross-section view on Fig. 4B. The size of the sub-tunnels in width was between 2.4–4.5 μm with the angle between 35 to 65° is from sub-tunnels and original surface. The creation of sub-tunnels could be affected by several factors as diameter of water droplets, spraying angle (10°) of water jet used during water jet treatment and preposition of materials to formation of cleavage facets and fracture. The sub-tunnels were observed more often than cracks (Fig. 5A).

In the next stage of material removal the propagation and linking of cracks at the end of sub-tunnels probably starts. This process continues as new cracks nucleation, their propagation and coalescence of cracks resulting in chipping-off material of sample. The similar formation of erosion damage through sub-tunnels is mentioned by Kamkar for rolled Ti–6Al–4V alloy [9]. However, the cracks observed on the cross-section view for Ti–6Al–4V alloy have preferential inclination around 20° relative to the rolling direction [9]. Any preferential inclinations of cracks have not been observed in sample Fe-Si (this study). Formation of tunnels more likely corresponds with spraying angle of PWJ and formations of cleavage fracture facets. Also formation of sub-surface cracks (Fig. 5) agrees with erosion character due to cleavage fracture.

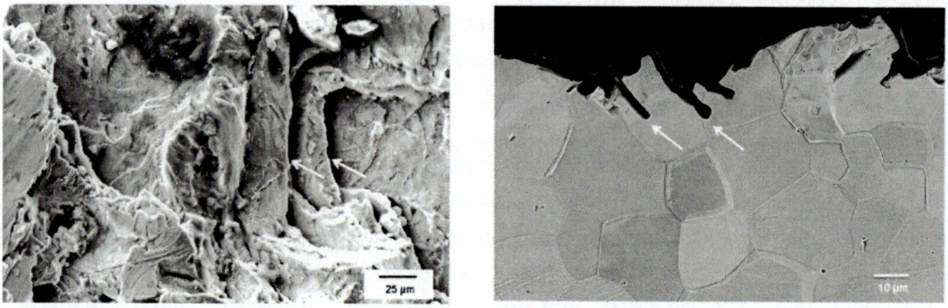

A B

Fig. 4. Evolution of sub-tunnels: A) two narrow pass separated by each other by material of sample, B) formation of sub-tunnels on cross-section view.

Figure 5 shows that propagation of crack is not limited only to grain boundary. The cross-section view on the sample showed formation of surface and sub-surface cracks. Transgranular crack propagation mode was observed as illustrated in Fig. 5B. Also several cases of crack propagation through the merging of isolated nucleated cracks were noted. The cracks merge together and it leads to detachment of a large fragment of material. The combination of surfaces and sub-surfaces cracks together with sub-tunnels leads to propagation of damage in a larger scale. The similar effects as surface and sub-surface cracking, secondary pits and asperity formation were observed by Gujba [16] for titanium alloy. Titanium alloy Ti-6Al-4V showed development of fatigue striation at advanced stages [16]. For different type of titanium alloy (Ti64) ductile slip bands were noted and plastic deformation under continuous water droplet impacts [8]. The slip bands were observed in primary α-grain on the surface Ti-6Al-4V during both incubation stage as well as onset of material removal [17]. Also for titanium alloy, mainly due to hydraulic penetration over pre-existing cracks and pits, ductile fracture mechanism was observed [18]. Although, the evolution of damage the surface of silicon steel was similar to damage observed for titanium alloy the final erosion character is different. Titanium alloys noted ductile fracture mechanism but silicon steel has erosion character by cleavage fracture.

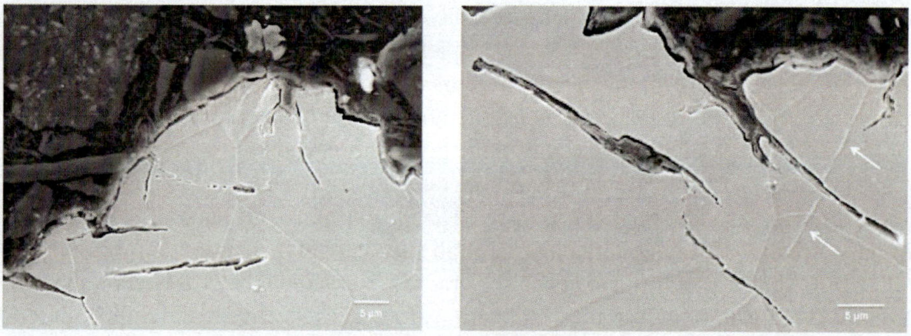

A B

Fig. 5. A) Details surface and subsurface cracks B) prolapse on materials.

In some cases a special feature (marked by arrows on Fig. 5B) were documented on cross-section view of materials treated by Water jet. The prolapses of materials were noted in grains on the edge of sample after metallographic preparation (grinding, polishing and mild chemical etching) before microscopy observation. The focused ion beam (FIB) technique was used to solve a problem with identification of this formation. The Fig. 6A shows prolapse on the sample before covering the prolapse by protective layer of platinum. Darker part of the right side of FIB ditch (marked by dashed line) is metallographic resin. The arrow on the Fig. 6B indicates the place of prolapse on the surface. With high probability the prolapse could be a very thin slip band emerged after chemical etching. The trace of a slip band can be spotted as a thin line of slightly different contrast located in small distance above the white dashed line. This formation can be created as a result of release of stress induced by pulsating water jet, because in the material before treatment have not been observed yet.

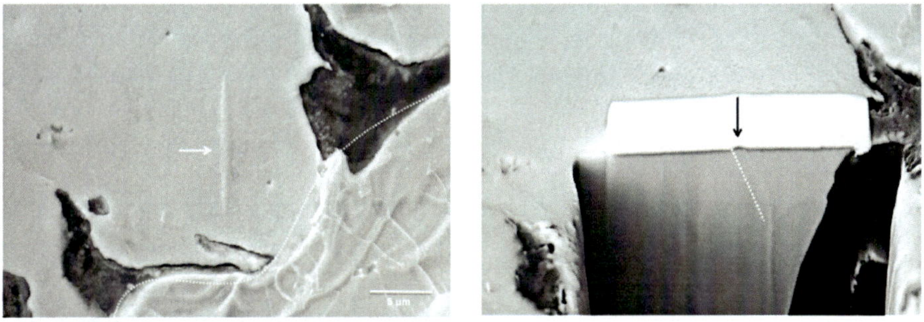

A B

Fig. 6. A) The prolapse of material in grain close to the surface. B) FIB cut going inside to material shows details of subsurface microstructure related to surface prolapse (BSE detector).

The cross-section of the sample was tested by method Electron backscatter diffraction (EBSD). The special interest was focused to the place close the surface, where the prolapses were observed. The EBSD analysis confirmed twins with stressed plane {112} with <−1−11> direction and misorientation angle ~60° (Fig. 7). The twins with comparable parameters were mentioned in the work of Christian [19]. The parameters

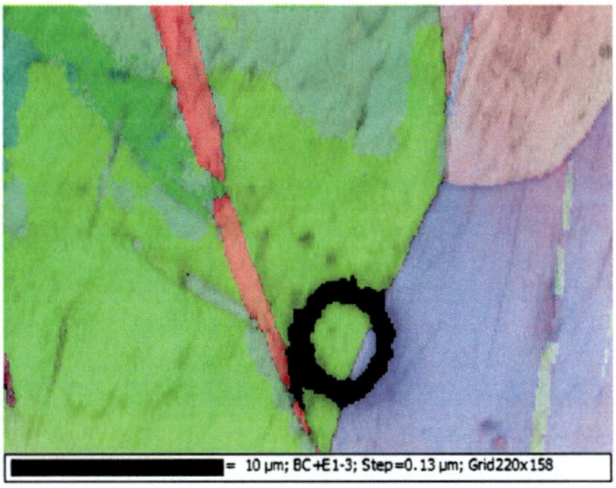

Fig. 7. The twins imaged by electron backscatter diffraction.

confirmed to the presence of deformation twins, which is not so much common in bcc metals.

The comparison of the size and shape of grains in materials before and after PWJ treatment at the cross-section view does not shown any visible changes. Only changes present in samples treated by PWJ are results of impacts the water jet (crack, sub-tunnels, prolapses) mentioned before.

3.3 Microhardness

The subsurface hardness was measured using Vickers method on the cross-sections of specimens by LECO LM 247 AT. The standard ISO 6507-1 has been followed using a load of 0.098 N for 10 s. The average from three microhardness values for the different depths starting from 10 μm beneath the jet impinged surface until a far distance of 1200 μm was plotted to the graph (Fig. 8A). The microhardness results of the sample measured machined under the above mentioned conditions show that the maximum value of microhardness is in the zone of deformation strengthening. The depth of treated layer, where microhardness improvement is detectable, is approximately 400 μm.

Hardness of rolled silicon steel is approximately 300 (HV0.2). This value correspond to the rolled Ti–6Al–4V alloy with hardness of ~330 HV 0.2) [20] or to pure Titanium with hardness of ~290 (HV 0.2) [20]. As was mentioned before, these materials also show similar erosion damage induced by water droplet impingement as studied silicon steel.

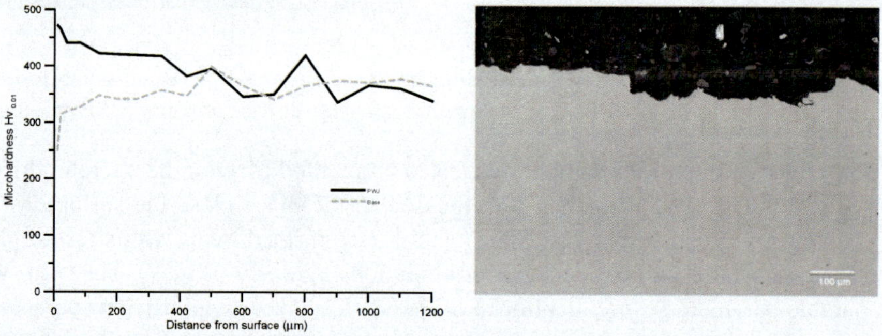

A B

Fig. 8. A) Comparison of microhardness trend for sample after water jet treatment (black line) and original material (grey dashed line). B) SEM image of sample after water jet with Vickers microhardness indents.

4 Conclusions

The method of pulsating water jet (PWJ) can be used as one of the methods for mechanical surface treatment. The application of a pulsing water jet method to silicon steel led to changes in microstructure. On the surface of the sample were observed various stages of erosion. The small pits with asperity on the edge of craters and small micro-cracks were characteristic for the initial stages of damage silicon of steel. The increasing number of water droplets caused fusion of pits to larger craters. After many passes of water jet the final structure exhibits the strong presence of features typical for cleavage fracture.

Observation on cross-section view revealed the presence of surface and sub-surface cracks and sub-tunnels. In the grains close to the surface in presence of prolapses was recorded. These prolapses were identified as deformation twins. Also increase of values of hardness was noticed in the close subsurface region.

Acknowledgement. The authors are grateful for financial support of the project CEITEC - Central European Institute of Technology with research infrastructure supported by the project CZ.1.05/1.1.00/02.0068 financed from the European Regional Development Fund.

References

1. Boud, F., Loo, L.F., Kinnell, P.K.: The impact of plain waterjet machining on the surface integrity of aluminium 7475. In: Axinte, D., Clare, A., Kinnell, P. (eds.) 2nd CIRP Conference on Surface Integrity (CSI), Procedia CIRP, vol. 13, pp. 382–386. Elsevier (2014)
2. Hloch, S., Srivastava, M., Krolczyk, J.B., Chattopadhyaya, S., Lehocká, D., Simkulet, V., Krolczyk, G.M.: Strengthening effect after disintegration of stainless steel using pulsating water jet. Tehnički Vjesnik **25**(4), 1075–1079 (2018)
3. Foldyna, J., Klich, J., Hlaváček, P., Zeleňák, M., Ščučka, J.: Erosion of metals by pulsating water jet. Tehnički Vjesnik **19**(2), 381–386 (2012)
4. Foldyna, J., Sitek, L., Ščučka, J., Martinec, P., Valiček, J., Páleníkova, K.: Effects of pulsating water jet impact on aluminium surface. J. Mater. Process. Technol. **209**(20), 6174–6180 (2009)
5. Nastic, A., Yandouzi, M., Tieu, A., Vijay, M., Jodoin, B.: Forced pulsed water peening potential of VAR 300 M steel for aerospace applications in comparison to conventional glass bead peening. In: Conference Proceeding 24th International Conference on WATER JETTING 2018, BHR Group, Granfield, UK (2018)
6. Lehocka, D., Klich, J., Botko, F., Simkulet, V., Foldyna, J., Krejci, L., Storkan, Z., Kepic, J., Hatala, M.: Comparison of ultrasonically enhanced pulsating water jet erosion efficiency on mechanical surface treatment on the surface of aluminum alloy and stainless steel. Int. J. Adv. Manuf. Technol. **103**(5–8), 1647–1656 (2019)
7. Kong, M.C., Axinte, D., Voice, W.: Aspects of material removal mechanism in plain waterjet milling on gamma titanium aluminide. J. Mater. Process. Technol. **210**(3), 573–584 (2010)
8. Ma, D., Mostafa, A., Kevorkov, D., Jedrzejowski, P., Pugh, M., Medraj, M.: Water impingement erosion of deep-rolled Ti64. Metals **5**(3), 1462–1486 (2015)
9. Kamkar, N., Bridier, F., Bocher, P., Jedrzejowski, P.: Water droplet erosion mechanisms in rolled Ti-6Al-4 V. Wear **301**(1–2), 442–448 (2013)
10. Heymann, F.J.: Erosion by liquids. Mach. Des. **10**, 118–124 (1970)
11. Ruiz, D., Ros-Yáñez, T., Vandenbossche, L., Dupré, L., Vandenberghe, R.E., Houbaert, Y.: Influence of atomic order on magnetic properties of Fe-Si alloys. J. Magn. Magn. Mater. **290–291**(2), 1423–1426 (2005)
12. Ros, T., Houbaetr, Y., Fischer, O., Schneider, J.: Thermomechanical processing of high Si-steel (up to 6.3% Si). IEEE Trans. Magn. **37**(4), 2321–2324 (2001)
13. Ros-Yañez, T., Ruiz, D., Barrros, J., Houbaert, Y.: Advances in production of high-silicon electrical steel by thermomechanical processing and by immersion and diffusion annealing. J. Alloys Compd. **369**(1–2), 125–130 (2004)

14. Stanciu, C.D., Popa, F., Chicinas, I., Isnard, O.: Synthesis of the Fe-6.5% wt. Si alloy by mechanical alloying. Adv. Eng. Forum **13**, 109–113 (2015)
15. Švábenská, E., Roupcová, P., Vondráček, M., Lashin, A.R., Schneeweis, O.: High temperature corrosion of Fe-6wt% Si steel in various atmospheres. Mater. Corros. **70**(4), 593–604 (2019)
16. Gujba, A.K., Hackel, L., Kevorkov, D., Medraj, M.: Water droplet erosion behaviour of Ti-6Al-4 V and mechanisms of material damage at the early and advanced stages. Wear **358–359**, 109–122 (2016)
17. Kamkar Zahmatkesh, N.: Water droplet erosion mechanisms of Ti-6Al-4 V. Thesis (D.Eng.) - Ecole de Technologie Superieure, Canada (2014)
18. Huang, L., Folkes, J., Kinnell, P., Shipway, P.H.: Mechanisms of damage initiation in a titanium alloy subjected to water droplet impact during ultra-high pressure plain waterjet erosion. J. Mater. Process. Technol. **212**(9), 1906–1915 (2012)
19. Christian, J.W., Mahajant, S.: Deformation twinning. Prog. Mater Sci. **39**(1–2), 1–157 (1995)
20. Poondla, N., Srivatsan, T.S., Patnaik, A., Petraroli, M.: A study of the microstructure and hardness of two titanium alloys: commercially pure and Ti-6Al-4V. J. Alloys Compd. **486**(1–2), 162–167 (2009)

Evaluation of Cut Quality on AlMg4.5Mn Material Using AWJ Containing Recycled Abrasives

Miroslava Ťavodová[1], Nataša Náprstková[2(✉)], Pavel Kraus[2], and Ingrid Görögová[3]

[1] Technical University in Zvolen, TF, Zvolen, Slovakia
[2] Jan Evangelista Purkyně University in Ústí nad Labem, FME, Ústí nad Labem, Czechia
natasa.naprstkova@ujep.cz
[3] Slovak University of Technology in Bratislava, FMST in Trnava, Trnava, Slovakia

Abstract. Recycling the abrasives used in Abrasive Water Jet cutting is a theme that touches multiple areas. First of all, it is the economic aspect (increasing the price of abrasives on world markets, its availability and ensuring timely supplies, as well as the environmental aspect of abrasives extraction and transport). For used abrasive material recycling the sorting lines are used, which consist of sorting and drying equipment. However, recycled material should achieve sufficient cutting efficiency when reused. To determine whether it achieves recycled abrasive material is sufficiently effective, an experiment was performed. The assessment was evaluated through selected roughness parameters *Ra*, *Rq*, *Rz* and *Rt* and Abbott-Firestone material fraction curve for AlMg4.5Mn aluminum alloy samples (EN AW 5083). The results of the experiment can contribute to the widespread use of recycled abrasive material.

Keywords: Abrasive water jet machining · Recycled abrasives surface roughness parameters · Environment · AlMg4.5Mn

1 Introduction

Abrasive Water Jet (AWJ) technology makes it possible to cut and divide actually all construction materials. The main cuttings is an abrasive particle, which is a solid corpuscular component of the high-speed hydroabrasive jet. In particular, it is required to have high resistance to cleavage and abrasion, high hardness, isothermal habitus and chemical resistance. The choice of the most suitable abrasives depends, in addition to the technical requirements for quality and performance, on the price and environmental characteristics of the abrasives. Garnet (Indian, Bengal, Australian), corundum, olivine and the like are the most commonly used abrasive materials [1]. For several reasons, the most serious increase in abrasives prices, there has recently been an attempt to separate used abrasives, i.e. to recycle used ones. The abrasives are then returned to the AWJ material by separation process after regeneration.

D. Klichová et al. (Eds.): Water Jet 2019, LNME, pp. 229–241, 2021.
https://doi.org/10.1007/978-3-030-53491-2_24

The recycling or reuse of material in production becomes as a saving task for sustainable and environmental competitive production. Sustainability means not only achieving cost efficient process but also eco efficient. The design of the abrasive recycling system is desirable because of the growing growth and development of the abrasive water jet cutting industry. The use of recycled abrasives greatly reduces production costs, and hence the price of products resulting from such processing and increases competitiveness [2].

2 Materials and Methods

There is a large space for wide application of cutting technology with abrasive water jet, and thus the application of abrasive recycling [2].

The basic principle of AWJ cutting in terms of tribology is the interaction of two media, where the removal of material takes place as a result of wear by the erosion process. The erosion phenomenon is a wear process and is associated with the impact of a solid particle on the surface of a solid material. The surface after separation of AWJ material is characterized by two types of texture in basic terms. The material surface after AWJ cutting is characterized by two types of texture in basic terms. From the upper part the cutting surface is smooth to slightly roughened, in the second half the water jet creates a grooving. The first, smooth area is the result of cutting wear and the second, grooved, is called the deformation wear in the cutting zone. The cause is the characteristic feature of WJ and AWJ and it is the water jet kinematics itself. The water jet entering the material loses its kinetic energy, starts to deflect and creates a typical grooves [1, 3, 4].

The authors [5] researched the effect of traverse speed on abrasive waterjet machining of Ti-6Al-4V alloy. The experimental results their research indicate that the traverse speed of the jet is a significant parameter on the surface morphology and the widths and features of different regions formed in the cutting surface change according to the traverse speed. They were also observed that the ratio of conical taper and surface roughness increase with increasing traverse speed in chosen conditions.

In the work [6] the authors introduced a study on the recycling of a local garnet (origin: Southern India). The authors noted that the reusability with particles larger than 90 μm is 81, 49, 26 and 15% after the first, the second, the third and the fourth recycling.

Perec [7] based on his research indicates that the biggest recycling potential, near 50% characterized garnet, recycling potential of corundum is from 47 ÷ 67% range, but abrasive wear of focusing tube is 60 times bigger then garnet and maximal working time of focusing tube with corundum abrasive particles not exceed 20 min.

The authors in [8] investigated the recycling and recharging of Supreme Garnet in Abrasive Waterjet Machining. They researched that the reusability of Supreme garnet is higher than that of GMA garnet. For example, with the particle size of >90 μm, the reusability of Supreme garnet is 58.86%, while it is only 53.64% for GMA garnet. The cutting performance of recycled and recharged Supreme abrasives is slightly higher than that of the new Supreme garnet mash 80. The optimum particle size for the cutting performance is larger than 90 μm for both the recycling and the recharging. The authors in [9] state that the recycled abrasive material no longer contains inclusions and cracks in grains as it was already used in the process of cutting. It thus eliminates fracturing of grains during the process of waterjet mixing and focusing. Similarly, the applied coating covers the abrasive grain and penetrates into cracks. The grain is "glued together" and further fracturing is avoided.

Possibilities of abrasives recycling under water flow and the most important properties of recycled abrasives are:

- particle size of abrasives and its distribution,
- fragmentation is more pronounced in the mixing chamber and focusing nozzle than after cutting and it is higher with fresh abrasives than recycled abrasives,
- recycling if continued beyond three cycles, discontinuous cutting will result due to blockage of the flow channel with finer abrasive particles,
- depth of cut with abrasives,
- improved surface finish is obtained with test sample than mesh size 80,
- the usability of abrasives can be approximately 80% reused after the first, 50% after the second cut, $25 \div 30\%$ after the third cut, and 15% after the fourth cut [10].

According to [11], abrasives around the world is recycled through several methods, such as magnetic separation, gravity, sedimentation, filtration, flotation, mechanical recycling (WATING system), automated recycling (WARD system), etc. The process of mechanical recycling consists of two basic steps - sorting and drying.

The sorting is divided into coarse and fine. For coarse sorting, the waste container (Fig. 1a) generated by the AWJM is poured into the container. This falls on the vibrating sieve (Fig. 1b). What does not pass through the sieve sorter is a waste that can no longer be treated (Fig. 2a). The fallen particles are transported by an inclined screw conveyor for further separation (Fig. 2b, 3a).

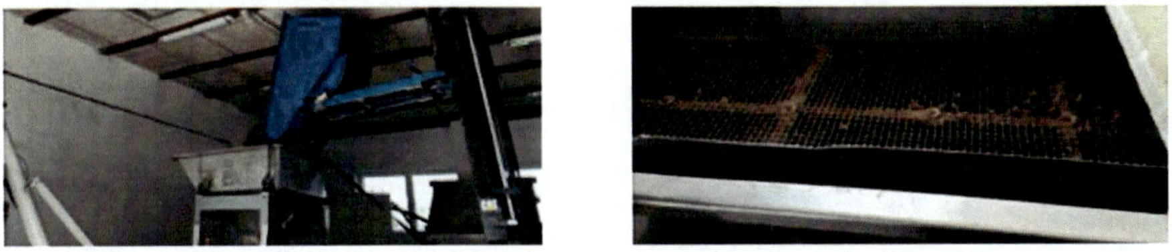

a) b)

Fig. 1. Sorting process of recycling abrasives - part 1.

a) b)

Fig. 2. Sorting process of recycling abrasives - part 2.

a) b)

Fig. 3. Sorting process of recycling abrasives - part 3.

Figure 3b shows the area of the fine sorting process. On the first sorting vibrating sieve, the separated material is mixed with water (Fig. 4a) and passed through another sieve (particles of about 3.0 mm). Coarse particles and dust go to waste. The wet recyclate, which by its size corresponds to the separated abrasives (up to mesh 100), accumulates in the transported Big Bags (Fig. 4b).

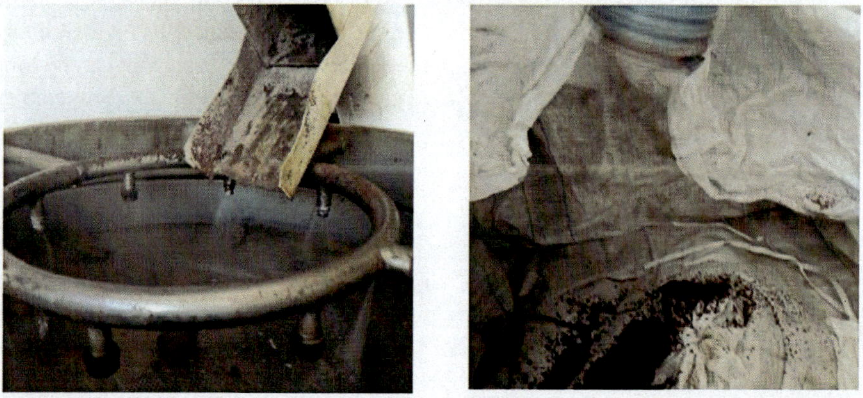

a) b)

Fig. 4. Sorting process of recycling abrasives - part 4.

The second step is drying. The recycle is poured into the container (Fig. 5a). By means of a screw conveyor it moves into the heating device. As a source of the heat for drying and heating spirals. These will ensure that the recycled material is dried. The water was evaporated. The output sorter separates the particles (Fig. 5b). Particles larger than 0.2 mm are coarse waste (above mesh 100), dust (below mesh 80), see Fig. 6a. The final output is dry recycled abrasives (Fig. 6b). The recycled material is transported to the container by a turbocharger at a flow rate of 3 L per hour (Fig. 7a). In order to save energy during recycled drying, the wet recycled material in the Big Bags is allowed to dry freely in the air, preferably in sunny places (Fig. 7b).

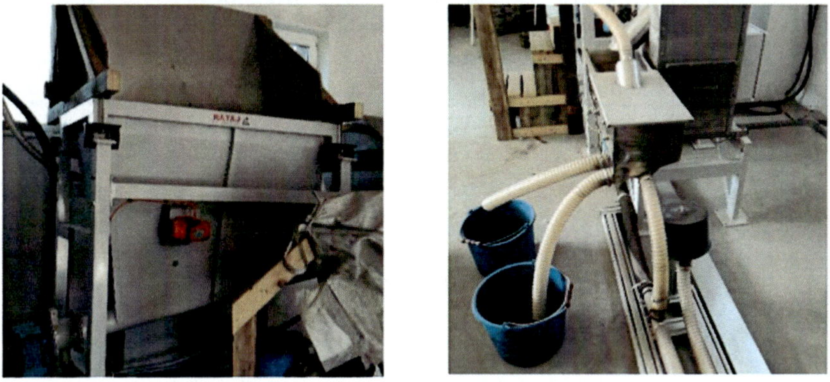

a) b)

Fig. 5. Drying process of recycling abrasives – part 1.

a) b)

Fig. 6. Drying process of recycling abrasives – part 2.

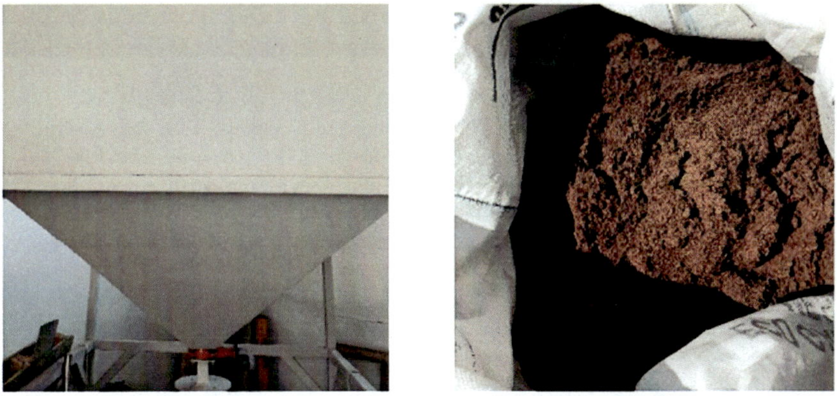

a) b)

Fig. 7. Drying process of recycling abrasives – part 3.

The reason for the abrasives reuse, in addition to the abovementioned problems on the abrasives market, is its applicability at first use. This is around 60%. The abrasive particles in the water jet only participates in the cutting that comes into contact with the material to be cut. What is inside the water jet remains actually unused. Therefore, it is assumed that the efficiency of the recycled material could be about 50%. As reported by the authors [11], the abrasives used and separated is not used alone in practice when returning to the manufacturing process, as it has lost its cutting ability by 40 ÷ 60% and is therefore mixed with the new abrasives in different proportions. The percentage of recycled abrasives mixing is shown in Fig. 8.

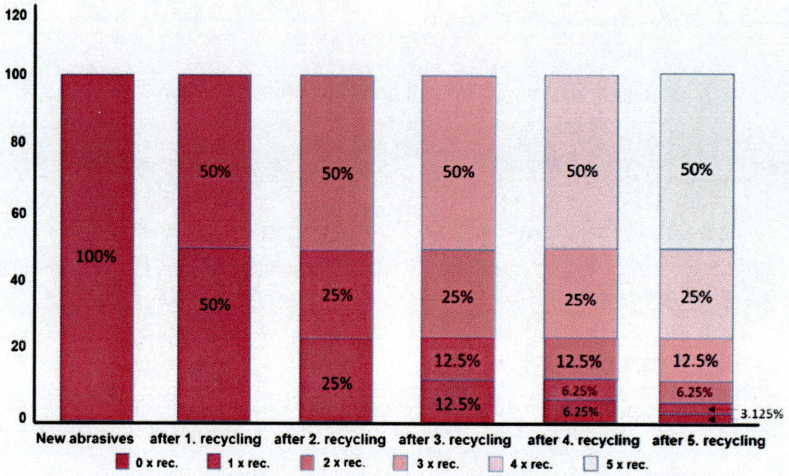

Fig. 8. Percent utilization of new and recycled abrasives [11].

Figure 9 shows new abrasives (Fresh abrasives) and after use in the AWJ process (Used abrasives). As can be seen, the recycled abrasives do not have the original sharpness that provides micro-cutting and micro-engraving in the cutting process. The abrasives could be successfully recycled in particle sizes down to 100 μm, or 140 mesh. When the abrasives collected from the cutting operations were recycled and once more used, it was found that the tested cutting area was smoother. The actual values of grain magnitude were a function of the nozzle design. It allows to expressing the opinion that recycled abrasives are more suitable to cutting process [12, 13].

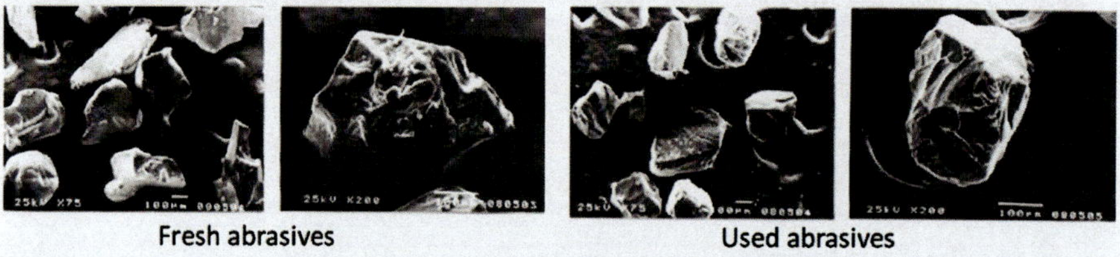

Fig. 9. Abrasive material before and after utilization in water jet cutting process [12].

Garnet (silicon dioxide SiO_4) is a sharp-edged inert abrasive, red-brown color. It is suitable as a substitute for silica sand. In addition to being used as abrasives in AWJ technology, when the process ensures maximum precision, quality and high efficiency and achieves maximum machining efficiency, it is also used to remove scale, paint but mainly to blast metal materials before further surface treatment, chrome plating, zinc plating, etc. In the last time garnet, as a suitable abrasive alternative, is increasingly being used for lace-up of stone, concrete, and other building materials, including facades and tiles, for example also in blasting technology (mobile sandblasting) in the perception of antigraffiti [1, 4, 13].

Assessment of the surface quality of the material is carried out in the technical practice by measuring the roughness. Appropriate characteristics and parameters are considered to be relevated surface roughness criteria [4, 14].

It is a technical variable that is decisive in terms of the coefficient of friction of the surfaces in contact. It is the sum of the unevenness of the surface with a relatively small distance that arises during production or its influence. The measurement of surface roughness parameters and characteristics is based on a unit of length l (μm). Surface roughness parameters are defined in several standards. These mainly include ISO 4287: 1999 Geometrical Product Specifications (GPS), Surface character: Profile method - Terms, definitions and surface character parameters [14]. When evaluating a surface in a tactile manner using the profile method, the machined surface profile, the primary profile P, is scanned after filtering out other roughness components such as waviness and shape. The roughness profile R is thus obtained. It is defined by several surface roughness parameters as *Ra*, *Rz*, *Rq* and the like.

Abbott-Fireston curve is used in addition to the surface roughness profile (Fig. 10). This curve is a graphical representation of the dependence of the relative material share on the profile cut position [15].

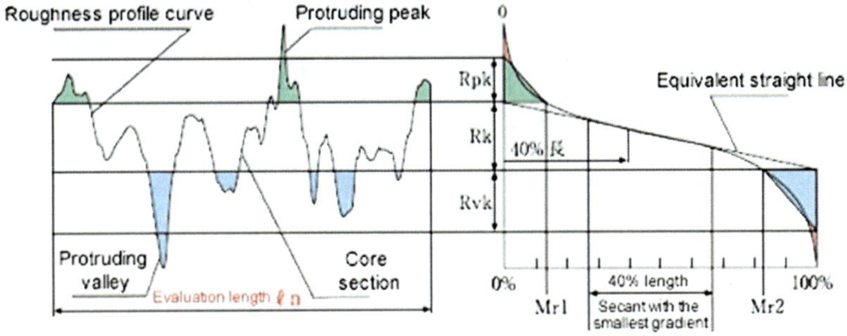

Fig. 10. Abbott-Firestone curve [15].

3 Experiment

In order to evaluate the efficiency of the recycled abrasive, samples of aluminum alloy EN AW 5083 (AlMg4.5Mn) were prepared on the machine UniJet (uses AWJT). Experimental material is a commonly used material. It has a wide range of applications where good quality is expected in terms of strength and corrosion and seawater resistance. It is used for engineering, shipbuilding, used for welded constructions, foaming molds, production of devices, production of various parts of vehicles and the like. The chemical composition and mechanical properties of EN AW 5083 are given in Table 1 in according with standards EN 573-3 and EN 485-2.

Table 1. Chemical composition and material properties of EN AW 5083 [16].

Chemical composition							
Si	Fe	Cu	Mn	Mg	Cr	Zn	Ti
0.4	04	0.1	0.4 ÷ 1	4 ÷ 4.9	0.05 ÷ 0.25	0.25	0.15

Mechanical properties			
Value name	Value signification	Units	Value
Tensile strength	Rm	$N{\cdot}mm^{-2}$	275 ÷ 350
Yield point	Rp0.2	$N{\cdot}mm^{-2}$	125 ÷ 190
Elongation	A5/A50	%	-/15
Brinell hardness	HB	–	75

Sample 1 (Fig. 11a) was cut using a new abrasives and Sample 2 (Fig. 11b) with 100% recycled abrasives. The type of abrasive grains was Australian Garnet. The sample thickness was 20.00 mm. During cutting, the parameters listed in Table 2 were set.

Table 2. Cutting parameters at AWJM.

Value name	Value signification	Units	Value
Cutting speed	v_c	$mm{\cdot}min^{-1}$	180
Abrasive flow rate	V	$g{\cdot}min^{-1}$	400
Grain size	MESH	MESH	80
Water pressure	p	MPa	400
Diameter of water nozzle	d_0	mm	0.13
Diameter of focusing tube	d_f	mm	0.91
Length of focusing tube	l_f	mm	76

Roughness parameters *Ra*, *Rq*, *Rz* and *Rt* were used to evaluate the resulting surface, with the samples distributed over six lines (Fig. 11). Measurements were performed five times on each line and for this purpose was used CNC measuring station SURFCOM 5000.

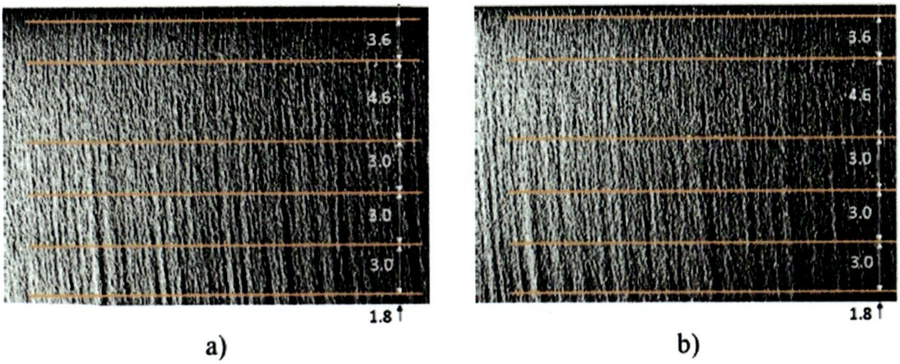

Fig. 11. Samples with marked roughness measurement lines - sample no. 1 - new abrasives a), sample no. 2 - recycled abrasives b).

The roughness parameters were measured with a measuring device SURFCOM 5000. The accuracy of the measuring device is $X = \pm (0.2 + L/1000)$ μm, where *L* is the length of the measured section in mm. For the measurement, a sensor was used whose arm is made of high-strength, very stiff and light carbon fiber tubes. The radius of the conical diamond tip was 2 μm with a 60° apex angle [17].

4 Results and Discussion

Figure 12 and 13 show the dependences of the individual surface roughness parameters on the distance at which they were measured (when average values were used for their alignment).

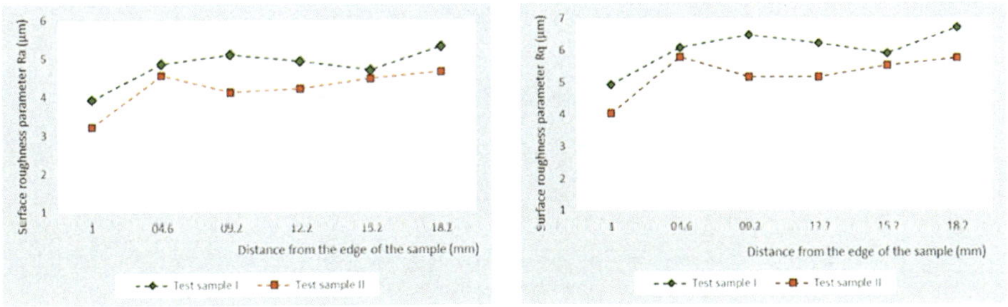

Fig. 12. Dependence of *Ra* and *Rq* on the measurement distance from the edge.

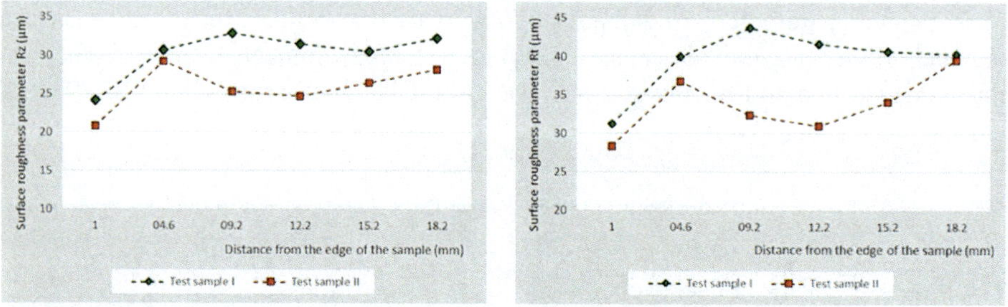

Fig. 13. Dependence of *Rz* and *Rt* on the measurement distance from the edge.

The above graphs show that the surface quality of sample 2 (cut by recycled abrasive) is comparable, even slightly better than sample 1. It can therefore be stated that for sample 2 compared to sample 1:

- the value of arithmetic deviation profile *Ra* improved from 4.7% to 19.3%,
- the value of square deviation profile *Rq* improved from 4.9% to 20%,
- the value of the highest height profile *Rz* improved from 4.8% to 22.8%,
- the value of overall height profile *Rt* improved from 1.8% to 26.1%.

All major improvements were observed on the third measured line (9.2 mm from the edge), i.e. at the point where the smooth zone ends, the water jet loses its kinetic energy and the grooved zone begins.

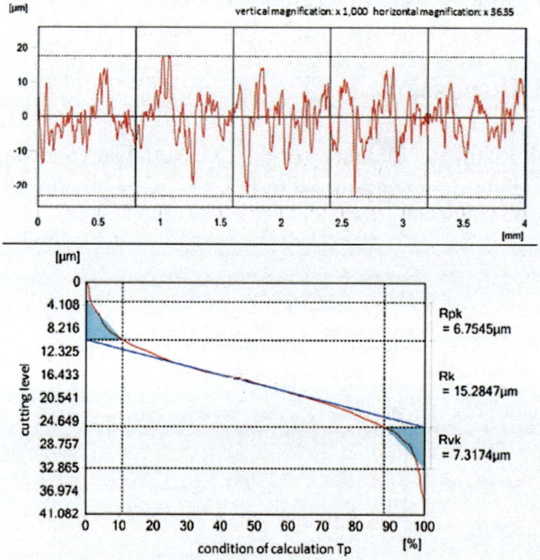

Fig. 14. Roughness profile and Abbott-Fireston curve at 9.2 mm from edge for Sample 1.

For further documentation of the cut quality, Fig. 14 and 15 show the surface roughness profile and the Abbot-Fireston curve, which serves to graphically depict the dependence of the relative material share values on the profile cut position. Also on the basis of these parameters (*Rpk* – reduced profile projection height, *Rk* – roughness profile core height, *Rvk* – reduced profile projection depth) it can be said that the sample 2 improved the surface quality.

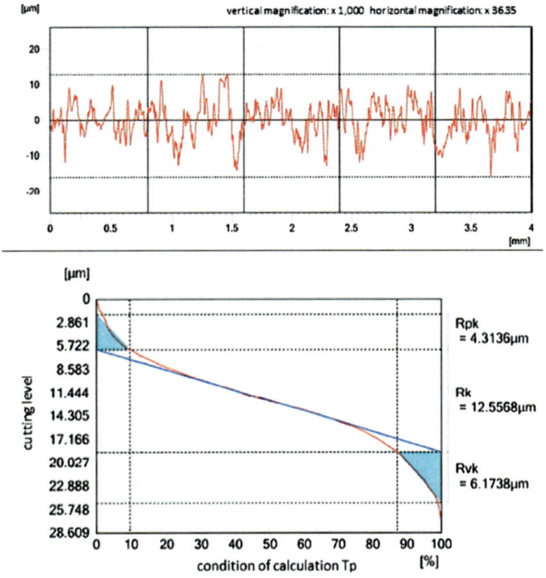

Fig. 15. Roughness profile and Abbott-Fireston curve at 9.2 mm from edge for Sample 2.

As the authors introduce in their work [13] appropriate species of abrasive material to the cutting process depends on a number of factors. One of the most important factors is the availability and price. For abrasives quality is often pays a high price, but this speaks volumes about its quality and purity. It is important to avoid abrasives containing silica (causing silicosis) and thus increase the risk for operator. Its consumption ranges from 0.1 to 1.0 $kg.min^{-1}$. The whole interface is associated with the type of pump and nozzle at WJ technology. The reusabilities of abrasives are determined as 81.77%, 57.50%, 34.37% and 17.72% after the first, second, third and fourth cuttings, respectively. Additionally, it is determined that recycling must be restricted three times due to the excessive disintegration of abrasives with further recycling. Moreover, it is concluded that cutting depth, kerf width and surface roughness decreases with recycling [18].

We can state that the results of the experiment correlate with the results of the research, which are presented in some scientific papers. According to the authors [2, 6, 8, 9], the efficiency of material cutting by WJ with recycled abrasives is high. Abrasive efficiency is higher than 53%. The authors wrote [2, 10] that recycled abrasives were more suitable to cutting process. The recycled abrasive material no longer contains inclusions and cracks in grains as it was already used in the process of cutting. It thus eliminates fracturing of grains during the process of waterjet mixing and focusing.

Similarly, the applied coating covers the abrasive grain and penetrates into cracks. The grain is "glued together" and further fracturing is avoided.

5 Conclusion

The global environmental trends lead to utilizing of so called "clean technologies" with minimum environmental impact. In case of high-speed continual waterjets application it is possible to expect substantial raise of products manufacturing efficiency [9, 16]. The experiment showed the legitimacy of use the recycled abrasives from the point of view of sufficient material cutting efficiency, evaluated through parameters *Ra*, *Rq*, *Rz* and *Rt*. Based on the results of the experiment, we can conclude that the use of recycled abrasives has a high degree of justification. The measured values of roughness characteristics showed better results in all cases, ie lower values of roughness parameters. This fact can be explained by the process that takes place in the abrasive water jet. In the jet, only the abrasives which comes into contact with the cut material, i.e. moves at the edge of the water column coming from the nozzle, is involved in the separation. What is inside the jet remains virtually unused. Therefore, as confirmed by the results of the experiment, when recycled abrasives is mixed with a new abrasive, it has sufficient cutting efficiency.

The question remains to what extent it is economically efficient to procure a recycling unit. Input costs are not low, and operating costs should also be considered. However, if several companies use recycled abrasives, the investment is expected to return to the owner within a reasonable time.

Acknowledgement. The article was supported by projects No. CZ.1.05/4.1.00/11.0260 EDIMARE, SGS UJEP 2019 and UJEP-IGA-TC-2019-48-01-2.

References

1. Hloch, S., Valíček, J.: Influence of factors on topography of surfaces created by hydroabrasive cutting (Vplyv faktorov na topografiu povrchov vytvorených hydroabrazívnym delením). FVT TU, Prešov (2008)
2. Čierna, H., Ťavodová, M.: Using the design of experiment method to evaluate quality of cuts after cutting aluminium alloy by AWJ. Manuf. Technol. **13**(3), 303–307 (2013)
3. Ťavodová, M.: The surface quality of materials after cutting by abrasive water jet evaluated by selected methods. Manuf. Technol. **13**(2), 236–241 (2013)
4. Hascalik, A., Çaydaş, U., Gürün, H.: Effect of traverse speed on abrasive waterjet machining of Ti–6Al–4V alloy. Mater. Des. **28**, 1953–1957 (2007)
5. Kantha Babu, M., Chetty, O.V.K.: A study on recycling of abrasives in abrasive water jet machining. Wear **254**, 763–773 (2003)
6. Perec, A.: Disintegration and recycling possibility of selected abrasives for water jet cutting. DYNA **84**(203), 249–256 (2017)
7. Pi, V.N., Hung, T.Q.: Recycling and recharging of supreme garnet in abrasive waterjet machining. Abrasive Technology, IntechOpen (2018)
8. Sitek, L., Martinec, P.: Abrasives and possibilities of increase in efficiency of abrasive waterjets. MM Sci. J. **3**, 877–881 (2016)

9. Kmec, J., Bičejová, Ľ., Gombár, M., Vagaská, A., Sobotová, L.: Wating abrasive recyclacion system after hydroabrasive erosion process. Int. J. Eng. **3**, 415–418 (2012)
10. Karková, M., Sobotová, L.: Quality, use and recycling of abrasives used in AWJM (Kvalita, využívanie a recyklácia abrazíva používaného pri AWJM). Transf. inovácií **30**, 54–57 (2014)
11. Sobotová, L., Badida, M., Karková, M.: Waste material recyclation in abrasive water jet technologies. Acta Mech. Slovaca **19**(4), 50–55 (2015)
12. ISO 4287:1997 Geometrical Product Specifications (GPS): Surface texture: profile method - terms, definitions and surface texture parameters
13. Roughness (3D) Parameter. https://www.olympus-ims.com/en/knowledge/metrology/roughness/3d_parameter/
14. EN AW-5083 (AlMg4.5Mn). http://www.cnc.at/media/pdf/material/Datenblatt_AlMg4-5Mn_EN_AW-5083.pdf. Accessed 11 June 2019
15. Chetty, O.V.K, Babu, N.R.: Some investigations on abrasives in abrasive waterjet machining. In: 10th American Waterjet Conference, August 14–17, Houston, Texas, USA (1999)
16. Duspara, M., Palatinuš, T., Maric, D., Samardžić, I., Ivandić, Ž., Stoić, A.: New Approach of Recycling of Abrasives for Water Jet Cutting. In: Advances in Manufacturing Engineering and Materials, January 2019 (2019). https://doi.org/10.1007/978-3-319-99353-9_4
17. Surface Measuring System – Surfcom 5000, Zeiss. https://datasheets.globalspec.com/ds/4248/CarlZeissIMT/1ADDE0CC-4852-4723-B397-F763C823CDF7. Accessed 11 Apr 2019
18. Aydin, G.: Performance of recycling abrasives in rock cutting by abrasive water jet. J. Cent. South Univ. **22**(3), 1055–1061 (2015). https://doi.org/10.1007/s11771-015-2616-5

Check for updates

Monitoring of Abrasive Waterjet Cutting and Drilling

Martin Tyč, Irena M. Hlaváčová(✉), and Jiří Kozelský

VSB - Technical University of Ostrava, Ostrava, Czechia
irena.hlavacova@vsb.cz

Abstract. In order to find a suitable way of on-line monitoring of the drilling and cutting process, the cutting head vibrations during drilling through various metal materials were measured with piezo-electric accelerometers attached around the cutting head. Three different types of experimental designs were carried on: drilling of different metals with the same thickness, operation with different abrasive mass flow rates without any workpiece and drilling of the same material with different types of abrasives. The most promising was the finding that inner dimension of the focusing tube is the most affecting parameter for cutting head vibrations. Continuous measurement of cutting head vibrations may therefore serve as an indicator of worn focusing tube. From the point of view of the on-line monitoring of drilling or cutting efficiency the sensors attached on the cutting head did not bring substantial benefit. No direct connection between dominant frequencies in vibration spectra and machined material was proved. The RMS of the vibration signal grows with the increasing mass flow rate and changes with different types of abrasives, but no significant direct relationship between abrasive characteristics and measured signal was discovered. Therefore, accelerometer placement on the workpiece appears to be much more promising from the point of view of drilling on-line monitoring.

Keywords: Acoustic emissions · Waterjet control · Cutting head vibrations

1 Introduction

The global character of modern civilization accelerates not only our personal lives but much more the industrial world. There is intensive competition involving great efforts to implement new more effective manufacturing processes - more flexible, faster and more accurate, more energy efficient and more environmental friendly. One of the production processes with great potential is abrasive water jet milling. Some of its numerous advantages are: negligible thermal distortion, flexibility, low cutting forces, and applicability to machine difficult to cut materials. Nevertheless, its wider application requires more accurate control of the material removal process. It is really challenging to develop proper solutions for process monitoring. Application of optical monitoring is hindered by tenderness of optical devices, therefore measurement of acoustic emissions (AE) was considered to be the most promising way of on-line control from the very beginning of

D. Klichová et al. (Eds.): Water Jet 2019, LNME, pp. 242–251, 2021.
https://doi.org/10.1007/978-3-030-53491-2_25

the technology spreading [1]. The feasibility of using root mean square acoustic emission (AE-RMS) to monitor the AWJ drilling depth on-line was investigated as early as the last decade of the past millennium [2]. However, it soon came out that there are too many factors entering the milling process, some of them arising from the tool itself, the others from the workpiece and so it is really difficult, to differentiate their influence on the acoustic signal. The methods of processing and interpretation of the AE signal are different with different authors.

An acoustic emission signal (AE) is a high frequency stress wave originating from deformation zones, generated by rapid release of strain energy. There are a lot of different sources of AE during metal cutting, connected with the cut material, nevertheless, jet flow turbulence, cavitation and interaction between individual abrasive particles represent considerable sources of AE as well [3]. Most of authors dealing with AWJ measure AE signals on the processed workpiece during cutting. AE detected by sensors placed on the surface of the workpiece enabled to monitor energy transfer rate during machining in [4]. Another research group installed the sensors on the sizes of workpiece and detected AE with frequencies above 100 kHz together with vertical forces measurement [5]. They coupled energy delivered to the workpiece by the jet and energy presented in the abraded volume and finally came to the conclusion that root mean square of the AE energy increases linearly with an increase in depth of cut. This concept, however, was contradicted by other authors [3, 6–9].

Another method generally used for monitoring of either tool or process condition in many areas of machining is vibration diagnostics. The studied vibrations can be subdivided in external, or exogenous, and intrinsic, or endogenous [10]. In case of abrasive water jets (AWJ) the exogenous vibrations are those measured on the workpiece, which are caused by the jet interaction with the workpiece, the equipment and the water in the catcher. Some of these vibrations may have identic origin as AE, but their frequencies are lower, they belong to the sound part of spectrum. The inherent vibrations of the jet caused by the cutting head, by the movement system, by the intensifier system and by the nozzle, are endogenous. Although this classification seems to help understanding to the process, the identification of individual sources of measured signal is far from being simple and determined. The correct diagnostics depends much on the experience of the researcher. Vibrations may be measured either on the workpiece [11] and classified as exogenous or on the cutting head [10, 12] involving endogenous ones or on both places [13]. The results of measurement were aimed to cutting quality, control of cutting parameters or tool diagnostics.

Although a lot scientific papers were published dealing with AWJ monitoring and most of authors state that their results are simply applicable in practice, we did not find in literature a method which should really help to distinguish on-line, whether cutting through is successful or not, or precast the depth of drilling on the basis of recorded acoustic signal. With the aim to solve these practical problems we tried to search proper solution. Our present paper is dealing with measurement and evaluation of the cutting head vibration during materials disintegration by AWJ, but influence of the exogenous vibrations was considered as well. The main aspect of our rating was availability of the method for application in real cutting and drilling tasks. There were two main reasons to start with such experimental design: sensors for vibration detection – accelerometers were available and measurement on the focusing tube seemed to be promising [9] and more attentive to measuring equipment.

2 Experiments

All the experiments were carried on in the Department of Physics Laboratory of the Liquid Jet at the Technical University of Ostrava with high pressure pump PTV 19/60 (Flow HSQ 5X). The maximum pressure is 415 MPa (working pressure 380 MPa), the waterjet nozzle used throughout this work has inside diameter of 0.25 mm, the focusing tube has an inside diameter 1.02 mm (a new one) and length of 76 mm. The cutting head vibrations during drilling through various metal materials were measured with three piezo-electric accelerometers PCB Model 352C33 attached counter-clockwise around the cutting head (Fig. 1). Accelerometer number 3 was placed opposite to the mouth of the abrasive feeding hose; the other two were approximately 120° apart from it. Positioning of the samples and cutting head was realized through the cutting X-Y CNC table WJ1020-1Z-EKO. The data collection was controlled by PC with measuring card NI PCI-6251 M series DAQ and evaluated by the National Instrument (NI) program Signal Express.

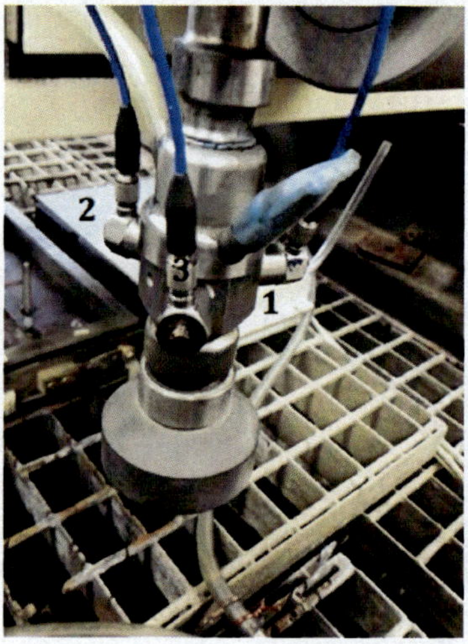

Fig. 1. Accelerometers attached to the cutting head.

Although the research is aimed to on-line monitoring of drilling and cutting operation, three preliminary stages of measurement were realized in the first round in order to exclude ostensible connections and misinterpretations. Detailed analysis of some references having dealt with similar topic led to correction of their consideration and explained discrepancy of our results.

2.1 Influence of Material Type

Seven different steel samples, a copper sample and two aluminium ones (Table 1) were drilled under the same working conditions. All the samples were 10 mm thick. Abrasive waterjet was operating at the same position on the x-y table with identical operating parameters, namely mass flow rate 226 g per minute. Abrasive used was Australian garnet 80 mesh, pump operating pressure was 380 MPa.

Table 1. Key characteristics of metal materials used for drilling [14].

Material type	Density	Tensile strength	Hardness
WRN (DIN) standard	kg·m^{-3}	MPa	Vickers
1.057 (St 52-3)	7772	443	127
1.0503 (C 45)	7651	623	165
1.7131 (16 MnCr 5)	7746	670	260
1.7225 (42 CrMo 4)	7879	651	190
1.4541 (X6 CrNiTi 18 10)	7521	523	165
1.2436 (X210 CrW 12)	7674	630	243
3.2315 (AlMgSi1Mn)	2660	327	106
3.1325 (AlCu4MgSi)	2767	419	122
3.4345 (AlZn5Mg3Cu)	2738	556	184
2.0060 (E-Cu57)	8434	211	41
HARDOX 500 PLATE	7856	1679	512

2.2 Influence of Position and Abrasive Mass Flow Rate

In order to analyse vibration signal properly, it is necessary to distinguish individual sources of vibration. One of the possible interfering factors was supposed to be the cutting *x*-*y* table. Both *x* and *y* beams may have different (probably higher) stiffness at their edges compared to their central part. Therefore frequency of vibrations may be higher when the cutting head operates at the smaller *x* or *y* coordinates.

Another non-willing source of vibrations may be the interaction with splashes of water from the catcher. This influence could be eliminated by lowering the catcher water level but such solution increases significantly the operating noise and therefore impairs the operating environment, it may also increase RMS on hearing frequencies. It was decided to measure vibrations in five *x*-different coordinates, namely (in mm) 200, 400, 600, 800 and 1000, and two different *y* coordinates (in mm) 100 and 400. The opening of the abrasive focusing tube was situated 233 mm above water level, in the second round of measurement it was put up to 293 mm.

Two different abrasive mass flow rates 226 and 393 g·min^{-1} were applied during these experiments; no workpiece was present under the cutting head.

2.3 Influence of Abrasive Type; Worn Focusing Tube Pattern

In order to gain sufficiently long non-spoiled (see Results and discussion subchapter) signal, 30 mm thick steel sample was chosen to be drilled. Five different types of abrasives were used for drilling, one of them in 3 different particle size modifications: Australian garnet, particle sizes 50, 80, 120 mesh, sorted Indian and Ukrainian garnet 80 mesh, quartz sand and silicon carbide (both non-sorted). Mass flow rate of all abrasives was individually calibrated and set to be 250 $g{\cdot}min^{-1}$.

Based on own previous experience as well as reports of other authors [15, 16] silicon carbide was expected to cause excessive wear of the focusing tube, therefore another (already worn) focusing tube was used for his type of abrasive. The inner dimension of this tube was greater and it was no more exactly circular. Unfortunately its wear could not be evaluated numerically and linked with measured vibration signals, because we were not able to measure the angle of rotation between the tube and the three accelerometers.

3 Results and Discussion

The accelerometers used for measurement allow measurement of frequencies up to 10 kHz. This fact substantially limited the range of our measurement. Measured data were filtered by second order Butterworth high pass filter so that the most important disturbing factors could be eliminated. The noise signal coming from stepper motors or the water catcher waves have low frequencies, mostly below 1 kHz and this way of processing enabled us to get rid of them.

A serious problem appeared during the measurement. At the beginning of each measurement a high single peak, reaching value of about 8 *g*, appeared. Within a short time interval lasting few seconds a smaller negative peak followed it (Fig. 2). Sometimes it lasted almost 10 s after the drilling start till the signal stabilized itself. Unfortunately we failed in solving the problem and finding the source of the undesired signal during the course of measuring. We had to cut off the initial part of all the signals and start evaluation after a certain delay. Aluminum 3.2315 (AlMgSi1Mn) appeared to be the less resistant material, in each trial it was drilled through before the signal became stable, it had to be excluded. Therefore, ten samples were processed.

Having measured the time signal, RMS and maximum amplitude were evaluated by the program and the frequency spectrum was gained using Fast Fourier Transform. The results for RMS values are summarized in graph in Fig. 3.

The other parameters of frequency spectra of vibrations, namely diminished plots of FFT spectra for accelerometers 1 and 3 (the spectral schemes of the accelerometers 1 and 2 were very similar, just no. 1 was selected for illustration), the main frequency peak, RMS and maximum value for all three accelerometers are presented in Table 2. Some expected correlations appeared: both measured aluminum samples, which are well machinable by AWJ than other materials, have significantly smaller values of the RMS and at the same time the acoustic energy for these samples is more uniformly divided between more frequencies. On the other hand the steel 1.2436 is the worst machinable among measured materials and all its RMS values are the highest ones. Nevertheless, other materials, e.g. copper, Hardox or other types of steel do not prove such specific trends, moreover, the results from individual accelerometers differ for

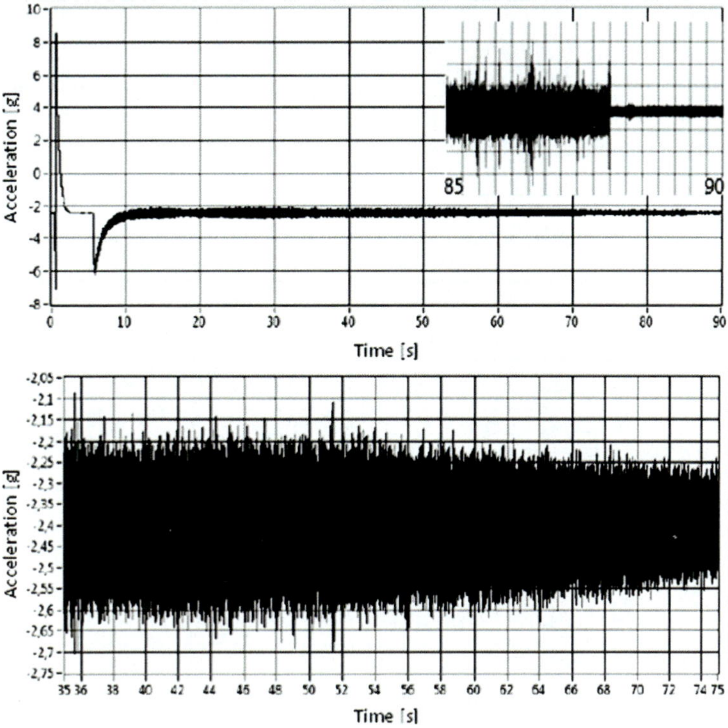

Fig. 2. Measured time running of acceleration with two undesirable initial peaks; later on it was found out that these peaks were probably induced by mechanical vibration caused during opening of pneumatically controlled valve. Detailed examination of the problem is in progress.

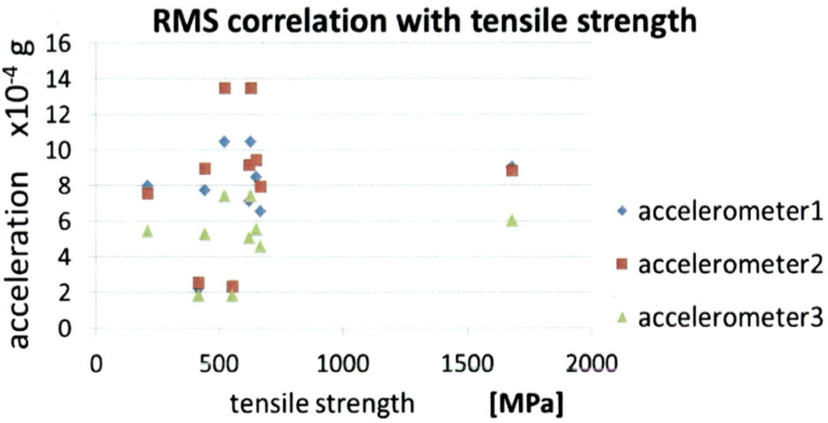

Fig. 3. Measured values of vibration RMS compared with tensile strength of materials.

them. More detailed analysis of signals will be therefore necessary. Some surprising results have to be verified by repeated measurement involving more materials so that discrepancies are explained. Supplementary measurement with sensors placed on the workpiece should be carried on, to prove or disprove the theory about machinability of material being detectable by analysis of cutting head vibrations.

Table 2. Signal processing of metal materials drilling.

material type	f_1	RMS_1	max_1	f_2	RMS_2	max_2	f_3	RMS_3	max_3
WRN standard	kHz	$\times 10^{-4}$	$\times 10^{-4}$	kHz	$\times 10^{-4}$	$\times 10^{-4}$	kHz	$\times 10^{-4}$	$\times 10^{-4}$
1.057	8.158	7.72	51.4	8.379	8.97	63.5	8.742	5.26	51.8
1.0503	7.375	7.12	61.2	8.087	9.17	78.6	8.719	5.07	35.0
1.7131	7.503	6.57	52.9	8.201	7.95	62.5	8.681	4.58	39.7
1.7225	7.522	8.47	69.2	7.881	9.44	72.2	7.794	5.53	41.0
1.4541	7.452	8.82	75.4	8.173	9.32	76.0	7.702	5.92	50.8
1.2436	7.378	10.47	83.4	8.090	13.47	97.0	8.693	7.41	57.7
3.1325	7.619	2.20	12.7	8.350	2.54	17.7	8.418	1.82	12.8
3.4345	7.676	2.20	14.6	8,333	2,36	14.3	8.685	1.83	13.2
2.0060	7.514	7.96	71.3	8.129	7.58	55.7	7.588	5.43	43.4
HARDOX	7.474	9.02	74.5	8.028	8.83	62.5	7.714	6.02	47.0

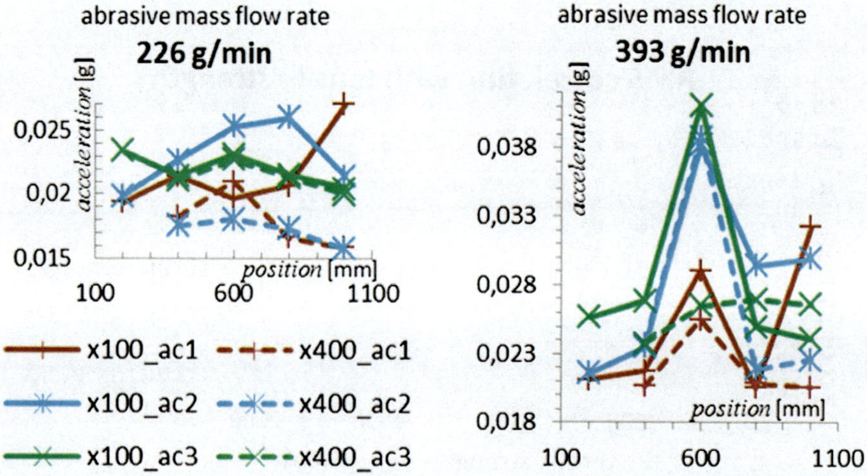

Fig. 4. The measured values of vibration RMS for various cutting head displacements and two different abrasive mass flow rates: left graph - mass flow rate 226 g·min^{-1}; right graph - mass flow rate 393 g·min^{-1}. Operation near the edge region of the cutting table is drawn in full line.

The idling drilling performance was performed as a supplementary measurement to test influence of device stiffness. The values of RMS are divided into two groups according to *y*-position of head and presented in Fig. 4. As the jet was travelling through a long path in air, the operation was much noisier and measured RMS values are more than 20 times higher than in the real drilling cases.

In almost all cases the RMS value increased with mass flow rate of abrasive – this finding corresponds with the fact that higher mass flow rate represents more impacts from individual particles but it was in contradiction with results published in [11]. Deeper analysis of the paper led to conclusion that considerations of the authors were based on wrong calculation of the number of particles involved. Our sensors were placed on the abrasive head, therefore, they might display aspects, not considered at all: disintegration of the particles in the abrasive chamber, their velocity and shape distribution. Detailed examination of these factors is planned in future experiments.

The RMS did not decrease for higher position of cutting head, though elimination of water waves was expected in this position. The results of measurement therefore confirmed that wave influence is sufficiently excluded by low frequency filtering.

The frequencies were not higher near the edge region of the table, where higher stiffness was expected; no significant influence of head position was demonstrated.

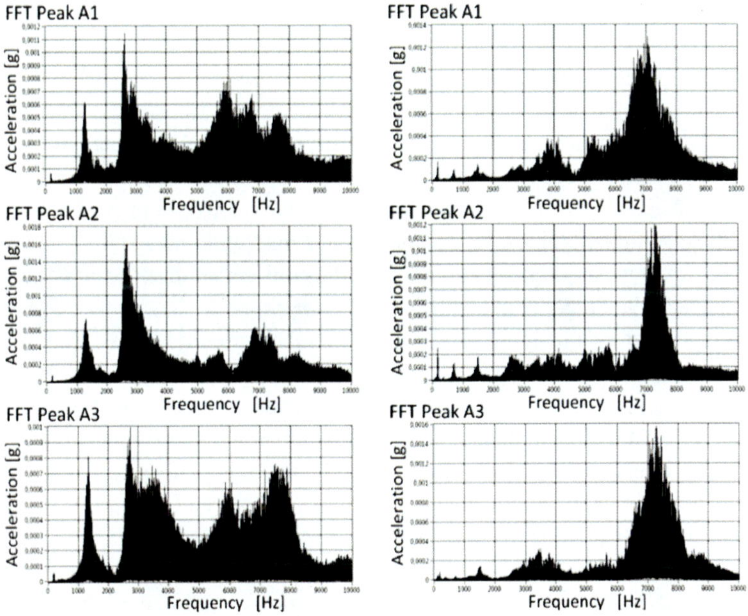

Fig. 5. The frequency spectrum of worn tube used with silicon carbide (left) is visibly different from other measured spectra represented by non-sorted Ukraine garnet on the right.

The results of thick steel sample drilling performed by different types of abrasive were very similar, there was a huge bundle of frequencies occurring between 6 kHz and 8 kHz, while in case of the accelerometer number 1 this peak was usually wider starting at 5 kHz and ending at nearly 9 kHz. The Ukraine garnet spectrum is presented in right part of the Fig. 5 to demonstrate this finding.

The spectrum of silicon carbide is completely different. The value of RMS was by about 30% higher than with other abrasives, but the energy spread was more uniformly along all frequencies between 2.5 kHz and 8 kHz with moderate minimum at 5 kHz. The two groups of abrasives certainly cannot be compared as different tubes were used, but based on the finding, that there were just mild variations observed among various types of abrasives, the fundamental difference between the two groups should be derived from the change of diameter of the focusing tube.

4 Conclusions

Accelerometer signals were measured on the cutting head of the AWJ device during drilling operation in three different experimental set ups. The signals were processed by Signal Express program and evaluated. The parameters included into evaluation were: acoustic signal RMS, frequency spectrum, peak frequency and maximum value. The changing parameters were in the three experimental sets successively: material type, abrasive mass flow rate and operating position, type of abrasive and diameter of the focusing tube. Evaluation of the results led to the following conclusions.

The basic signal parameters (RMS and characteristic frequencies) are determined neither by the mass flow rate of the abrasive nor by material characteristics. Although the mass flow rate was the same for all the materials, the results differed, and similar materials provided different results. We did not prove any direct connection between dominant frequencies in the vibration spectra and machined material.

The second experimental set up confirmed that the vibrations are not substantially influenced by displacement of the cutting head throughout the working area of the device. The RMS of the vibration signal grows with the increasing mass flow rate.

The third experimental configuration proved that the vibrations of the cutting head depend on the type of abrasive. The presented research, however, did not discover significant direct relationship between abrasive characteristics and measured signal. The inner dimension of focusing tube was found to be the most affecting parameter for cutting head vibrations. Continuous measurement of cutting head vibrations may therefore serve as an indicator of worn focusing tube.

From the point of view of the on-line monitoring of drilling or cutting efficiency the sensors attached on the cutting head did not bring substantial benefit, although detailed analysis of signals may bring more satisfactory results. The time cost of such detailed analysis, however, would not be compensated by eventual benefit as it should be applied for special configuration only without possibility of generalization. The accelerometers placement on the workpiece appears to be more promising.

References

1. Mohan, R.S., Momber, A.W., Kovacevic, R.: On-line monitoring of depth of AWJ penetration, using acoustic emission technique. In: 12th International Conference on Jet Cutting Technology, BHR, Cranfield, UK, pp. 649–664 (1994)
2. Kovacevic, R., Kwak, H.S., Mohan, R.S.: Acoustic emission sensing as a tool for understanding the mechanisms of abrasive water jet drilling of difficult to machine materials. Proc. Inst. Mech. Eng. Part B – J. Eng. **212**, 45–58 (1998)
3. Mikler, J.: On use of acoustic emission in monitoring of under and over abrasion during a water jet milling process. J. Mach. Eng. **142**, 104–115 (2014)
4. Rabani, A., Marinescu, I., Axinte, D.: Acoustic emission energy transfer rate: a method for monitoring abrasive waterjet milling. Int. J. Mach. Tools Manuf **61**, 80–89 (2012)
5. Hassan, A.I., Chen, C., Kovacevic, R.: On-line monitoring of depth of cut in AWJ cutting. Int. J. Mach. Tools Manuf **44**, 595–605 (2004)
6. Mohan, R.S., Momber, A.W., Kovacevic, R.: Energy dissipation control in hydro-abrasive machining using quantitative acoustic emission. Int. J. Adv. Manuf. Technol. **20**, 397–406 (2002)
7. Matsuoka, K., Forrest, D., Tse, M.K.: On-line wear monitoring using acoustic emission. Wear **162**, 605–610 (1993)
8. Deam, R.T., Lemma, E., Ahmed, D.H.: Modelling of the abrasive water jet cutting process. Wear **257**, 877–891 (2004)
9. Hloch, S., Valicek, J., Kozak, D., Tozan, H., Chattopadhyaya, S., Adamčík, P.: Analysis of acoustic emission emerging during hydroabrasive cutting and options for indirect quality control. Int. J. Adv. Manuf. Technol. **66**, 45–58 (2013)
10. Monno, M., Ravasio, C.: The effect of cutting head vibrations on the surfaces generated by waterjet cutting. Int. J. Mach. Tools Manuf **45**, 355–363 (2005)
11. Hreha, P., Radvanská, A., Hloch, S., Peržel, V., Królczyk, G., Monková, K.: Determination of vibration frequency depending on abrasive mass flow rate during abrasive water jet cutting. Int. J. Adv. Manuf. Technol. **1–4**(77), 763–774 (2015)
12. Krenický, T., Rimár, M.: Monitoring of vibrations in the technology of AWJ. Key Eng. Mater. **2012**(496), 229–234 (2011)
13. Radvanská, A., Hreha, P., Knapčíková, L., Petrů, J., Zlámal, T., Monka, P.: Influence of focusing tube wear on vibrations and surface roughness. Tehnički vjesnik **2**(22), 271–278 (2015)
14. Strnadel, B., Hlaváč, L.M., Gembalová, L.: Effect of steel structure on the declination angle in AWJ cutting. Int. J. Mach. Tools Manuf **64**, 12–19 (2013)
15. Axinte, D.A., Srinivasu, D.S., Kong, M.C., Butler-Smith, P.W.: Abrasive waterjet cutting of polycrystalline diamond: a preliminary investigation. Int. J. Mach. Tools Manuf **49**, 797–803 (2009)
16. Aydin, G., Kaya, S., Karakurt, I.: Effect of abrasive type on marble cutting performance of abrasive waterjet. Arab. J. Geosci. **11**(12), 357 (2019)

Effects of the Cutting Angle on the Kerf Formation During Near-Net-Shape Fabrication with the Abrasive Water Jet

Eckart Uhlmann and Constantin Männel(✉)

Technische Universität Berlin, Institute for Machine Tools and Factory Management (IWF), Berlin, Germany
maennel@iwf.tu-berlin.de

Abstract. The abrasive water jet (AWJ) is a suitable tool to cut all kinds of materials, including difficult to machine materials, with constant and low tool wear. Thus, the process is a promising method to reduce the manufacturing costs for these materials. In order to expand the 2.5D cutting abilities of the AWJ to manufacture 3D parts, milling operations have been investigated. In this paper, an AWJ milling approach is investigated, applying only two kerfs along the parts' boundaries, from different sides of the material. By intersecting the two kerfs, a solid piece of material can be removed from the part and thus, the approach enables a higher material removal rate (MRR) aggregated. The aim of this study is to expand the capabilities of this milling process by the angle of cut, orthogonal to the feed speed, to enable broader manufacturing range. For this purpose, the angle of cut was investigated along with other parameters, using design of experiments to obtain the effects and limitations of the milling operation. In order to understand cutting behavior in depth, further tests with single passes of the AWJ were conducted. The results reveal the capabilities of the milling operation for a broad range of the angle of cut. Furthermore, the application of the operation was demonstrated on a turbine blade's fir tree connection.

Keywords: Abrasive water jet · Milling · Titanium aluminide

1 Introduction

AWJ is an unconventional manufacturing technology that allows the cutting of all kinds of materials including brittle, super-hard and non-homogenous materials [1]. Furthermore, the wear of the AWJ is, unlike conventional cutting, not influenced by the material properties and the surface structure. Hence, the AWJ is a suitable technology to machine difficult to cut materials [2]. One of those materials is titanium aluminide, which offers a high strength to density ratio and good thermal stability up to 700 °C. The material can be used to replace nickel-base alloys for jet engines and reduce the overall weight [3]. Due to the specific properties, this material causes high inserts' wear during conventional chipping, due to the hardness and the low heat absorption of the material. Under

D. Klichová et al. (Eds.): Water Jet 2019, LNME, pp. 252–261, 2021.
https://doi.org/10.1007/978-3-030-53491-2_26

these premises, the application of AWJ, with material independent wear, seems to be a promising alternative manufacturing method. However, since the surface of all AWJ operations need further finishing operations after the cutting, the process is considered a near-net-shape fabrication operation.

The most common AWJ operation is cutting through the material, Fig. 1a. This operation is comparatively simple, because the remaining jet energy is absorbed by a catcher. The challenge of AWJ machining is to control the material removal when all of the jet's energy is induced in the material. However, this operation is necessary to manufacture 3D parts, Fig. 1b. These AWJ milling operations have been investigated by e.g. Hashish [4], Fowler et al. [5], Axinte et al. [6] and others. The investigations revealed the feasibility of the approach if a parameter set with low water pressure, high feed speed and multiple passes is applied. The material removal rate is generally low for these operations. A way to compensate the lower material removal rate is to intersect two kerfs, see Fig. 1c, and remove a solid piece of material, Fig. 1e, instead of chipping all the material. The feasibility and potential of this operation was demonstrated by Faltin [6] and in previous investigations [7, 8]. The investigations show that the effects of the jet deflection have a strong influence on the kerf formation and therefore it is most likely that the jet deflection influences the kerf formation for varying angle of cuts as well.

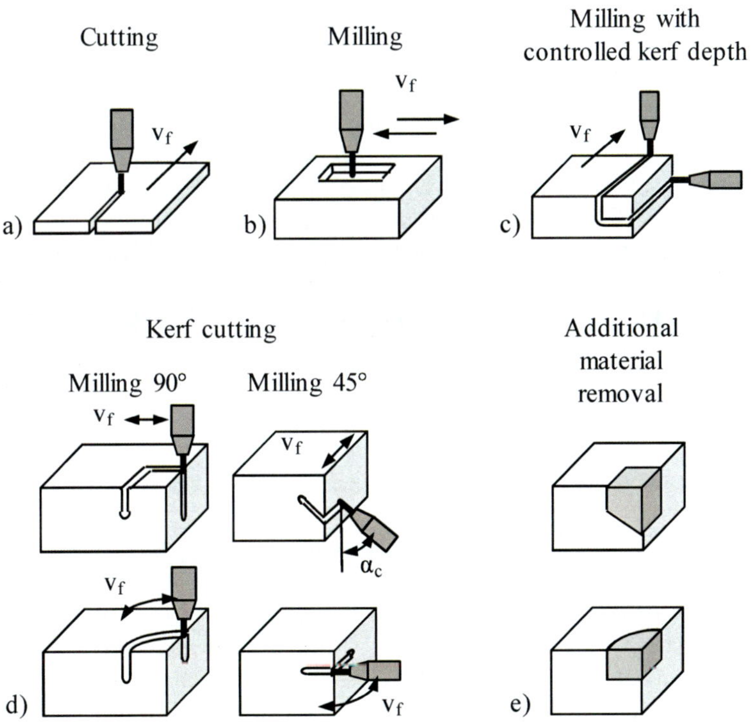

Fig. 1. AWJ; a) Cutting; b) Milling; c) Milling with controlled kerf depth; d) Efficient near-net-shape fabrication by AWJ; e) Additional material removal.

The fundamental cutting behavior of loose accelerated abrasive particles was described by Finnie [9] and Bitter [10]. They introduce the two material removal mechanisms cutting and deformation depending on the angle of cut. Furthermore, they describe the superposition of both effects on ductile material resulting in strongly varying material removal depending on the angle of cut. This curve starts with a strong increase of the material removal (erosion) for ductile materials at low cutting angles of about $\alpha_c = 10°$ [10]. This increase results in a peak around $\alpha_c = 15°$. A further increase of the angle of cut α_c causes a decrease of the material removal until a high material removal for a cutting angle of $\alpha_c = 90°$. These effects are expected to strongly affect the AWJ milling operation with varying angles of cut α_c.

This paper focuses on the extension of the milling operation with controlled kerf depth through the investigation of the effects of the angle of cut, orthogonal to the feed speed, Fig. 1d. In a first step the kerf formation is investigated and discussed. Secondly, an additional test plan is carried out to investigate the local cutting behavior, in order to answer the questions raised by the kerf formation. In addition, a search is conducted for the specific angle of cut at which the cutting behavior changes. Finally, the understanding of the process is used to apply the process on a turbine blade's fir tree connection to demonstrate the capabilities of the manufacturing method.

2 Experimental Setup for the Investigation of the Kerf Formation

In order to acquire the necessary understanding about the AWJ operation, a test plan was constructed, including the angle of cut α_c, orthogonal to the feed speed, Fig. 2a, the water pressure p, the feed speed v_f, the number of passes n and the abrasive mass flow rate $\dot{m}_a$. Fixed parameters were the stand of distance $l_s = 2$ mm, the orifice diameter $d_o = 0.25$ mm, the focus tube diameter $d_f = 0.76$ mm as well as the focus tube length $l_f = 76$ mm. Garnet mesh 120 by GMA Garnet (Europe) GmbH, Hamburg, Germany, was used as abrasive material. A full factorial test plan, Table 1, was applied with four factor levels for the angle of cut and two levels for all other parameters. The test plan contained two repetitions of each point. In addition, the center points for each angle of cut were considered. Thus, the test plan comprises 204 tests in total. The experiments were conducted on gamma titanium aluminide Ti 43.5Al 4Nb 1Mo 0.1B (TNM-B1) on a specimen with a workpiece length of $l_w = 60$ mm. All experiments were conducted using a robot guided water jet cutting machine, type HRX 160 L by Maximator Jet GmbH, Schweinfurt, Germany, Fig. 2b.

The effects of the parameters on the kerf depth d_K, kerf angle α_K, kerf width w_K and the kerf ground waviness w_t were measured using a light transmitting measurement projector type PJN 322 by Mitutoyo Deutschland GmbH, Neuss, Germany. The surface characteristics, surface waviness w_s and surface roughness Rz were measured with optical measurement device microProf MPR 100 by FRT GmbH, Bergisch Gladbach, Germany. All profiles were analyzed using an evaluation length of $l_n = 12.5$ mm, a sampling length of $l_r = 2.5$ mm and a cut-off wavelength of $\lambda_c = 2.5$ mm. Furthermore, the definition of the target values and the positions for the surface measurement are given in Fig. 2c.

Table 1. Design of Experiments (DoE).

Parameter		Levels			
		–	–	+	++
Water pressure p	MPa		100	200	
Feed speed v_f	mm/min		3000	5000	
Abrasive flow rate $\dot{m}_a$	g/min		150	250	
Number of passes n	–		200	400	
Angle of cut α_c	°	22.5	45	67.5	90
Focus tube diameter d_f	mm	0.76			
Orifice diameter d_o	mm	0.25			
Focus tube length l_f	mm	76			
Stand of distance l_s	mm	2			
Workpiece length l_w	mm	60			

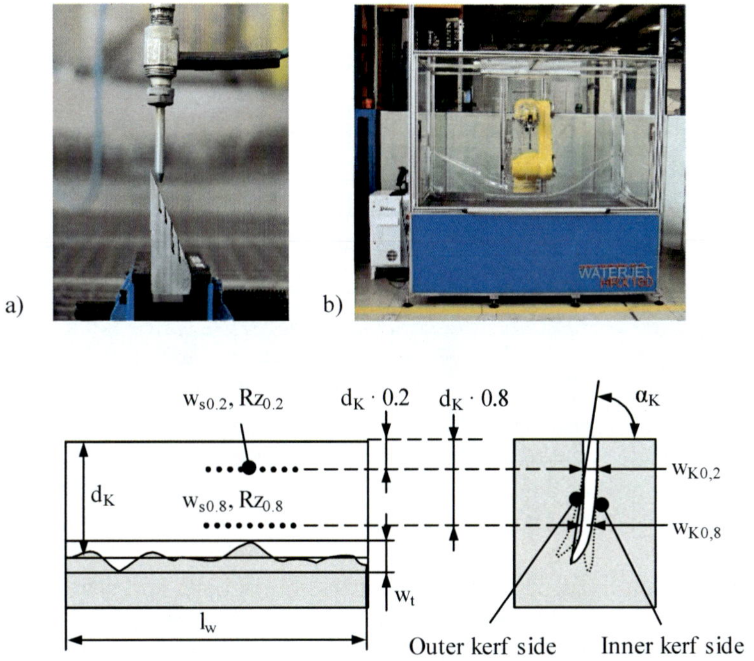

d_K	Kerf depth	$Rz_{0.2}$	Roughness Rz at 20 % of the d_c
w_t	Kerf ground waviness	$w_{s0.2}$	Surface waviness at 20 % of the d_c
$w_{K0,2}$	Kerf width at 20 % of d_K	$Rz_{0.8}$	Roughness Rz at 80 % of the d_c
$w_{K0,8}$	Kerf width at 80 % of d_K	$w_{s0.8}$	Surface waviness at 80 % of the d_c
c) α_K	Kerf angle		

Fig. 2. Experimental set up; a) Angle of cut α_c, orthogonal to the feed speed; b) Robot guided AWJ machine; c) Target values.

3 Results and Discussion of the Kerf Formation

Figure 3 states the main effects of characteristic kerf values, which were defined above in Fig. 2c. The diagram shows a slight decrease for the main effect of the kerf depth d_K, when the angle of cut is changed from $\alpha_c = 90°$ towards $\alpha_c = 45°$. A further reduction of the angle of cut to $\alpha_c = 22.5°$ causes a 40% drop of the kerf depth d_K. The widths of the kerf show a similar but reversed behavior to the kerf depth d_K. At the angle of cut of $\alpha_c = 22.5°$, the width increases strongly of about 200%. Furthermore, it should be noted that the width at the bottom of the kerf is wider than at the kerf entrance. The kerf ground waviness w_t shows a smaller change. The value is the highest for $a_c = 90°$. The Fig. 3 reveals that, in average, lower angle of cut α_c cause lower kerf ground waviness w_t.

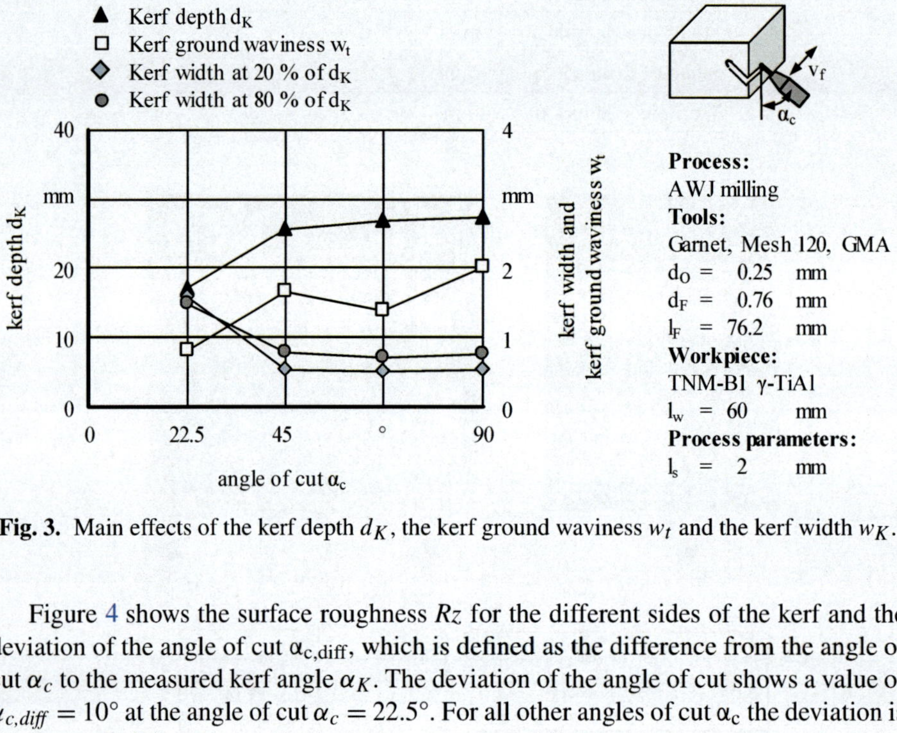

Fig. 3. Main effects of the kerf depth d_K, the kerf ground waviness w_t and the kerf width w_K.

Figure 4 shows the surface roughness Rz for the different sides of the kerf and the deviation of the angle of cut $\alpha_{c,diff}$, which is defined as the difference from the angle of cut α_c to the measured kerf angle α_K. The deviation of the angle of cut shows a value of $\alpha_{c,diff} = 10°$ at the angle of cut $\alpha_c = 22.5°$. For all other angles of cut α_c the deviation is almost negligible $\alpha_{c,diff} < 1°$. Furthermore, a clear difference between the outer and the inner kerf can be observed. The surface roughness of the outer kerf side, Fig. 2c, also shows a strong increase at the angle of cut $\alpha_c = 22.5°$. This surface is inconsistent and full of deep craters. The surface roughness of the inner kerf side is not affected by this behavior and is very smooth and thus low in surface waviness and roughness. Thus, the outer surface waviness and roughness are in average around eight times higher than on the inner side.

The results reveal that a decreasing angle of cut causes at first very little change in the kerf parameters. However, once a critical value has been exceeded, between the angle of cut of $\alpha_c = 45°$ and $\alpha_c = 22.5°$, a majority of the kerf parameters change drastically. Due to the strong change of the kerf parameters it can be assumed that the cutting mechanisms change between these angles. The fundamental cutting behavior was revealed above in the introduction. The behavior of the erosion described by Bitter [10] should be related to the kerf depth d_K, because similar effects occur during the process. However, both behaviors have no similarities. For example, the kerf depth d_K maximum is at $\alpha_c = 90°$ while the material removal described by Bitter [10] peaks around $\alpha_c = 15°$.

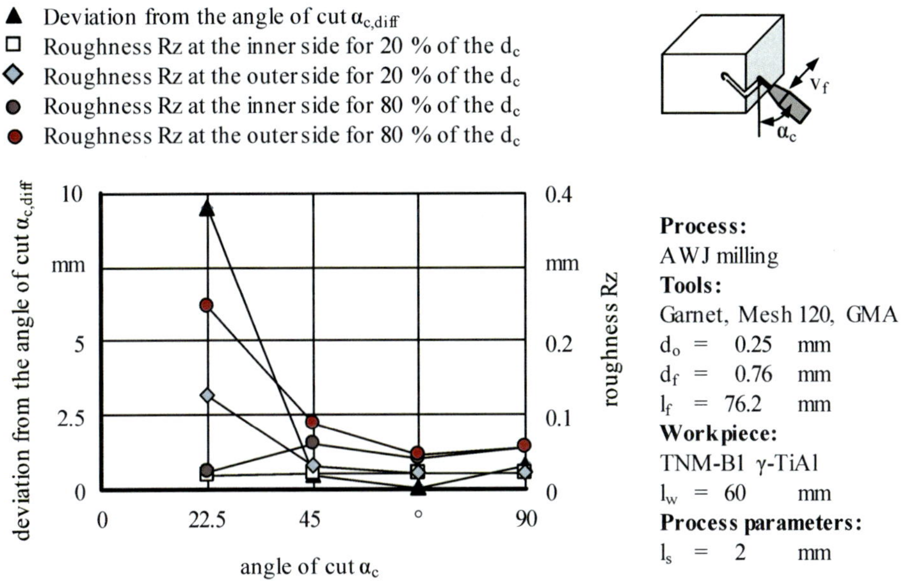

Fig. 4. Main effects of the deviation of the angle of cut $\alpha_{c,diff}$ and the surface roughness Rz.

In previous works the jet deflection, or secondary jet, was found to have a crucial influence on the kerf formation [6]. Thus, a detailed analysis of the local cutting behavior seems to be necessary. In order to achieve a better understanding of the local cutting behavior a further investigation on the material removal mechanism of AWJ was carried out.

4 Experimental Setup for the Investigation of the Local Cutting Behavior

To investigate the local cutting behavior, the material removal mechanism of AWJ cutting were examined in a test plan containing the same parameter settings given in Table 1, except the number of passes n which has been changed to $n = 1$. In addition to the main (primary) target material (TM) another piece of material (secondary) was added to measure the effect of the secondary jet. This piece was placed orthogonal on the primary material, Fig. 5. The material loss was measured on both the primary and secondary TM for all parameter settings.

5 Results and Discussion of the Local Cutting Behavior

The resulting material loss is given in Fig. 5. The diagram shows that the water jet cuts most material on its first impact under the angle of cut of $\alpha_c = 45°$. In general, the behavior of this primary material removal is similar to the one described by Bitter [10], but with a shifted peak. The secondary material removal increases strongly with decreasing angles of cut α_c. At the angle of cut $\alpha_c = 22.5°$ both values are almost the same. The combined material removal of both jet impacts correlates well with the erosion described by Bitter [10]. As a consequence, it can be assumed that the material loss of Bitter [10] comprehends both primary and secondary material removal. However, in order to understand the kerf formation of AWJ milling under the angle of cut, the two material removals must be differentiated.

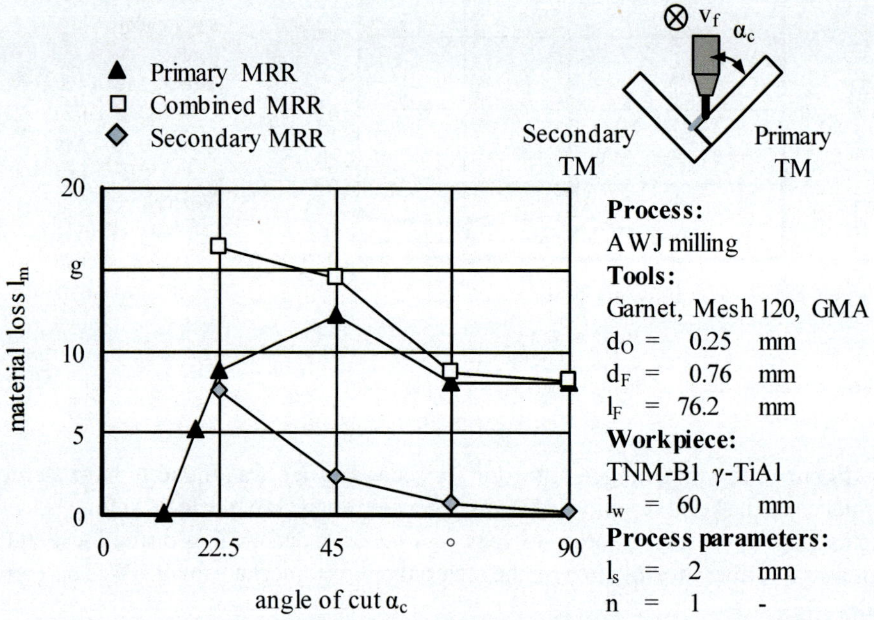

Fig. 5. Main effects of the material loss during single passes.

The above described kerf forms can be explained with the following deliberation for the local cutting situation. At a angle of cut $\alpha_c = 90°$ the jet impacts the material orthogonal. During the impact of the waterjet, the surface changes slightly to a lower cutting angle, Fig. 6a. Thus, the secondary jet is directed along the kerf. The secondary jet further increases the depth of the kerf. Additionally, the primary material removal is strong and the secondary material removal to the sides is neglectable, Fig. 6c. This behavior is repeated over the passes, since the new surface conditions are similar to the start conditions.

At the angle of cut of $\alpha_c = 45°$ the primary material removal is high and the second material removal is low, see Fig. 5. Furthermore, the second jet is orthogonal to the material surface and is thus directed to one side of the kerf, Fig. 6b and Fig. 6e. As a result, a high amount of material is removed and the kerf is very wide after the first pass. During the next passes the jet impacts a kerf with an angle of cut $\alpha_c \neq 45°$. The local cutting angle changes towards $\alpha_c = 90°$ during the next passes. Thus, the local cutting behavior changes back to the state described above. During the running-in stage more material is removed. This additional material removal appears on the kerf sides, increasing the kerf's width. However, the progression in the z direction is limited, because the secondary jet does not deepen the kerf. Therefore, the kerf depth d_K at the end of a process is a bit lower for the angle of cut $\alpha_c = 45°$. The effect is likely to appear in the same way, but with a lower intensity, for the angle of cut of $\alpha_c = 67.5°$, Fig. 6d.

If the jet impacts the surface at the angle of $\alpha_c = 22.5°$, the primary jet cuts a wide and shallow kerf at first. In addition, the jet is deflected and directed orthogonal to the material surface. This secondary jet causes approxemately the same amount of material removal as the first impact, Fig. 6f. Hence, the material removal towards the side of the kerf is as strong as the first impact. Due to the strong secondary jet which effects one

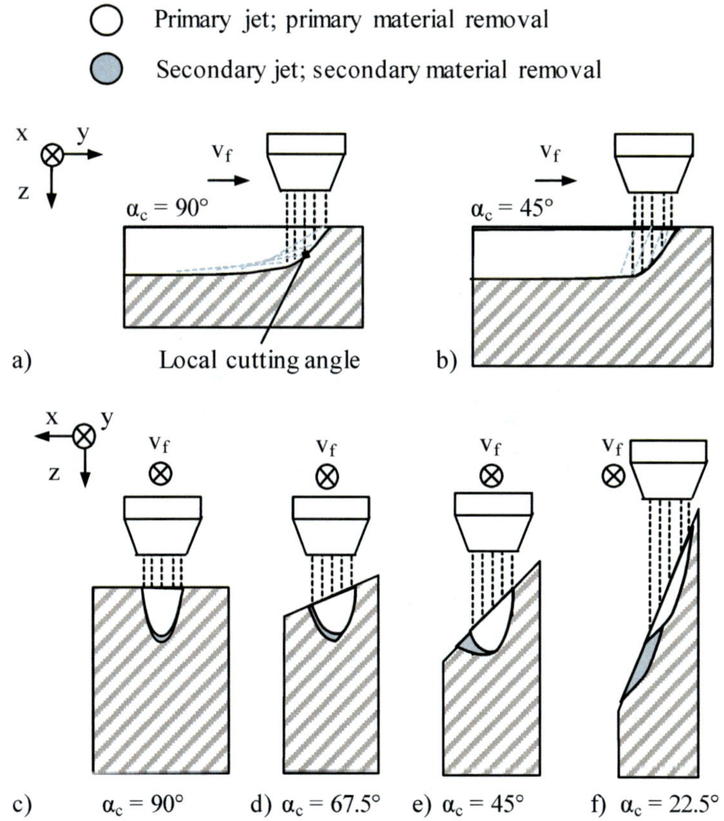

Fig. 6. Kerf formation; a) Local cutting angle with secondary jet for $\alpha_c = 90°$; b) Secondary jet for $\alpha_c = 45°$; c) Kerf form after the first pass for $\alpha_c = 90°$; d) Kerf form after the first pass for $\alpha_c = 67.5°$; e) Kerf form after the first pass for $\alpha_c = 45°$; f) Kerf form after the first pass for $\alpha_c = 22.5°$.

side of the kerf, the angle of cut remains around the previous value of $\alpha_c = 22.5°$. Thus, this behavior is being repeated for every pass. This behavior results in the much wider kerf width, the reduced kerf depth d_K, the high deviation from the angle of cut and the differences in the surface roughness.

6 Finalization and Application of the Investigation for the Angle Of Cut

In order to complete the investigation about AWJ milling with various angles of cut α_c, an additional test was carried out to find the exact angle of cut $\alpha_{c,min}$ where the kerf formation changes. The test revealed that until the angle of cut of $\alpha_{c,min} = 39°$ the material removal follows the stable mechanisms of the cutting at $\alpha_c = 90°$. For lower angles of cut the mechanisms change to the cutting described for angle of cut of $\alpha_c = 22.5°$.

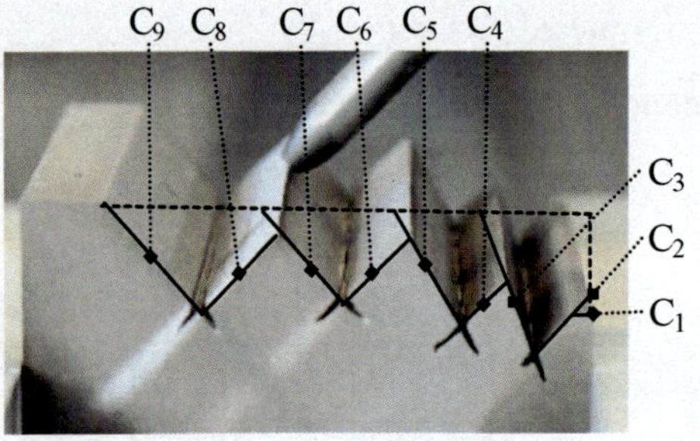

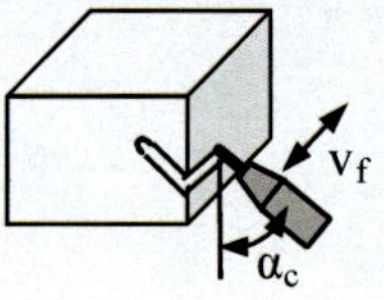

Process:
AWJ milling
Workpiece:
TNM-B1 γ-TiAl
l_w = 60 mm

Varied process parameters:

n_1 = 100	-	α_{c1} =	90°
n_2 = 245	-	α_{c2} =	45°
n_3 = 470	-	α_{c3} =	70°
n_4 = 190	-	α_{c4} =	70°
n_5 = 430	-	α_{c5} =	60°
n_6 = 285	-	α_{c6} =	75°
n_7 = 480	-	α_{c7} =	45°
n_8 = 345	-	α_{c8} =	90°
n_9 = 440	-	α_{c9} =	45°

Tools:
Garnet, Mesh 120, GMA
d_o = 0.25 mm
d_f = 0.76 mm
l_f = 76.2 mm

Fixed process parameters:
l_s = 2 mm
$\dot{m}_a$ = 150 g/min
p = 100 MPa
v_f = 5000 mm/min

Fig. 7. AWJ milled fir tree connection.

In the last step, the obtained understanding about AWJ milling with angles of cut α_c is used to manufacture a turbine blade's fir tree connection. Therefore, nine cuts C1 to C9 were performed under the angle of cut of α_c and the number of passes n stated in Fig. 7. All applied parameters as well as the resulting part are given in Fig. 7.

7 Conclusion

The results show that AWJ milling with controlled kerf depth can be used efficiently within a large range of the angle of cut α_c, orthogonal to the feed speed. A critical angle of cut was found at $\alpha_c = 39°$. Above this angle the kerf depth d_K changes only very little, due to different running-in behaviors. This change can easily be implanted in process models such as MLR. Below the cutting angle of $\alpha_c = 39°$ a different cutting behavior was observed. The consideration of the local jet impact and knowledge about the primary and secondary material removal have revealed the differences in the cutting behaviors during one pass of the water jet. Most notably is that the primary and secondary jet cause the same amount of material removal at the angle of cut $\alpha_c = 22.5°$. This observation and the deliberation carried out, provides a good explanation for the differences of the final kerf. Finally, the results allow the near-net-shape fabrication using AWJ milling for difficult to machine materials, demonstrated on a turbine blade's fir tree connection of TNM B1.

Acknowledgements. This paper is based on results acquired in the projects DFG UH 100/165-3 and DFG UH 100/206-2, which are kindly supported by the DEUTSCHE FORSCHUNGSGEMEINSCHAFT (DFG).

References

1. Mohankumar, V., Kanthababu, M., Raveendran, R.: Review on machining aspects in metal matrix and ceramic matrix composites using abrasive waterjet. Appl. Mech. Mat. **766–767**, 643–648 (2015)
2. Hashish, M.: Wear in abrasive-waterjet cutting systems. In: Wear of Materials, pp. 769–776 (1987)
3. Kothari, K., Radhakrishnan, R., Wereley, N.M.: Advances in gamma titanium aluminides and their manufacturing techniques. Prog. Aerosp. Sci. **55**, 1–16 (2012)
4. Hashish, M., Turning, M.: Drilling with abrasive-waterjet. In: 9th International Symposium on Jet Cutting Technology, vol. 2, pp. 113–131 (1988)
5. Fowler, G., Shipway, P., Pashby, I.: Abrasive water-jet controlled depth milling of Ti6Al4V alloy - an investigation of the role of jet-workpiece traverse speed and abrasive grit size on the characteristics of the milled material. J. Mater. Process. Technol. **161**(3), 407–414 (2005)
6. Faltin, F.: Endkonturnahe Schruppbearbeitung von Titanaluminid mittels Wasserabrasivstrahlen mit kontrollierter Schnitttiefe. Berichte aus dem Produktionstechnischen Zentrum Berlin. In: Uhlmann, E (ed.). Dissertation, Technische Universität Berlin. Stuttgart: Fraunhofer IRB (2018)
7. Uhlmann, E., Männel, C.: 3D-Vorkonturierung mittels Wasserabrasivstrahl. Zeitschrift für wirtschaftlichen Fabrikbetrieb **113**(7–8), 479–483 (2018)
8. Uhlmann, E., Männel, C.: Modelling of abrasive water jet cutting with controlled depth for near net shape fabrication. Procedia CIRP **81**, 920–925 (2019)
9. Finnie, I.: The Mechanism of Erosion of ductile Metals. Proceedings of the third U. S. National Congress, pp. 527–532 (1958)
10. Bitter, J.G.A.: A study of erosion phenomina. Part II. Wear **6**, 169–190 (1963)

Author Index

D. Klichová et al. (Eds.): Water Jet 2019, LNME, pp. 263–264, 2021.
https://doi.org/10.1007/978-3-030-53491-2